**Successful Trouble Shooting for Process Engineers**

*D. R. Woods*

## *Further Titles of Interest:*

Büchel, K. H., Moretto, H.-H., Woditsch, P.

**Industrial Inorganic Chemistry**
**Second, Completely Revised Edition**

2000
ISBN 3-527-29849-5

Weissermel, K., Arpe, H.-J.

**Industrial Organic Chemistry**
**Fourth, Completely Revised Edition**

2003
ISBN 3-527-305789-5

Mollet, H., Grubenmann, A.

**Formulation Technology**
**Emulsions, Suspensions, Solid Forms**

2001
ISBN 3-527-30201-8

Sundmacher, K., Kienle, A. (Eds.)

**Reactive Distillation**
**Status and Future Directions**

2003
ISBN 3-527-30579-3

Rauch, J. (Ed.)

**Multipurpose Plants**

2003
ISBN 3-527-29570-4

Elias, H. G.

**An Introduction to Plastics**
**Second, Completely Revised Edition**

2003
ISBN 3-527-29602-6

Hattwig, M., Steen, H. (Eds.)

**Handbook of Explosion Prevention**

2004
ISBN 3-527-30718-4

Oetjen, G.-W., Haseley, P.

**Freeze-Drying**
**Second, Completely Revised and Extended Edition**

2004
ISBN 3-537-30620-X

Hagen, J.

**Industrial Catalysis**
**A Practical Approach**

1999
ISBN 3-527-29528-3

Jakobi, R.

**Marketing and Sales in the Chemical Industry**
**Second, Completely Revised Edition**

2002
ISBN 3-527-30625-0

Bamfield, P.

**Research and Development Management In the Chemical and Pharmaceutical Industry**
**Second, Completely Revised and Extended Edition**

2003
ISBN 3-527-30667-6

*Donald R. Woods*

# Successful Trouble Shooting for Process Engineers

## A Complete Course in Case Studies

WILEY-VCH

WILEY-VCH Verlag GmbH & Co. KGaA

**Author**

***Prof. Donald R. Woods***
Chemical Engineering Department
McMaster University
Hamilton
Ontario
Canada, L8S 4L7

1st Edition 2006
1st Reprint 2008



**Library of Congress Card No.:** applied for

**British Library Cataloguing-in-Publication Data**
A catalogue record for this book is available from the British Library.

**Bibliographic information published by the Deutsche Nationalbibliothek**
The Deutsche Nationalbibliothek lists this publication in the Deutsche Nationalbibliografie; detailed bibliographic data are available in the Internet at http://dnb.d-nb.de.



Printed in the Federal Republic of Germany.
Printed on acid-free paper.

**Typesetting** Kühn & Weyh, Satz und Medien, Freiburg
**Printing** betz-druck GmbH, Darmstadt
**Bookbinding** Litges & Dopf GmbH, Heppenheim8

**ISBN-13:** 978-3-527-31163-7
**ISBN-10:** 3-527-31163-7

# Contents



*Successful Trouble Shooting for Process Engineers.* Don Woods

ISBN: 3-527-31163-7

**Appendix I**



**Appendix J**

# Preface

My McMaster University colleague Tom Marlin describes trouble shooting as "the bread and butter" activity of engineering. Indeed, the financial health of a process unit depends so much on the skill of the engineers to trouble shoot problems promptly, safely and effectively.

Training in trouble shooting should be part of every undergraduate engineer's education. Yet, it rarely is, even though the introduction of trouble-shooting examples receives a warm welcome by the students. As Scott Lynn of UC Berkeley reports, "Our experience was that most students really got into the spirit of the thing and trouble shooting was one of the most popular parts of the course." Perhaps some of the reasons why the development of trouble-shooting skill is not introduced are the need for excellent problem-solving skills, the lack of a variety of industrial problems and, perhaps most significantly, the student's lack of a rich set of practical experience and understanding of equipment. There may also be a lack of the faculty's confidence in using such open-ended experiences. Whatever the reason, I have designed this book to overcome these shortcomings. I hope that trouble-shooting skill development becomes part of every undergraduate experience.

Training in trouble shooting in industry tends to occur from the school of hard knocks, by trial and error and gradually from the experience of solving problems as they occur, with no well-designed program of instruction. This is relatively inefficient and it does little to develop confidence. This book is designed to improve skill and confidence of process engineers and engineering students.

This book is based on my experience developing trouble-shooting skills in undergraduate engineering programs, in short courses in industry and in courses presented at conferences.

This book is designed to help develop your skill and confidence. This book is tailored to help you improve your skill no matter where you are in your journey to become an outstanding trouble shooter.

A number of excellent books have been published about trouble shooting. Liberman ("Trouble-shooting process Operations") describes a wide range of problems that he encountered, the fault that he discovered and the corrective action. His personal approach to trouble shooting is illustrated. Saletan ("Creative Trouble Shooting in the Chemical Process Industries") provides interesting examples to illustrate different components in the trouble-shooting process. However, no specific educa-

*Successful Trouble Shooting for Process Engineers.* Don Woods

ISBN: 3-527-31163-7

tional plan is apparent. No activities, with feedback, are provided for skill development. Branan's "Rules of Thumb for Chemical Engineers" is mainly an excellent collection of rules of thumb. In addition he has chapters on trouble shooting and plant startup. He includes material from a range of topics and resources but he does not presented a synthesis of this material. The focus in these books tends to be a personalized approach of how the author solved trouble-shooting problems. Not everyone will or should follow Lieberman's (1985), Saletan's (1994), Gans's (1983), Kister's (1979) or my style in trouble shooting. The key is to identify your style and develop confidence in using it.

Developing your style and skill requires that we draw on the extensive research about the trouble-shooting process. A skill development program should give you a chance to solve a wide range of trouble-shooting problems, to think about how you solved them and to set goals for improvement. The central core of this book is 52 trouble-shooting cases that are presented in a unique format that allows you to select the process you will use to solve the problem. Feedback is given to help you assess your approach. Target skills used by successful trouble shooters are given; structured activities provided, and feedback is supplied. This book is unique in its coverage, ease in use, focus on skill development using proven methods, self-selection and inclusion of activities that are challenging but fun to do. Included are a range of self-assessment tools.

Here are the details:

Chapter 1 outlines four types of trouble-shooting problems and summarizes the five key skill areas needed in trouble shooting: skill in problem solving, practical knowledge about a range of process equipment, specific knowledge about safety, hazards, systems thinking and people skills. A *self-test* is included to help identify which of the five key skill areas might be of most interest to you. Five trouble-shooting cases are posed from a variety of industries and unit operations: distillation, heat exchange, pumps, adsorption and crystallization and that pose a range of difficulty. The results of the self-tests can be used to guide you as to how best to use the remaining chapters and appendix material.

The focus of this book is on developing skill in the mental process used to solve trouble-shooting problems. Chapter 2 summarizes the research evidence of what skilled trouble shooters do, provides a *Trouble-Shooter's Worksheet* (and illustrates its application) and a feedback form to help focus attention on the problem-solving, synthesis, data-handling and decision-making activities used. This gives you a chance to compare the processes used with those used by skilled trouble shooters, and hence improve your skill and confidence in trouble shooting.

To illustrate the application of these skills five scripts are provided in Chapter 4 of trouble shooters tackling the trouble-shooting problems **Cases #3–7**. These are *real* problems taken from industrial experience; only the names of the trouble shooters have been changed. Each of the five scripts consists of about three parts with each part concluding with a few questions for you to consider. This reflective break was introduced to give you a chance to reflect on how you would have handled the case, and to decide what you should do next. I recommend that, as you read each script, you *play the game*. An assessment is given of the problem-solving processes used by each of the trouble shooters. Other examples of the process are given in Appendix C and scattered as activities throughout most of the Chapters. **Case #3** in Chapter 6; **Case #6**, in Chapter 6; **Case #8**, in Chapters 2 and 6; **Cases #9 and 10**, in Chapters 5 and 6; **Case #11** in Chapter 6; **Cases #12–18** in Chapter 7.

The central activity of the book is in Chapter 8. Here, trouble-shooting problems are posed so as to help you develop your skill. The activity asks you to select an action or question to take for each selected case (from about 30 possible actions). A coded answer to each action is given in Appendix D. By posing a series of actions you will gather evidence until you have "solved the problem". Feedback about the process is given in Appendix E where an answer is given and key elements of the process used by an experienced trouble shooter are listed. These problems are sequenced and classified so that you can start with easy and familiar Cases and build up your confidence gradually. The classification notes the degree of difficulty, the type of equipment involved, and the chemicals/process technology involved. Some of the Cases relate to similar processes. For example, six cases relate to the depropanizer-debutanizer system. Two, to the ethylene process; five, to the ammonia-reformer.

Since the trouble-shooting cases require the use of the five key skills, the rest of the book provides skill-development activities for each of these five skills.

**Skill #1: problem solving.** The development of *problem-solving skills* is the theme of Chapters 5 and 6. In Chapter 5 the focus is on awareness, strategies, exploring the problem, creativity and self-assessment. Target skills are given, activities are introduced, a range of tasks are given (in Chapter 5 and Appendix F) and feedback is provided. Chapter 6 provides activities to develop skill in gathering data, checking hypotheses and critical thinking. For the various skills being developed, the process is illustrated (in the context of a trouble-shooting case), and tasks are given. This activity-based, workshop-style approach has been proven to be extremely effective; the proof is given in the award-winning paper" Developing problem-solving skills: the McMaster Problem Solving program, "Journal of Engineering Education", April, vol 86, no 2, pp. 75–91, 1997. A wide range of tasks are provided from which you can select those most pertinent to your experience with feedback available in Appendix G.

**Skill #2: knowledge of process equipment.** Chapter 3 provides a convenient summary of the *practical aspects about equipment* needed for trouble shooters of over 50 different types of process equipment. For most types of process equipment the following information is given: overall fundamentals, guidelines for good operation, and trouble shooting. For trouble shooting, typical symptoms are given together with a prioritized list of typical causes. Some will want to keep this text handy for just this summary of practical know-how. More details are given in Appendices A, H and I.

**Skill #3: process safety and properties of materials.** Chapter 3 also gives some succinct rules of thumb related to *safety and hazard identification* in Section 3.12.

**Skill #4: systems thinking.** Guides to and activities to help develop "A*systems thinking*" are given in Chapter 3 in Sections 3.1 and 3.11 and Appendix B.

**Skill #5: people skills.** Chapter 7 addresses *interpersonal skills* and looks at the factors influencing personal performance. More is given in Appendices C, F and G.

Chapter 9 offers ideas of what to do next.

This book would not have been possible without the help of many. In the seven companies for whom I worked before coming to McMaster University, I was fortunate to have worked with a variety of excellent trouble shooters who patiently helped me polish my skill, Don Ormston and Ted Tyler of Distiller's Company Ltd, Saltend, UK; Stan Chodkiewicz, Polysar, Sarnia, J. Mike F. Drake, British Geon Ltd, Barry, South Wales.

I thank Tom Marlin, Adam Warren, Iryna Bilovous, the late R.B. Anderson, Archie Hamielec, Terry Hoffman, Cam Crowe, John Vlachopoulos, Raja Ghosh, Douglas Dick, Dave Cowden and Lisa Crossley, McMaster University; Peter Silveston, University of Waterloo; Jud King and Scott Lynn, University of California, Berkeley; Ian Doig, University of New South Wales; Frank Bajc; Pierre Cote, Zenon Environmental, Douglas R. Winter and Robert French, Universal Gravo-plast, inc, Toronto, my students, my alumni who sent back problems (and answers), participants in the industrial workshops and in the conference workshops and Esso Chemicals, Nova Corporation, Prices of Bromborough, Unilever who generously provided me with problems and gave me permission to use them.

McMaster Alumni who sent me Cases (and answers) include Bill Taylor (B Eng '66), Ian Shaw (B. Eng. '67), John Gates (B. Eng. '68), Don Fox (B. Eng. '73), R.J. Farrell (B. Eng. '74), Jim Sweetman (B. Eng. '77), Mike Dudzic (B Eng '80), Mark Argentino (B. Eng. '81), Vic Stanilawczik (B. Eng. '83), Gary Mitchell (B. Eng. '83), David Goad (B Eng and Mgt, '91), Kyle Bouchard (B Eng '93), Doug Coene (B. Eng. '97) and Jonathan Yip (B. Eng. '97).

I have learned much from the cases solved and the approaches taken by Norman Lieberman, David Saletan and Henry Kister that they published in their books and articles.

I thank Tom Marlin, and Brendan J. Hyland (B Eng and Society, '97). With financial support from McMaster University Instructional Development program, they produced detailed versions of over 40 cases, some of which were used in this book.

I am especially indebted to Luis J. Rodriguez, Downstream Oil Company, Waterdown; Douglas C. Pearson, Technical Support Consultant, Parry Sound and Tom Marlin, McMaster University, who gave me feedback and detailed suggestions on the case problems.

Many colleagues supplied me with interesting trouble-shooting cases and information about the cause and perhaps some details about the TS process used to solve the problem. However, in writing up the interactive cases, I had to provide additional information to flesh out the case, provide some red herrings and address a broad range of possible hypotheses so that the fault is not immediately obvious. I have done my best, and any errors in this elaboration are mine.

Waterdown, September 2005 *Don Woods*

# 1
# What is Trouble Shooting?

Process plants operate about 28 days of the month to cover costs. The remaining days in the month they operate to make a profit. If the process is down for five days, then the company cannot cover costs and no profit has been made. Engineers must quickly and successfully solve any troublesome problems that occur. Sometimes the problems occur during startup; sometimes, just after a maintenance turn-around; and sometimes unexpectedly during usual operation.

A trouble-shooting (TS) problem is one where something occurs that is unexpected to such an extent that it is perceived that some corrective action may be needed. The trouble occurs somewhere in a system that consists of various pieces of interacting equipment run by people. The TS "corrective" action required may be:

- to initiate emergency shut-down procedures,
- to forget the situation; it will eventually correct itself,
- to return the situation to "safe-park" and identify and correct the cause and try to prevent a reoccurrence,
- to identify and correct the cause while the process continues to operate under current conditions.

Here are two example TS problems.

Example **Case #1**:
*"During the startup of the ammonia synthesis reactors, the inlet and outlet valves to the startup heater were opened. The pressure in the synthesis loop was equalized. The valves to the high-pressure stage of the synthesis gas compressor were opened and the firing on the start-up heater was increased. However, we experienced difficulty getting the fuel-gas pressure greater than 75 kPa; indeed a rumbling noise is heard if we try to increase the pressure. The process gas temperature is only 65 °C. What do you do?"*

Example **Case #2**:
*"The pipe on the exit line from our ammonia storage tank burst between the vessel and the valve. An uncontrolled jet of –33 °C ammonia is streaming out onto the ground. What do you do?"*

*Successful Trouble Shooting for Process Engineers.* Don Woods

ISBN: 3-527-31163-7

Trouble shooting is the process used to diagnose the fault safely and efficiently, decide on corrective action and prevent the fault from reoccurring. In this chapter we summarize the characteristics of a trouble-shooting problem, give an overview of the trouble-shooting process and "systems" thinking used to correct the fault and present an overview of this book.

## 1.1 Characteristics of a Trouble-Shooting Problem

TS problems share four common characteristics; TS problems differ in their seriousness and when they occur. Here are the details for each.

### 1.1.1 Similarities among TS Problems

TS problems share the following four characteristics: a) exhibit symptoms of deviations from the expected, b) have tight time constraints, c) are constrained by the physical plant layout and d) involve people.

a) Trouble-shooting situations present *symptoms*. The symptoms may suggest faults on the plant or they might be caused by trouble upstream or downstream. The symptoms may be false and misleading because they result from faulty instruments or incorrect sampling. The symptoms might not reflect the *real* problem. For example, in **Example Case #1** the cause is not that the fuel-gas pressure is too low. Instead, the suction pressure of the synthesis gas compressor was lower than normal, the alarms on the cold bypass "low flow" meter had been disarmed and the *real* problem was that there was insufficient process gas flow through the heater.
b) The *time* constraints relate to safety and to economics. Is the symptom indicative of a potential explosion or leak of toxic gas? Should we initiate immediate shutdown and emergency procedures? The release of ammonia, in **Example Case #2**, causes an immediate safety hazard. Time is also an economic constraint. Profit is lost for every minute when off-specification or no product is made.
c) The *process configuration* constrains a trouble shooter. The process is fabricated in a given way. The valves, lines and instruments are in fixed locations. We may want to measure or sample, but no easy way is available. We have to work within the existing process system.
d) Sometimes the cause of the problem is *people*. Someone may not have followed the expected procedure and was unwilling to admit error. Someone may have opened the bypass valve in the belief that "the process operates better that way." As in **Case #1**, the alarm may have been turned off. The orifice plate may have been put in backwards. Someone may have left his lunch in the line during the construction. Instructions may have been misinterpreted.

## 1.1.2
## Differences between TS Problems

Here are four ways that TS problems differ. Some TS problems pose a) safety and health hazards. TS problems can arise b) during startup, c) after a shutdown for maintenance or after a change has been made and d) during usual operations.

## 1.2
## Characteristics of the Process Used to Solve Trouble-Shooting Problems

The TS process or strategy used differs depending on the *type* of TS problem. Yet, the TS process has five common key elements.

### 1.2.1
### How the Type of Problem Guides the TS Process or Strategy

The four different *types* of TS problems (described in Section 1.1.2) call for different TS strategies.

- Handling trouble that poses a hazard

At the design stage engineers should anticipate causes of potentially unsafe and dangerous operation (through such analyses as HAZOP and fault tree) and prevent hazardous conditions from ever occurring. They should include the four elements of control: the usual control, alarms, system interlock shutdown, SIS, and shutdown/relief. However, despite best efforts trouble can occur – such as in Example **Case #2**.

The TS strategy is to recognize unsafe conditions and initiate emergency measures or, where possible, to return the operation to "safe-park" conditions where operation is safe until the trouble is solved.

- Handling trouble during the startup of a new process

When we start up a process or new approach for the first time, we may encounter trouble-shooting problems. However, because these are "first-day" problems they have characteristics that differ from the usual trouble that can occur on an existing process. Hence, a different set of information or experience, and sometimes approach, can be useful. In particular, four events could cause trouble:

1. garbage or stuff left in the lines or equipment,
2. incorrect installation, for example, a pipe hooked up to the wrong vessel,
3. during startup, there are often many people around to get things going correctly – this can interfere with the lines of communication,
4. residual water or air left in process vessels and lines.

Furthermore, although we have theory and often computer simulations to provide ideas about how the plant or process *should* be operating; we have no actual data. Example **Case #1** is a startup problem.

The TS strategy is to focus on the basic underlying principles and create hypotheses about how the process and operations should function.

The financial penalty is usually higher for delays during startup. The penalties include penalties written into the contract for delays, insurance costs and government regulation costs.

- Handling trouble that occurs after a maintenance turnaround or a change.

Changes that can cause faulty operation include

1. equipment is taken apart for maintenance,
2. processing conditions change because, for example, the feedstock is changed,
3. there is a change in operating personnel.

In these examples, we have information about performance *before* and *after* the change.

The TS strategy is to identify the change that seems to have triggered the fault.

- Handling trouble that occurs during usual operation or when conditions change gradually.

Sometimes we encounter trouble when the process is operating "normally" or when we gradually increase the production rate.

The TS strategy is to focus on the basic underlying fundamentals of how the process works, create hypotheses that are consistent with the evidence and use tests to confirm the hypothesis.

### 1.2.2
### Five Key Elements Common to the TS Process

Skill in trouble shooting depends on five key elements: 1) skill in problem solving, 2) knowledge about a range of process equipment, 3) knowledge about the properties, safety and unique characteristics of the specific chemicals and process conditions where the trouble occurs, 4) "system" thinking and 5) people skills. Here are some details about each.

For *general problem solving,* one of the most important skills is in identifying which evidence is significant and how the evidence relates to appropriate hypotheses and conclusions.

Concerning the importance of *knowledge about process equipment,* the differences between skilled and unskilled trouble shooters are more in their repertory of their experiences than in differences in general problem-solving skills. In other words, it is the knowledge about process equipment, common faults, typical symptoms and their frequency that is of vital importance. A trouble-shooter's effectiveness depends primarily on the quality of the knowledge that relates i) symptom to cause; and ii) the relative frequencies of the symptoms and the likelihood of causes.

*Specific knowledge about the chemicals* and equipment configuration must be known to handle safety and emergencies. For example, if knowledge of the hazards of ammonia is not known, then Example **Case** #2 is not treated with the urgency required.

Trouble occurs in a process "*system*" even though it might initially appear as though it is in an isolated piece of equipment. Equipment interacts; people interact with the equipment. Viewing the trouble-shooting problem in the context of a "system" is vital.

*Interpersonal skills* are needed. The interpersonal skills needed between the trouble shooter and the people with whom he/she must interact include good communication and listening skills, building and maintaining trust and understanding how biases, prejudice, and preferences lead to interpersonal differences in style.

## 1.3 Self-Test and Reflections

Reflect on your trouble-shooting skills based on the five common key elements described in Section 1.2.2. Rate yourself-on the five or six elements in each category and then set goals to improve. A rating of 0 means that *nothing is known*. The maximum scale is 10. Descriptions are given for ratings of 1, 5 and 10.

(1) Problem-solving skill as applied to trouble shooting

– Monitoring, being organized and focusing on accuracy: rate: ______

1 = aware that it's important when problem solving. 5 = monitor about once per 5 minutes, use a personal "strategy", tend to let time pressures dominate. 10 = monitor about once per minute, use an evidence-based strategy flexibly and effectively, focus on accuracy, check and double check frequently.

– Data handling, collecting, evaluating and drawing conclusions: rate: ______

1 = think of a variety of data to be collected. 5 = systematically collect data that seem to test the hypotheses, unclear of accuracy of data, unaware of common faults in reasoning, emphasis on opinions. 10 = systematically decide on data to collect and correctly identifies its usefulness; aware of the errors in measurements; use valid reasoning, focus on facts, aware of own biases in collecting data.

– Synthesis: creating and working with hypotheses as to the cause: rate: ______

1 = aware that should have a hypothesis. 5 = can identify several working hypotheses that seem technically reasonable. 10 = can generate 5 to 7 technically reasonable hypotheses for any situation; willing to change hypotheses in the light of new data.

– Decision making: rate: ______

1 = use intuitive criteria. 5 = systematic, consider many options, unaware of any biases. 10 = use measurable *must* and *want* criteria explicitly, prioritize decisions and aware of personal biases and try to overcome these.

(2) Experience with process equipment

– Centrifugal pumps: rate: ______

1 = flow capacity and head, location of inlet and exit, principle of operation. 5 = NPSH and problems related to this, impact of reverse leads on the motor, correct location of the pressure gauge on the exit and the implications of shutting the exit valve, pumps operate on the head-capacity curve and the implications. 10 = implications of worn volute tongue and worn wear rings, lubrication, seals and glands.

– Shell and tube heat exchangers: rate: ______

1 = size area. 5 = size on area and $\Delta p$, baffle-window orientation, correction to MTD for multipass system, some options for control. 10 = tube vibrations, steam traps, nucleate versus film boiling and conditions, different causes of fouling, maldistribution issues and can use a variety of control options.

– Distillation columns: rate: ______

1 = estimate the number of trays, know impact of feed conditions, reflux ratio and bottoms and overhead composition. 5 = familiar with a variety of internals and can size/select, size downcomers, issues related to sealing downcomers, familiar with some control options, can describe the interaction between condenser and reboiler. 10 = jet versus downcomer flooding, surface tension positive vs negative, pump arounds, vapor recompression, wide variety of control options.

(3) Knowledge about safety and properties of material on the processes with which I work
I can list the conditions and species that pose:

– Flammable risk rate: ______

1 = can identify individual species and conditions for five chemicals that might produce “flammable risk”. 5 = can identify individual and combinations of species and conditions for over 30 chemicals and the process faults or failures that might produce “flammable risk”. 10 = can identify individual and combinations of species and conditions for over 100 chemicals and the process faults or failures that might produce “flammable risk”.

– Health risk rate: ______

1 = can identify individual species and conditions for five chemicals that might produce “health risk”. 5 = can identify individual and combinations of

species and conditions for over 30 chemicals and the process faults or failures that might produce "health risk". 10 = can identify individual and combinations of species and conditions for over 100 chemicals and the process faults or failures that might produce "health risk".

– Explosive risk rate: ____

1 = can identify individual species and conditions that might produce "explosive risk" for five chemicals. 5 = for over 30 chemicals and can identify individual and combinations of species and conditions and the process faults or failures that might produce "explosive risk". 10 = for over 100 chemicals and can identify individual and combinations of species and conditions and the process faults or failures that might produce "explosive risk".

– Mechanical risk rate: ____

1 = can identify pressure and moving equipment risk for about five types of equipment. 5 = can identify overpressure, thermal and moving equipment risk for a P&ID with 20 pieces of Main Plant Items, MPI. 10 = can identify overpressure, thermal, corrosive and moving equipment risk for a P&ID with 50 MPI.

– Unique physical and thermal properties: rate: ____

1 = can identify chemicals and conditions that have "unique properties" for one chemical. 5 = for 10 chemicals. 10 = for 30 chemicals.

(4) "Systems" thinking.

– Faulty operation of and carryover from/to upstream/downstream equipment: rate: ____

Can estimate/predict the effects of pulses, cycling, contamination on downstream equipment. Can predict potential sources of pulses, cycling and contamination from upstream equipment. 1 = for one piece of equipment. 5 = for a P&ID with 10 MPI. 10 = for a P&ID with 40 MPI.

– Impact of environmental conditions rate: ____

1 = can estimate the environmental impact for the atmosphere from about 10 main plant items. 5 = for about 20 MPI and atmospheric, aqueous and solid impact. 10 = for about 50 MPI and atmospheric, aqueous and solid impact.

– Pressure profile: rate: ____

1 = can calculate a pressure profile for one pipe from detailed calculations. 5 = can use rules of thumb to estimate the pressure profile for about five piping configurations. 10 = can estimate pressure profiles for a P&ID with interconnecting piping with 50 MPI.

– Process control: rate: ______

0 = Unable to identify and rationalize a process control system. 5 = For a P&ID with 10 MPI, can identify good and bad process control; can identify the presence and absence of four levels of process control (control, alarm, SIS, relief and shutdown). 10 = For a P&ID with 40 pieces of equipment, can identify good and bad process control; can identify on the P&ID the presence of and absence of four levels of process control (control, alarm, SIS, relief and shutdown).

(5) People skills

– Communication skills: rate: ______

1 = write or speak to tell them what you know, use acceptable grammar and follow expected format. 5 = correctly identifies single audience, answers needs and questions; includes some evidence related to conclusions, reasonably well organized with summary, coherent and interesting, defines jargon or unfamiliar words, grammatically correct and follows the expected format and style. Some misunderstanding occurs in some verbal or written instructions. 10 = correctly identify multiple audiences, answer their needs and questions; include evidence to support conclusions, well organized with summary and advanced organizers, coherent and interesting, defines jargon or unfamiliar words, grammatically correct and follows the expected format and style. Verbal and written instructions are carried out correctly.

– Listening skills: rate: ______

1 = listen intuitively. 5 = aware of some elements of listening and usually can demonstrate *attending*. 10 = aware of the characteristics and foibles of listening, skilled at opening conversations, attending, following and reflecting.

– Fundamentals of relationships: rate: ______

1 = handles relationships intuitively. 5 = aware of most of the fundamentals and unacceptable behavior. 10 = claims and respects fundamental rights and avoids using contempt, criticism, withdrawal and defensiveness.

– Developing and building trust: rate: ______

1 = knows a few principles for developing trust; 5 = understands how to develop trust. 10 = can develop mutual trust naturally.

– Building on another's personal preferences: rate: ______

1 = intuitively aware of own preferences and that others are different. 5 = explicitly aware of own preferred style and aware of uniqueness of others but not very effective in exploiting the differences positively. 10 = familiar with my uniqueness and those of my colleagues and use the differences to improve our work instead of promoting conflict.

Total your scores. Identify the areas with the lowest scores and set goals for yourself. For problem solving, see Chapters 2, 5 and 6. For experience with process equipment, see Chapter 3 and Appendix A. For knowledge about safety, see Chapter 3. For "systems thinking", see Chapter 3 and Appendix B. For people skills, see Chapter 7. If you have high scores in all areas, Congratulations. Go directly to Chapter 8 and enjoy!

## 1.4 Overview of the Book

This book is about improving your approach to trouble shooting. This book has basically five parts. Chapters 2 and 3 provide details about the mental process and practical knowledge of common symptoms and causes for a variety of process equipment. Chapter 4 gives some examples of trouble shooters in action as they work through a variety of problems. This is included to give you a chance to reflect on your approach. Chapters 5, 6 and 7 provide example training opportunities to polish your skill in trouble shooting in the areas of problem solving, critical thinking and testing hypotheses and interpersonal skills, respectively. Chapter 8 gives cases that you, the reader, can use to polish your skill. The final chapter suggests the next level of considerations to polish your skill further.

## 1.5 Summary

Trouble-shooting situations present *symptoms*, symptoms that may not reflect the real problem. Trouble shooters are constrained by time and the existing equipment layout. Trouble-shooting situations inevitably include people.

Solving a trouble-shooting problem uses the five elements: skill in problem solving, knowledge about equipment and about hazards, skill in *systems* thinking and people skills.

Problems occur that pose a hazard, when the process is started up for the first time, when the process is started up after change or maintenance or during usual operations or when we are trying to increase the capacity of the process. Slightly different TS strategies are used for the different types of TS problem.

## 1.6 Cases to Consider

Here are five cases. Consider each and write out the approach you would take to start each. For example, you might ask *What is the problem? What questions might I ask? What are the possible causes? What tests might I do? What samples might be taken for analysis?*

**Case #3:** The Case of the cycling column

The shut-down and annual maintenance on the iC4 column has just been completed. When the operators begin to bring the column back on-stream the level in the bottom of the column cycles madly, that is, the level rises slowly about 0.6 m above the normal operating level and then quickly drops to about 0.6 m below normal. The process then repeats. You have been called in as chief trouble shooter to correct this fault. It costs our company about $500/h when this plant is off-stream. Get this column working satisfactorily. The system is given in Figure 1-1.

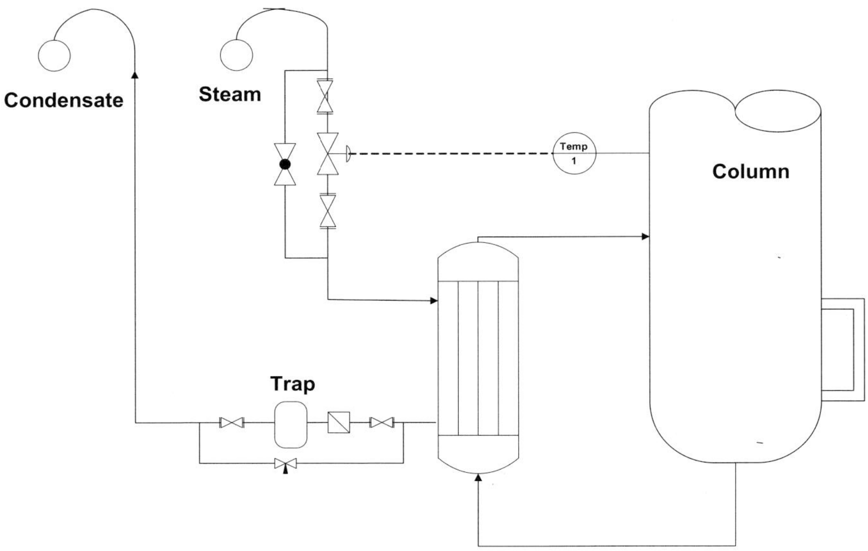


**Figure 1-1** A distillation column for Case #3.

**Case #4:** The case of the platformer fires

Heavy naphtha is converted into high octane gasoline in "Platforming". Byproducts of the reaction include low-pressure gas and hydrogen-rich gas containing 60–80% hydrogen. The products from the platformer reactor (at 4.8 MPa g and 500 °C) are heat exchanged with the feed naphtha to preheat the reactor feed. Figure 1-2 illustrates the layout. In the past three weeks since startup we have had four flash fires along the flanges of the stainless steel, shell and tube heat exchanger. The plant manager claims that because of the differential thermal expansion within the heat exchanger, because of the diameter of the exchanger (1 m), and because it's hydrogen, we're bound to have these flash fires. The board of directors and the factory manager, however, refuse to risk losing the $90 million plant. Although the loss in downtime is $10,000/h, they will not let the plant run under this flash-fire hazard condition. "Fix it!" says the technical manager. Maintenance have already broken six bolts trying to get the flange tighter, but they just can't get the flanges tight enough.

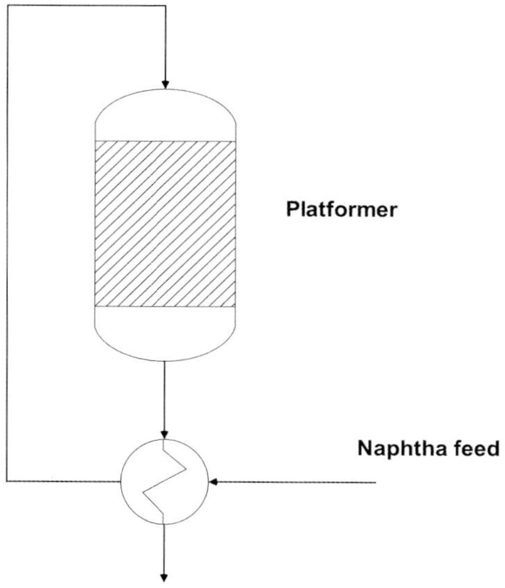


**Figure 1-2** The platformer for Case #4.

**Case #5:** The sulfuric acid pump problem

Dilute sulfuric acid is stored in a horizontal, cylindrical tank in the basement, as is shown in Figure 1-3. The tank diameter is 1.8 m; the length, 3.6 m. An exit line goes from the bottom of the tank and rises 3.6 m up through the ground floor to a centrifugal transfer pump that pumps the acid to a reservoir 7.5 m above the ground level.

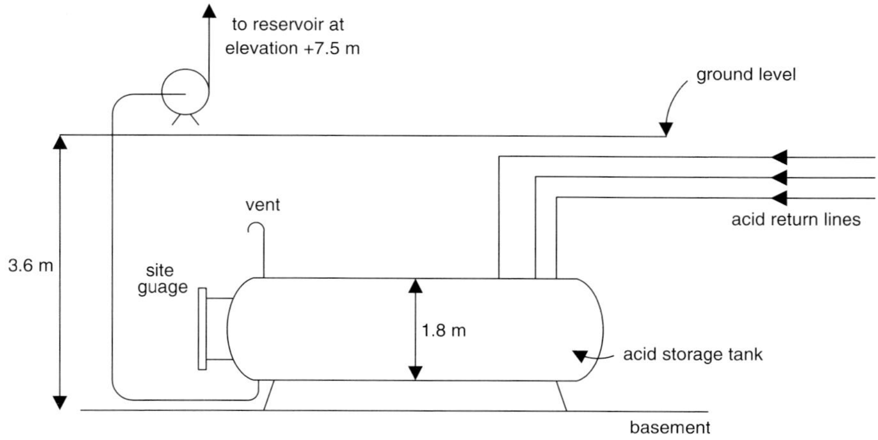


**Figure 1-3** The configuration for Case #5.

Acid is recycled to the tank from different parts of the process at such a rate that about every two hours the pump is activated to transfer the acid to the elevated reservoir. However, each time the transfer pump operates, the level gauge at the side of the tank shows that there is still about 0.7 m of acid in the bottom of the tank when the transfer pump makes a "crackling" noise that the operator says "sounds like cavitation". At this time the operator stops the pump. This means that the transfer pump has to operate more frequently than need be and that cavitation may be eroding the impeller. What do you suggest that I do to fix the problem?

**Case #6:** The case of the utility dryer (courtesy of C.J. King, University of California, Berkeley)
Our plant has a utility air drying unit, which dries all of the utility air used for the pneumatic instrument lines and other purposes. The air is compressed to about 550 kPa g and then passes through the drying unit, the flow diagram of which is attached. For this unit, two beds are hooked in series with the first bed being regenerated and the second bed drying. The first bed experiences two hours of regeneration with hot air followed by a one hour flow of cold incoming air to cool down the regenerated bed. The second bed drys the air for 3 hours. After 3 hours, the flows switch so that the regenerated bed becomes the drying bed and vice versa. The plant operators are following the vendor's instructions in setting the timer dials on the various valves: all the valves (the four way valves, V2 and V3, and the three-way valve, V1) are thrown every three hours. The 3-way valve, V1, is also thrown two hours after a cycle change to send fresh air to cool the regenerated bed. The hot air used to regenerate the bed is heated in a steam heater with the TRC-1 set at 175 °C. The present utility air flowrate through the dryer is 4000 Ndm$^3$/s or about 1/2 design flowrate. The proportionating valve is governed by pressure P3. At present, full pressure is kept on the valve; the valve is shut so that no air goes directly to the dryer bed. The diagram shows the valve settings for Bed A being regenerated and Bed B, drying. All the air flow goes, via the 3-way valve V1 to the steam heater for the regeneration phase of Bed A. The adsorbent is activated alumina with typically 0.14–0.22 kg water adsorbed/kg dry solid. Each bed contains 5000 kg of activated alumina. The available sample valves are labeled "S".

Now that it is winter we have been experiencing much colder nights, and we have encountered several instances where the instrument air lines have been freezing. This has been traced to the air coming out of the drying unit being too wet, on average. We estimate that this problem will cost us about $8,000 per day until we get it fixed. The job is yours – fix it.

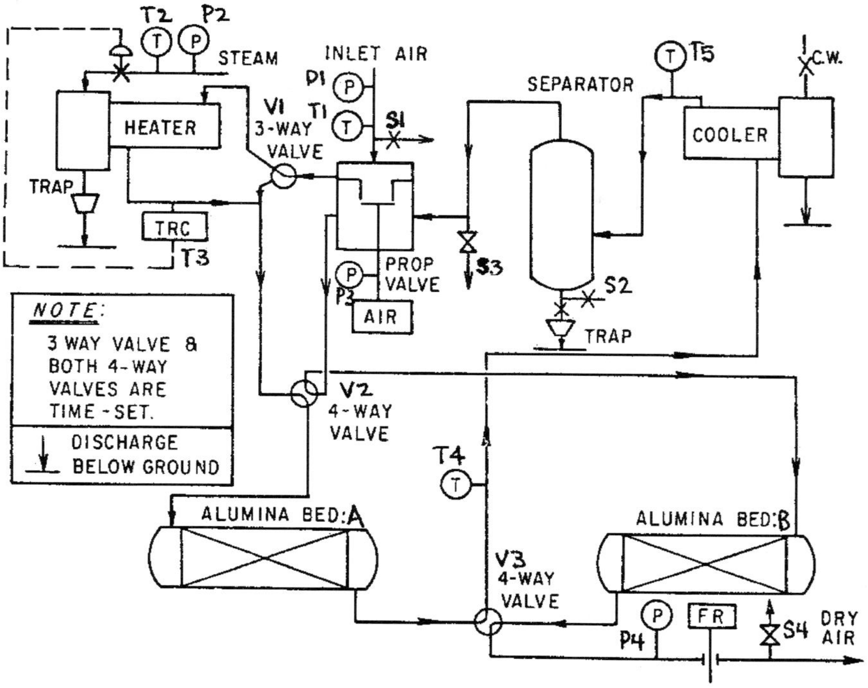


**Figure 1-4** The utility dryer for Case #6.

**Case #7:** The case of the reluctant crystallizer (the case is supplied by W.K. Taylor, B Eng. McMaster, 1966 and used with permission)
Process solution, at 55 °C, enters the vacuum crystallizer (VC) where it is concentrated and cooled to cause precipitation of the product.

Normally, the first and second stage ejectors are used to start syphoning feed solution into the VC until it is two-thirds to three-quarters full. The first hour of operation is done at 6.5 kPa absolute supplied by the first- and second-stage ejectors with city water to the interstage condenser. When the batch cools to 40 °C the booster ejector and the barometric condenser are turned on to give an absolute pressure of 2.5 kPa abs. The batch time is 8 hours during this time the liquid level in the VC slowly drops about 40 to 50 cm. The city water is much colder than the bay water and so to ensure that the temperatures in the barometric leg is less than 26 °C, city water can be used to supplement the bay water. If the booster ejector is turned on too soon, it will not hold but rather *kicks out*. This happens when the steam goes directly into the VC instead of through the ejector nozzle. This phenomena makes a recognizable sound.

Today, the plant operator phones, "The booster does not hold! After about half to one hour of operation it *kicks out*."

While the booster was holding, the liquid level dropped at such a "fantastic rate" that you could actually watch the level drop, whereas it would normally drop 40 to 50 cm over an 8-hour period.

Pressure gauge **P7** indicated a "wildly fluctuating pressure". The needle jumped back and forth from 140 to 550 kPa g while the booster was "holding".

All the other pressures and temperatures were normal. Here is a summary:

| | Pressure, kPa g | | | | | | | |
|---|---|---|---|---|---|---|---|---|
| | Steam | | | | | Water | | |
| | P1 | P2 | P3 | P4 | P8 | P5 | P6 | P7 |
| Normal readings | 550 | 550 | 550 | 725 | 550 | 0–35 | 310 | 205 |
| Today | 550 | 550 | 550 | 725 | 550 | 0–35 | 310 | 140–550 |

| | Temperature, °C | |
|---|---|---|
| | Barometric legs | |
| | T1 | T2 |
| Normal readings | < 27 | < 27 |
| Today | < 27 | < 27 |

Figure 1-5 illustrates the system.

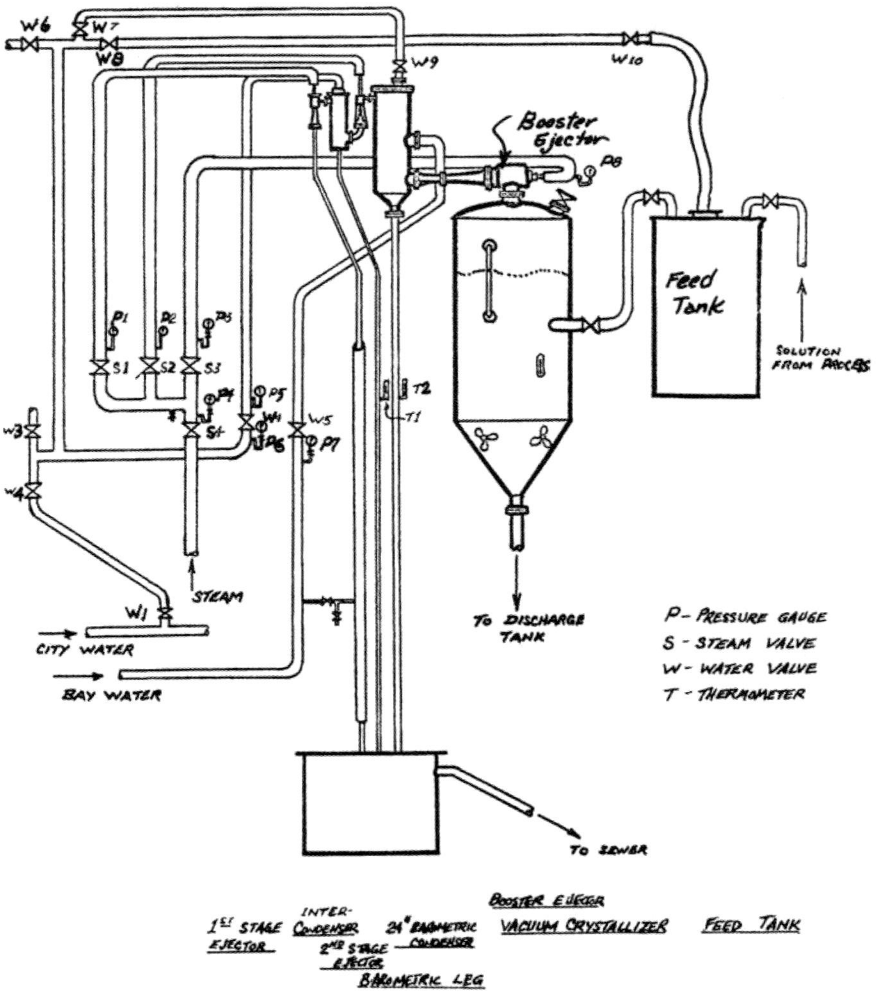


**Figure 1-5** The vacuum crystallizer for Case #7.

Feedback for these cases is provided in Chapter 4.

# 2
# The Mental Problem-Solving Process used in Trouble Shooting

A trouble-shooting problem occurs. As the trouble shooter processes the evidence, he/she mentally scans past experience to see if he/she has successfully solved anything like it before. If that past experience is limited; if there are no past examples that bear any relationship to the current troubling problem, then this is a "problem to solve". The process used will be called *"problem solving."* Problem solving is illustrated in Figure 2-1. Data are gathered and an internal, mental representation is created of the problem situation. That mental representation is compared with past experience to see if a problem similar to this has been solved successfully in the past. If not, then "it's a problem!". We systematically draw on our problem-solving skills, scan our bank of pertinent knowledge and combine our skills and knowledge to "solve the problem". We then elaborate and take time to encode and store that successful solution to the problem in our mental experience bank. In the future, when we encounter a similar problem we don't have to agonize through all of the problem-solving process. We simply recall a solution from our experience in a process we call *"exercise solving"*.

*"Exercise solving"* is illustrated in Figure 2-2. Data are gathered and an internal, mental representation is created of the problem situation. That mental representation is compared with past experience to see if a problem similar to this has been solved successfully in the past. If yes, then "it's an exercise!". The past solution is recalled from experience and modified to solve the current problem.

Beginning trouble shooters with limited experience start their journey as *problem solvers* and gradually build up experience. Experienced trouble shooters are primarily *exercise solvers* who draw on their knowledge and experience. Research evidence suggests that for experienced trouble shooters 95% of the situations they encounter will be "exercises". They still need problem-solving skill for 5% of the situations.

In Section 2.1 we summarize research about problem solving, in general. These characteristics of skilled problem solvers are used in solving any type of problem such as setting goals, making decisions, making a purchase and trouble shooting.

In Section 2.2 additional research evidence into the process of solving trouble-shooting problems is given. In Section 2.3 is given a worksheet or template for solving trouble-shooting problems. Also given is an assessment form to provide feedback about one's performance as a trouble shooter. An example use of the Worksheet is given in Section 2.4 for **Case #8**.

*Successful Trouble Shooting for Process Engineers.* Don Woods

ISBN: 3-527-31163-7

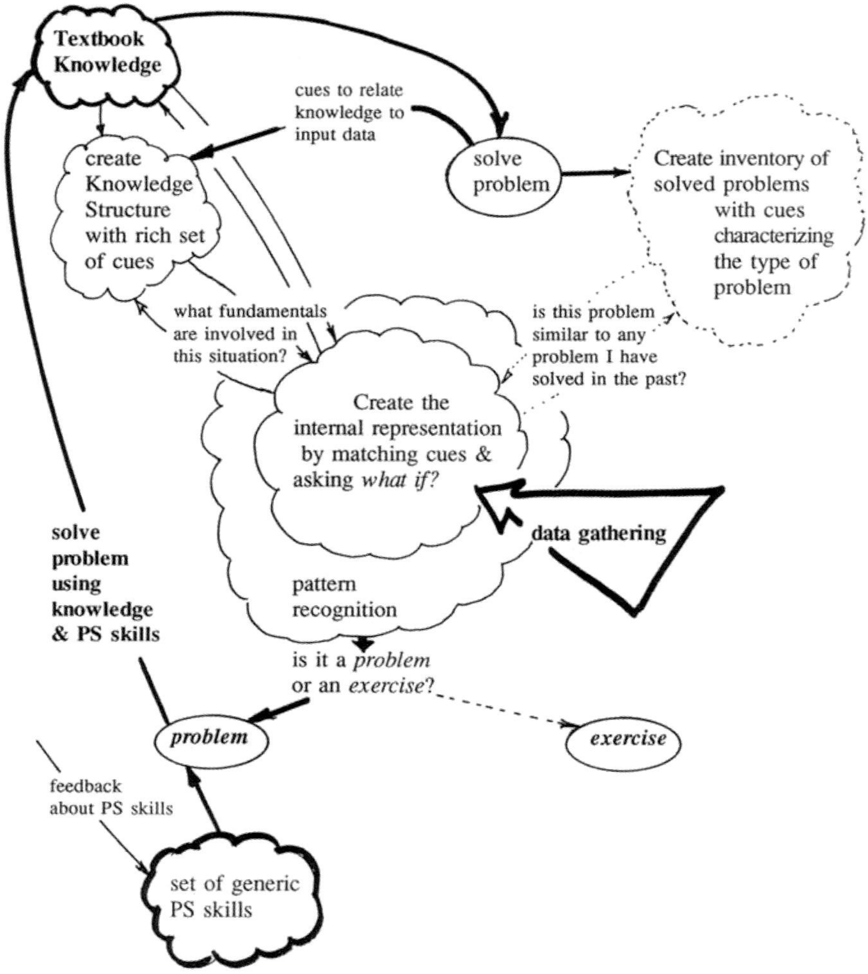


**Figure 2-1** Problem solving.

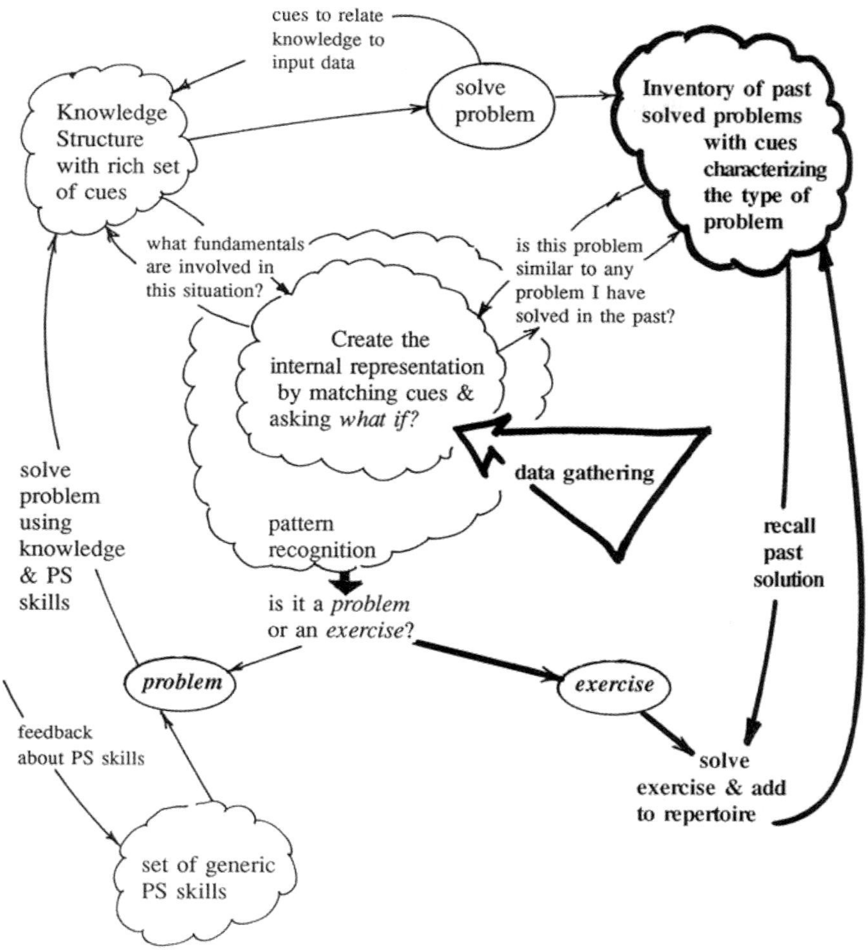


**Figure 2-2** Exercise solving.

## 2.1 Problem Solving

Here are eighteen characteristics of skilled problem solvers. The first eight could be called "the problem-solving process or how"; the second set of characteristics are called "synthesis". The third class is called "decision making". The other important characteristics, "data and analysis", are given in Section 2.2.3. Research has shown that attitudes and other related skills are also important.

The "problem-solving process, how":

1. *Be able to describe your thought processes as you solve problems.*
2. *Be organized and systematic.* An evidence-based strategy for solving problems, in general, is given in Figure 2-3. The six stages are as follows. 1. Engage with the problem or dilemma, listen, read carefully and manage your distress well. Say "I want to and I can!" 2. Analyze the data available and classify it: the "goal, the givens, the system, the constraints and the criteria". 3. Explore: build up a rich visual/mental picture of the problem and its environment; through simplifying assumptions explore the problem to see what is really important; identify the real problem. 4. Plan your approach to solving the problem. 5. Carry out the plan and 6. Check the accuracy and pertinence of your answer. Did it answer the problem? satisfy the criteria? Reflect on the problem-solving process used to discover new insights about problem solving. Elaborate on the answer and the problem situation to discover answers to other problems, to extend the solution to other situations and relate this problem experience to other technical problems you have solved in the past. Cue this experience into memory. This systematic approach is *not* sequential. Skilled problem solvers bounce back and forth between the stages. A typical approach would be engage, analyze, engage, explore, engage, explore, analyze, engage, explore, plan, engage and so on.
3. *Focus on accuracy* instead of speed.
4. *Actively write things down.* Make charts, draw diagrams, write down goals, list measurable criteria and record ideas from brainstorming.
5. *Monitor and reflect.* Mentally keep track of the problem-solving process and monitor about once per minute. Typical monitoring thoughts are "*Have I finished this stage? What have I discovered so far? Why am I doing this: if I calculate* this, *what will* this *tell me? What do I do next? What seems to be the problem? Is this the real problem? Should I recheck the criteria?*" Typical reflections that look back on the process and attitudes used are: "*This didn't work, so what have I learned? Am I focusing on accuracy or am I letting the time pressures push me to make mistakes? Am I managing my stress? I can do this! Am I monitoring the process?* "
6. *Explore the "real" problem by creating a rich perspective of the problem.* During the explore stage, see it from many different points of view. Be willing to spend at least half the total available time defining the problem. Ask many *what if* questions. Try to bound the problem space. "*Swim with the data*" to see how it responds. Identify the *real* problem, by asking a series of *Why?* questions to generalize the situation and to see the problem in the context of a "system". This activity of identifying the real problem was called the Explore stage and is the heart of the problem-solving process.
7. *Identify the subcomponents of the problem, yet keep the problem in perspective.*
8. *Are skilled at creative and critical thinking.*

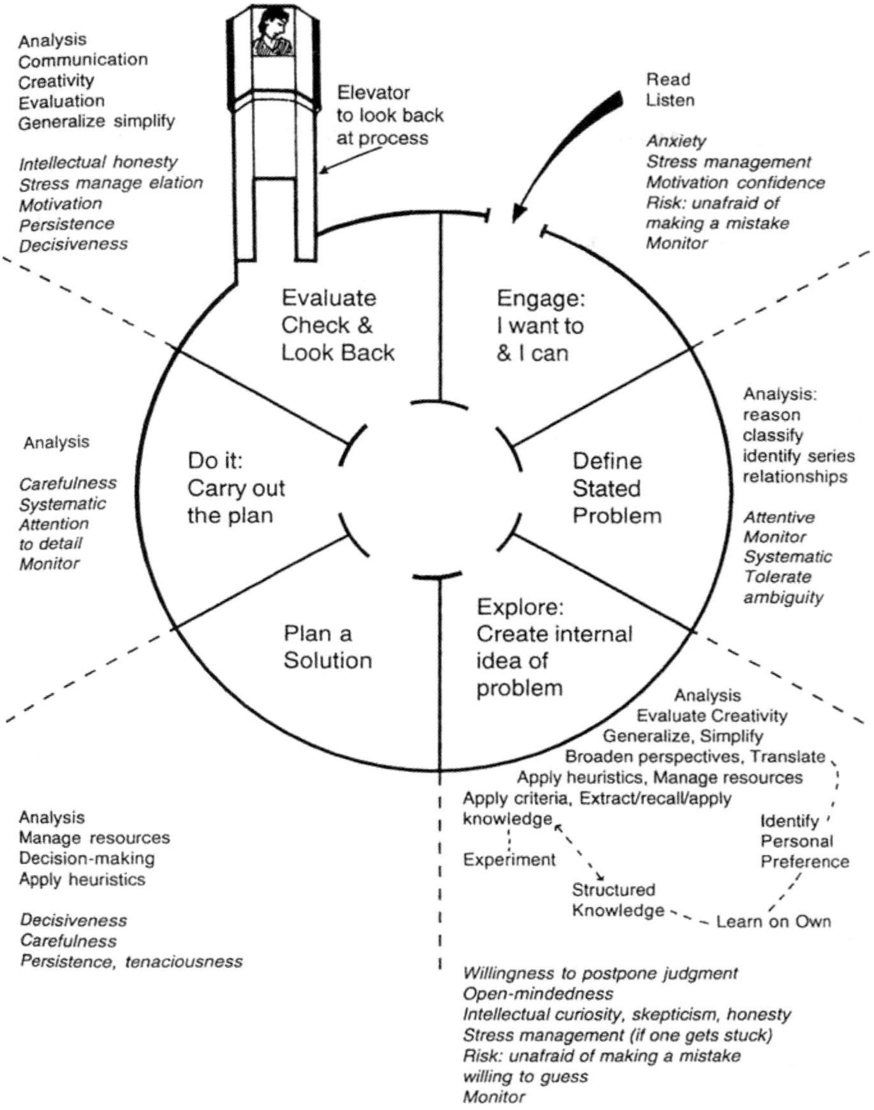


**Figure 2-3** The MPS Strategy for problem solving.

These first eight items we could call "Problem-solving process, how". Table 2-1 lists detracting and enriching behaviors. Activities to help develop these skills are given in Chapter 5.

The next two items are related to "Synthesis" with detracting and enriching behaviors listed in Table 2-1.

9. *Be flexible.*
10. *Keep at least five hypotheses active.* Do not quickly close on one hypothesis.

Issues related to "decision making" are next.

11. *Spend time where it benefits you the most.* Use Pareto's principle (80% of the results can be found from 20% of the effort). Find the key 20%.
12. *Be an effective decision maker.* Express the goal as *results* to be achieved rather than as *actions* to be taken. Make decisions based on criteria that are explicit and measurable. Distinguish between *must* criteria (the process must have an internal rate of return of 35%) and *want* criteria (the process might have the potential to be licensed). Reject options that do not meet the *must* criteria. Use a rating system to score the *want* criteria.

The remaining research evidence relates to "attitude toward problem solving" and some related skills.

13. *See challenges and failure as opportunities for new perspectives.*
14. *Be willing to risk.*
15. *Manage stress well.* Solving problems is stressful. When we initially encounter a problem we experience distress because of the uncertainty. Such stress tends to immobilize us. When we successfully solve a problem we experience the joy and exhilaration of stress (that distracts us from checking and double checking that our answer is the best). A certain level of stress motivates us. Excessive stress makes us make mistakes. Data suggest that operators with confidence and training working under high stress make 1 mistake in 10 actions. Operators with confidence and training who receive feedback about their actions and are under low stress make 1 mistake in 1000 actions. Although these data refer to plant operators, the same trends can be extended to suggest how stress, lack of reflection and feedback might interfere with engineering practice. High stress would be a rating of over 450 on the Holmes–Rahe scale (Holmes and Rahe, 1967).

    Ten suggested approaches to managing stress include: worry only about things over which you have control, include physical exercise as part of your routine, have hobbies and destimulating activities in which you can lose yourself, plan ahead, avoid negative self-talk, rename the events that are stressful to you, build a support system, be decisive, put the situation into perspective and use role models of others who have succeeded.
16. *Manage your time well.* Covey (1990) offers excellent suggestions on time management. Identify problems and decisions according to their importance and urgency. Shift the important situations to being non-urgent. Learn to say "No".
17. *Understand your strengths, limitations and preferred style.* See Section 6.1.3, part c.
18. *For problems involving people, use the 85/15 rule.* 85% of the problems occur because of rules and regulations; 15% of the problems are because of people.

## 2.2
## Trouble Shooting

The general problem-solving characteristics listed in Section 2.1 apply to trouble-shooting problems. However, unique to this type of problem are other research findings. Here we summarize the use of the strategy, in Section 2.2.1, general considerations, in Section 2.2.2 and hypothesis testing in Section 2.2.3.

### 2.2.1
### Considerations when Applying the Strategy to Solve Trouble-Shooting Problems

Two different types of ideas help us focus on our use of the general strategy: problems where there is an apparent change and where there is no apparent change.

**a. Using this Strategy for "Change" Problems**
The overall strategy, described in Section 2.1.1, is applied to identify the change that occurred to cause to trouble. The hypothesis is that the symptoms arise because of some change made to the system. Therefore the plan is to identify the change. The basis of the approach is to learn to ask the right questions. Kepner–Tregoe (1985) illustrate the application of this approach. The questions that usually are most helpful are those that help identify an obvious change.

- what is happening and what should be happening but is not, and "is this difference significant?"
- where is and where is not,
- who is and who is not,
- when is and when is not.

This approach is usually most helpful for "people problems", for problems that occur just after maintenance and for processes that have worked well in the past and now seem to be malfunctioning after a change has been made – in raw material, in operating procedure, in operators, in weather or in season.

**b. Using this Strategy for "Basics" Problems**
The same overall strategy, described in Section 2.1.1, is applied when a change is not obvious. The emphasis is different in that we focus on the basics instead of on a change. The conditions when this apply could be because 1) we are starting up a process for the first time, and we have no practical data of what should be happening or 2) something internal to the process changes and we have no simple way to identify that change. No one ordered the raw materials from a new supplier. No one repaired a pump. No one changed the temperature setting on the heater. Instead, inside the equipment a hunk of corroded metal fell into the liquid; or a truss weakened and gave way inside the vessel. The catalyst bed collapsed. We cannot easily identify the change because there is nothing to "see" from the outside. For this situation, we rely on our fundamental principles and knowledge of how the process and equipment *should* operate, we create hypotheses, check for consistency between the

hypotheses and the evidence and test the hypothesis. The questions that tend to be successful are as follows.

- What basically is going on in this process?
- What fundamentals are important?
- What are the key operating principles that guide the operation of the equipment?
- How are the fundamentals reflected in the observed data?

### 2.2.2 Problem-Solving Processes Used by Skilled Trouble Shooters

Here are the characteristics of skilled trouble shooters especially when we are trying to diagnose the cause when using a "basics" strategy.

1. Generate hypotheses early based on limited cues. Consider the most common hypothesis first.
2. Be systematic and organized: each piece of information requested should relate to an organized plan of attack.
3. Make hypotheses consistent with the evidence: no hypothesis should be more specific or more general than the evidence justifies.
4. Keep two to five competing hypotheses under consideration at any one time.
5. Explore the option of multiple causes especially when evidence suggests a single rare cause. Do not neglect the possibility of two common causes. If two or more common causes would produce disastrous results, and you cannot confirm or refute these causes; act as if both are the cause.
6. Whenever a new or revised hypothesis is generated, check the implications of previous cues.
7. Prioritize test procedures; use the simple, inexpensive ones first before exploring the high-cost option.
8. Do not guess. Use a systematic TS process.
9. Find the root cause; do not correct the symptoms.

Some example data (from the Medical literature, Elstein et al., 1978) are:

- total different bits of information sought/given: 200 per case,
- total bits accumulated before the first hypothesis was generated: 20,
- number of active hypotheses: 3,
- total number of different hypotheses throughout whole case: 6,
- number of cues acquired: 50
- number of critical cues acquired: 50

Table 2-1 describes detracting and enriching behaviors for these activities under the topic *synthesis*.

### 2.2.3
### Data Collection and Analysis: Approaches Used to Test Hypotheses

Successful trouble shooters

- assign a weighting of +++, ++, +, 0, –, – –, – – – to the cues/evidence and then select the hypothesis that maximizes the positive cues or that has the maximum difference between positive and negative cues.
- use Bayes' theorem if the probabilities of various causes are known.
- are sensitive to and try to overcome personal biases (related to *premature closure* and *anchoring*).
- consider the evidence with respect to *all* hypotheses (to overcome the most commonly encountered bias of *pseudodiagnosticity* or *overinterpretation*).
- gather data to disconfirm a hypothesis and are willing to discard a "favored" hypothesis (to overcome *confirmation bias*).
- consistently use fundamentals when analyzing the evidence-hypothesis link (to overcome *representativeness bias*).
- use diagrams, trees and tables to systematically chart hypotheses, cause and evidence (to overcome omitting cues and overcome the limitations of Short-Term Memory).
- restrain from creating a new hypothesis for each new clue and thereby generate excessive data and have trouble with closure.

Table 2-1 lists detracting and enriching behaviors for these activities under the two topics of *data analysis* and *decision making.* Ideas on how to improve this skill are given in Chapters 5 and 6.

## 2.3
## Overall Summary of Major Skills and a Worksheet

The research summarized in Sections 2.1 and 2.2 can be converted into a Trouble-Shooter's Worksheet to guide our approach and an assessment form, to give feedback for growth. These are discussed in turn.

### 2.3.1
### Getting Organized: the Use of a Trouble-Shooter's Worksheet

To help us to be systematic, we use the McMaster 6-step strategy in Figure 2-3 and convert this into a worksheet. A succinct version highlighting the key features is given in Trouble shooters Worksheet 2-1.

**Trouble-Shooter's Worksheet 2-1:** Succinct summary


1. **Engage:** Gather initial information.

- Establish if emergency priority: safety? damage? shut down or safe-park or continue?
- Describe what's going on.
- Manage panic: "*I want to and I can.*"
- Monitor: Have you finished this stage? Can you check? What next?

2. **Define the stated problem:** based on *given* information. If the information is not known at the stage, gather it later.

| | IS | IS NOT |
|---|---|---|
| WHAT | ______________ | (should be happening but it is not) |
| | ______________ | ______________ |
| WHEN | ?☐ ______________ | ?☐ ______________ |
| WHERE | ?☐ ______________ | ?☐ ______________ |
| WHO | ?☐ ______________ | ?☐ ______________ |

Identify situation as 1) startup new process; 2) startup after maintenance or change, 3) usual operation. *Monitor:* Have you finished this stage? Can you check? What next?

3. **Explore**: *Exercise*? or *problem*? Strategy for *change* or *basics*? Useful to broaden with *Why? Why? Why?*

Gather information. Perspectives: customers? suppliers? weather? changed economics? politics? environment?

- Prioritize: product quality, production rate or profit?
- Goal: safe-park? short term? long term? SMARTS$
- Data consistency? Pertinent fundamentals? Likelihood of problem type.
- Explore with *What if?*
- List changes made *and/or* list trouble-shooting experience: root causes based on symptom. (Chapter 3)
- Brainstorm hypotheses
- Hypotheses and evidence of symptoms:

Evidence of symptoms: a. ______________ b. ______________

c. ______________ d. ______________ e. ______________

| Working Hypotheses | Initial Evidence | | | | | Diagnostic Actions | | | |
|---|---|---|---|---|---|---|---|---|---|
| | a | b | c | d | e | A | B | C | D |
| 1 | | | | | | | | | |
| 2 | | | | | | | | | |
| 3 | | | | | | | | | |
| 4 | | | | | | | | | |
| 5 | | | | | | | | | |
| | | | | | | | | | |

**S** = supports; **D** = disproves, **N** = neutral

Diagnostic actions: A. ______________________ B. ____________________

C. __________________________ D. __________________________

4. **Plan**
5. **Do it**
6. **Look back**

**1. Engage**: Take in the evidence. Listen carefully to the phone call. Sense the evidence. Establish priority: Safety? Hazard containment? Equipment damage? If yes, then invoke emergency measures or alter conditions to *safe-park* the process in the interim. *For example, a distillation column might be "parked" by isolating it and continuing to operate under full reflux.* If emergency or safe-park options are not needed, continue. Take time to really understand the physical process. We find it useful to write down a word description of what happens in the process. If a diagram is available, trace around the lines and describe what is flowing in each line, how it is controlled and what should be happening. Once you feel that you have some understanding of the process, manage your distress by saying "*I want to and I can.* I have a strategy that works. Let's systematically follow it."

**2. Define the stated problem** Understand the given information. Classify and chart the information using IS and IS NOT for WHAT, WHERE, WHEN and WHO. List the symptoms as given. Don't guess! Don't overinterpret! Don't infer! Just systematically classify the given information. Some may not be known, so identify this as *information to be gathered.* Note whether this is startup of a new plant, startup after maintenance or change or usual operations. If this is not known, then identify this as *information to be gathered.*

**3. Explore**: Build a rich description of the situation. Gather *information to be gathered* noted in **Define.** See the situation from many different perspectives. Decide if this is an *exercise* or a *problem.* Decide if a more effective strategy might be to focus on *change* or *basics.*

Gain a rich perspective about the situation by considering the problem situation from the following points of view: a) viewpoint of fundamentals and practical operability, b) viewpoint of trouble-shooting rules of thumb: likelihood specific to type of situation or conditions, *example startup* (from Section 3.1), likelihood specific to equipment (Chapter 3 plus experience) c) viewpoint of controlling factors via *What if?* plus order-of-magnitude estimates to bound the behavior and identify key assumptions and variables, d) viewpoint of isolated equipment, equipment in the context of a subsystem, in the context of a mini-system, in the context of the plant site including utilities and in the context of the corporation; e) viewpoint in context of the weather, political, environmental, legal and economic environment; f) viewpoint of *trends* and time changes, g) perhaps broaden the context of the situation by asking *Why? Why? Why?* h) viewpoint of stakeholders: customers, plant manager, operators, vendors, process control, auditors, statisticians, instrument and control specialists, or unit operation specialist.

Then select the "real" goal. Decide priority based on product quality, production rate and profit. Describe the goal and criteria with the aid of the acronym SMARTS$, introduced by colleague Tom Marlin.

The Goal should describe results – not actions – and be expressed in **S**pecific and **M**easurable terms. The Goal should be **A**ttainable. The Goal should produce a **R**eliable and stable result. Use as Criteria: **T**imely: the problem should be solved quickly. Promptness is critical for emergency priorities in safety, containment of hazards and damage to equipment. This was considered in the **Engage** stage. The Goal should produce a **S**afe operation, safe product, safe startup and shutdown. Use as Criteria: **$** The downtime costs, the testing costs and the corrective and preventative costs should be minimized.

Chart the given evidence and symptoms.

Brainstorm hypotheses as to the root cause. Defer judgement. Don't be afraid to list crazy ideas and to spend an extra five minutes dreaming up and building on wild ideas. Remember, often the unique ideas are found in the midst of completely useless ideas.

List five to seven working hypotheses as to the cause. These should be expressed as potential *root* cause rather than *symptom* cause.

Systematically chart and check, based on experience and fundamentals, as to whether the evidence **S**upports, **D**isproves or is **N**eutral toward each hypothesis. Look for obvious flaws and inconsistencies. *For example, the measured data on a refrigeration unit are not consistent with refrigerant data on a pressure–enthalpy diagram.* or *On a pipe, the pressure gauge downstream reads higher than the upstream gauge.*

**4. Plan**

Use the criteria to select a sequence of diagnostic actions. Sometimes several actions can be combined. However, usually it is best to wait for the results from the first action before we recycle back to the **Explore** stage and relook at our hypotheses. The criteria in selecting an action include: will the action provide background information to ensure the problem is understood in context or is the action to test a hypothesis? will the action produce results that give the accuracy needed? is the action simple? inexpensive? safe? Stopping production to "inspect" or "change equipment" is usually very costly.

**5. Do it**

Carry out the first action in the plan. *Check. Monitor.*

**6. Look back**

Compare the results obtained with the hypotheses. Look back at the process used. Self-assess. Return to previous stages of **Engage, Explore, Plan** and continue.

An example of the use of this Trouble-Shooter's Worksheet is given in Section 2.4.

### 2.3.2
### Feedback about your Trouble Shooting

Based on the evidence presented in Sections 2.1 and 2.2, the four things to look for in an effective trouble shooter are: the overall approach to problem solving, data handling, synthesis, and decision making. Table 2-1 summarizes the detracting and enriching behaviors for each. Worksheet 2-2 provides a summary worksheet that can be used by you to self-assess your TS process or can be completed by an observer to give you feedback about your TS process.

**Table 2-1** Summary of detracting and enriching behaviors for trouble shooters.

*Problem solving in general: How you did it*

| Theme | Detracting behaviors | Enriching behaviors |
|---|---|---|
| **Monitors the thought process** | No assessment of potential gain from a question or action. | Asks "What will this get me?" |
| | Unclear of type and purpose of question asked; just asks what pops into mind. | Knows clearly the purpose: ask fishing or shooting questions; whether creating hypotheses or checking for change or gathering information for clarification. |
| | Does not monitor or ask questions as to Why? or implications. | Asks "Am I through?", "Am I finished with this task?", "Where is this leading me?", "This should tell me ..." |
| **Checks and double checks** | Assumes everything is OK. Does not check instruments, diagrams, hardware, procedures. | Checks and double checks instruments; checks if the equipment and lines are as on diagrams. Calibrates and recalibrates instruments. |
| **Is systematic** | Jumps all around, confused, and no apparent plan. | Identifies plan and follows it systematically yet flexibly. Uses tables or charts to keep track of idea flow. |

| | | |
|---|---|---|
| **Subproblems and perspective** | Keeps whole problem and does not identify subproblems. No identification of a strategy. | Breaks overall task into ones of situation clarification, or hypothesis testing and/or identify the change; into emergency action, cause identification, fault correction and future problem prevention. |
| | Confuses issues, factors, fault detection, solutions. | Identifies phases clearly and works through systematically. |
| | Solves a minor fault while the process explodes. | Keeps situation in perspective, does not get lost in a subproblem. |

*Data handling: What you did*

| Theme | Detracting behaviors | Enriching behaviors |
|---|---|---|
| **Data resolution** | Gathers data but does not know what it tells him/her. | Correctly identifies the usefulness of the data collected. |
| | Asks any old question. | Matches *hypotheses* with the *observed evidence* to see if the hypothesis is consistent with the evidence. |
| | Believes all he/she sees and hears; unclear of errors in information. | Explicitly states limitations of the instruments, measurements and checks these systematically. |
| | No data gathered explicitly. Jumps in making corrective action without stating possible hypothesis or cause. | Gathers data for problem clarification and hypothesis testing/or change rather than jumping in with corrective action without any data. |
| | Gathers data expensively, Takes process apart for everything. Overlooks simple ways of gathering information. | Gathers data easily through simple changes in operating procedure, puts controllers on manual. |
| | Asks for samples, but assumes that sample locations and procedures are as usual. | Is present when samples are taken, bottles labelled. |
| | Gives imprecise instructions: "Check out the instrument"; "Open up the exchanger". | Gives precise instructions. |
| **Actions based on fundamentals** | Based on intuition. | Based on fundamentals; estimates behavior based on fundamentals. |
| | | Does mass and energy balances with at least two independent measurements. |
| | | Does pressure profiles through units. |

| | | |
|---|---|---|
| **Reasoning** | Jumps to invalid conclusions. | Draws valid conclusions; tests both positive and negative: what is; what is not; if it does happen; if it does not. |
| | Error in inference: Confirmational bias. | |
| **Completeness** | Uses only part of the information. Doesn't check the design calculations, or data from startup or data from initial, clean fluid; didn't think of human error. | Uses all resources. |

*Synthesis: How you put it all together*

| Theme | Detracting behaviors | Enriching behaviors |
|---|---|---|
| **Hypotheses** | Becomes fixed, thinks of only or selects one hypotheses; selects one at the start and cannot become unfixed. | Keeps at least four working hypotheses; keeps options open as data are gathered. |
| | Makes everything complex. | Keeps it simple, especially if there is a "big failure". |
| | One view. | Many viewpoints: operators, design, human error, instruments, corrosion. |
| | Critical of ideas; limited brainstorming. | Defers judgement when appropriate. |
| **Flexibility** | Considers only a "basics" strategy or a "change" strategy and sticks with it regardless of the evidence. | Selects either a "basics" strategy or a "change" strategy. Shifts strategy when its warranted. |
| | Considers steady state only; considers only the facts | Considers unsteady state as well; considers the people too (the stress they might be under; the environment that allows open discussion; turf fights). |
| **Overall synthesis** | Cannot put all the ideas together into a reasonable story. Becomes fixed on one cause even when evidence points otherwise. | Can put the ideas together into a plausible explanation. |

*Decision making: How you put it all together*

| Theme | Detracting behaviors | Enriching behaviors |
|---|---|---|
| **Priorities** | No priorities for hypotheses; just start anywhere; indeed, may not even create a hypothesis! | Sets and uses priorities. Keeps track and moves from top priority to second. At least four working one hypothesis; keeps options open as data are gathered. |
| | No priorities for gathering evidence; just collects something No priorities about the urgency of the situation; diddles around while the plant explodes; keeps customers waiting until problem completely solved. | Prioritizes; gathers the easy and cheap tests first. Visits the site. Prioritizes urgency; willing to use a contingency plan to get things going safely and later corrects the real fault. |
| **Bias** | Biased, stacks the deck so the favorite fault will be selected even when the evidence refutes it. | Unbiased. Selects either a "basics" strategy or a "change" strategy. Shifts strategy when it is warranted. |
| | Biased: tests for only positive elements. | Tests for both positive and negative. Proves that hypothesis is correct and the other options are not correct. |
| **Overall process** | No criteria used, or if they are, they are not measurable. | Uses measurable criteria to make decisions. |

**Worksheet 2-2:** Summary observation form for feedback about Trouble Shooting.

TS name ______________________ Case _____ Initials ES _____ Obs ___

Rough work area:

**Process: how**
Monitoring ______________________
Checking ______________________
Systematic ______________________
Subs and perspective ______________

**Data/analysis: what**
Data resolution ______________
Fundamentals? ______________
Reasoning ______________
Completeness ______________

**Decision making: how**
Priorities ______________________
Bias ______________________

**Synthesis: what**
Hypotheses ______________
Flexibility ______________

Rating and Feedback

Clarity of Communication

None ☐ | Some ☐ ☐ ☐ | Most ☐ ☐ ☐ | All ☐

Process used:

None ☐ | Some ☐ ☐ ☐ | Most ☐ ☐ ☐ | All ☐

Data collections and analysis:

None ☐ | Some ☐ ☐ ☐ | Most ☐ ☐ ☐ | All ☐

Synthesis:

None ☐ | Some ☐ ☐ ☐ | Most ☐ ☐ ☐ | All ☐

Decision-making:

None ☐ | Some ☐ ☐ ☐ | Most ☐ ☐ ☐ | All ☐

Five Strengths:

______________________________________
______________________________________
______________________________________
______________________________________
______________________________________

Two areas for improvement

______________________________________
______________________________________

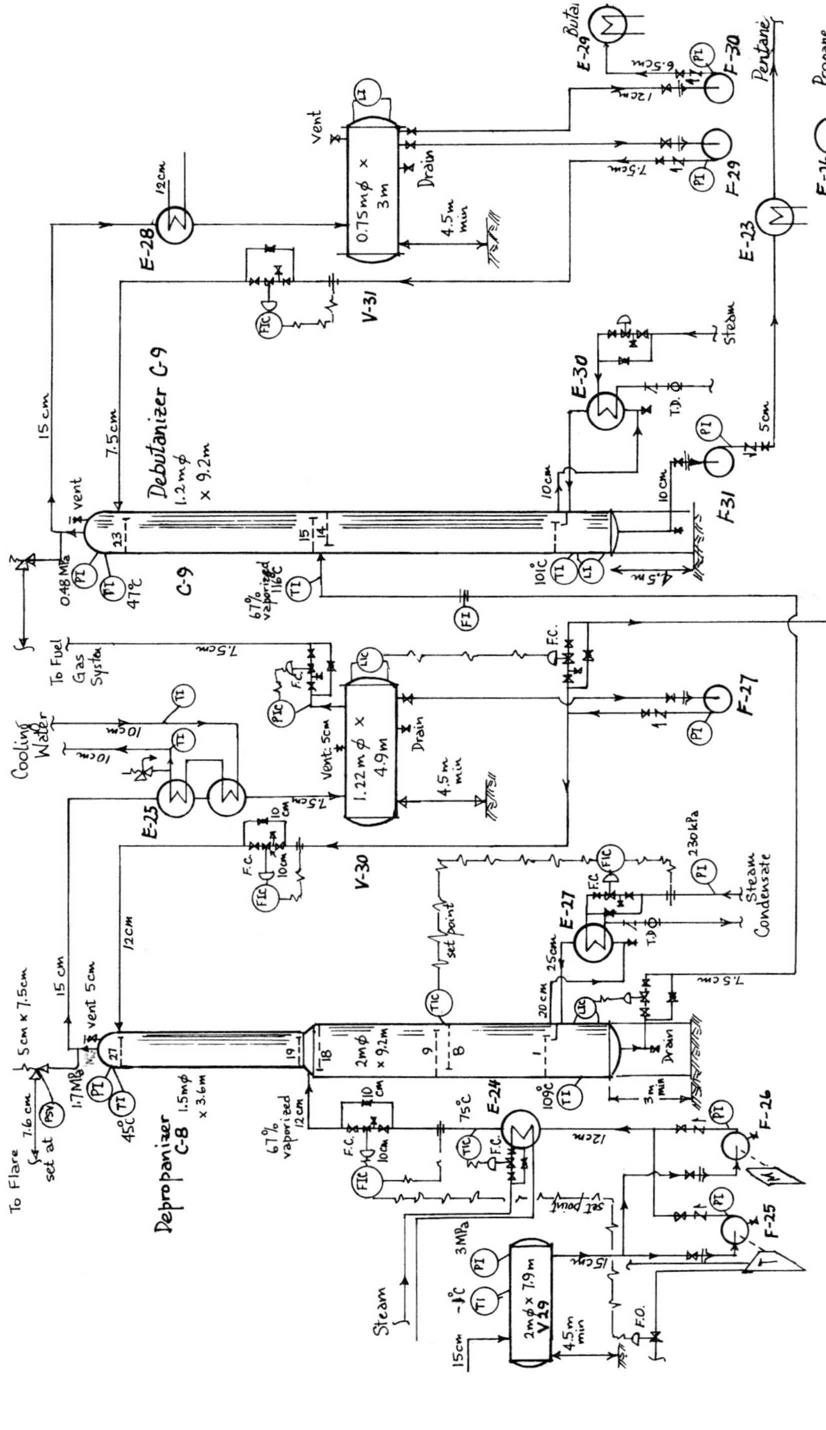


**Figure 2-4** P&ID for the depropanizer and debutanizer.

## 2.4 Example Use of the Trouble-Shooter's Worksheet

**Case #8 Depropanizer:** The temperatures go crazy (used courtesy of T.E. Marlin, McMaster University, Hamilton, Canada)

The process shown in Figure 2-4 is being started up for the first time. It has been running well for several shifts. Just an hour ago, a new operator came on duty and the operator changed the pressure at which the Depropanizer, C-8, is operated, raising the pressure by 0.1 MPa. About 10 minutes after the pressure was increased, the tray temperatures began to go crazy and the bottoms level started to decrease. Figure 2.4 shows the P and ID for this process.

**Trouble-Shooter's Worksheet 2-3: Case #8:** The depropanizer: the temperatures go crazy (© copyright 2003 Donald R. Woods and Thomas E. Marlin)

**1. Engage:** Write down what is said; what you sense, smell, hear. If someone is telling you, then use skilled reflective statements to ensure you accurately obtain the information.

- Emergency priority: Safety? Hazard? Equipment damage?
  shut down □; safe park □. If not □, then:
- Draw a sketch of the process and mark on values. Provide a description in words of what is going on. This is a simple distillation column but all the piping and instrumentation details make it look complex. On the LHS, feed from upstream processing enters drum V29. This feed is pumped (via either a steam driven or motor driven centrifugal pump, F25, 26) through a preheater, E24, and into the depropanizer at tray 18. As the name suggests, the purpose of this column is to take overhead "propane and all lighter species". Let's follow the overhead. The overhead is condensed in two condensers in series, E25, collected in overhead drum V-30 with the non-condensibles (such as methane and hydrogen) removed from the drum and vented to the fuel-gas system. The pressure on the column, C8, is controlled by the valve on the vent system, PV 10. Condensed propane is pumped, F-27, from the drum V-30 forward as product, through product cooler E-26, and returned to the column as reflux. The reflux is flow controlled. Following the bottoms: a thermosyphon reboiler is steam heated. The bottoms flows forward to the next column, the debutanizer. No pump is needed because of the pressure difference between the depropanizer, 1.7 MPa, and the debutanizer, 0.48 MPa. I'm not sure at this stage if this is a control "problem" so I won't elaborate further on the system at this time. I also will focus on the depropanizer, and not explore the debutanizer at this time.

- Manage any panic you might feel by saying "*I want to and I can*. I have a strategy that works. Let's systematically follow it."
- *Monitor:* Have you finished this stage? Can you check? What next? I've systematically followed my way around the flow diagram. I think I understand enough for now.

**2. Define the stated problem:** Systematically classify the *given* information using IS and IS NOT. If the information is not known at the stage check ?☐ to remind you to gather this information

| | IS | IS NOT |
|---|---|---|
| WHAT | tray temperatures "go crazy"; bottoms level decreases. | (should be happening but it's not) tray temperatures steady; bottoms level steady. |
| WHEN | ? ☐ 10 minutes after the pressure in column C8 increased by 0.1 MPa. | ? ☐ before the pressure increase; running well for several shifts. First plant startup. |
| WHERE | ? ☐ depropanizer, C8. | ? ☒ maybe upstream; no information about downstream debutanizer, yet! |
| WHO | ? ☐ new operator. | ? ☐ not with previous operators. |

– startup new process ☒ suggest use Basics
– startup after maintenance or change ☐ suggest use Change
– usual operation but changes made in operation but not in equipment ☐ suggest use Basics
– usual operation ☐ suggest use Basics.

- *Monitor:* Have you finished this stage? Can you check? What next? Yes. I think I've finished.

**3. Explore**: Gather *information to be gathered ?* ☐ in Define stage ☐. Perhaps questions about downstream effects. The decrease in liquid level could be because the flow has increased to the debutanizer (because the $\Delta p$) has increased.

*Exercise?* ☐ or a *problem?* ☒. I haven't seen anything like this before

Strategy: *change* ☐ or *basics* ☒.

Perspectives. *Why? Why? Why?* no information at this time that suggests this might be useful.

*Why?* ↑

*Why?* ↑

*Why?* ↑

---

*Why?* ↑

---

*Why?* ↑

Start → ____________________

- Prioritize: product quality ☒; production rate ☐; profit ☐
- **Goal:** safe-park? ☒: short term now with long term later ☒; long term now ☐

Action to be achieved: **S**pecific terms and **M**easurable:
level out the temperatures and the bottoms level

Attainable? total reflux is a start but I hope it's attainable
Reliable? depends on my short-term solution
Timely? will work on solving it quickly
Safe? cannot think of major hazard now
$ ____________________

- Check consistency of data/symptoms: inter-data consistency?
  OK ☐ no ☐
  data consistent with fundamentals?
  OK ☐ no ☐
- Type of problem: startup new process ☒ maybe mechanical electrical failure
  usual operation ☒: ambient temp? ☐ maybe fluids problems;
  high temperature? ☐ then maybe materials problems
  System? failure of heat exchanger ☐ > rotating equipment ☐ > vessels ☐ > towers ☐
- Identify key and *What if?*

*What if?* temperatures "going crazy" = temperature cycling
then focus on cycling symptoms / causes
*What if?* only temperature "cycling" and no decrease in level
then bottoms and tops temperature and pressures should be cycling too
*What if?* only bottoms level drop and no temperature "going crazy"
then root cause related to bottoms level drop
*What if?* column pressure increases
then condensation temperature at top increases; ΔT condenser increases and condensation should be easier; boiling temperature at bottoms increases; ΔT reboiler decreases so might shift from film to nucleate boiling giving higher heat flux, causing increased boilup or if nucleate to start with then insufficient area and boilup decreases.

- List changes made ☐ *and/or* trouble-shooting causes based on symptom ☒.

"Crazy temperatures" and decreasing bottoms level sounds like a *control* problem.

From Chapter 3, no symptoms listed for bottoms level dropping, but symptoms related to "oversized condenser" and "undersized reboiler" are:

*"Insufficient boilup"*: [fouling on process side]*/ condensate flooding, see steam trap malfunction, Section 3.5 including higher pressure in the condensate header/ inadequate heat supply, steam valve closed, superheated steam/ boiling point elevation of the bottoms/ inert blanketing/ film boiling/ increase in pressure for the process side/ feed richer in the higher boiling components/ undersized reboiler/ control system fault/ for distillation, overdesigned condenser.

But if there was "insufficient boilup", then the bottoms level should be increasing and not decreasing. This doesn't make sense??

For:

*"Cycling of column temperatures:"* controller fault/ [jet flooding]*/ [downcomer flooding]*/ [foaming]*/ [dry trays]* with each of the []* items listed as separate symptoms with their own root causes.

*[jet flooding]*:* excess loading/ fouled trays/ plugged holes in tray/ restricted transfer area/ poor vapor distribution/ wrong introduction of feed fluid/ [foaming]*/ feed temperature too low/ high boilup/ entrainment of liquid because of excessive vapor velocity through the trays/water in a hydrocarbon column.

*[downcomer flooding]**: excessive liquid load/ restrictions/ inward leaking of vapor into downcomer/ wrong feed introduction/ poor design of downcomers on bottom trays/ unsealed downcomers/ [foaming]*

*[foaming]**: surfactants present/ surface tension positive system/ operating too close to the critical temperature and pressure of the species/ dirt and corrosion solids.

*[Dry trays]*:* flooded above/ insufficient reflux/ low feedrate/ high boilup / feed temperature too high.

- Brainstorm root causes: summary of major ideas generated:

change in feed, too much overheads in feed, not enough feed, tray collapsed in stripping section, too much vaporized feed, pump F26 failure, pump F26 cavitates, increased pressure and $\Delta T$ increases, dry tray, flooded in rectification, insufficient reflux, low feedrate, feed temperature too high, boilup too high, too much feed to debutanizer, leak in bottoms, vaporizer flashes 90% (instead of 67%), failure of check-valve on idle pump outlet, boilup controller fault.

*Some of these are symptoms and not root cause, e.g. "not enough feed" "dry trays"*

- Hypotheses: list in Chart; Symptoms: code and list in chart; Analyze with **S** supports; **D** disproves and **N** neutral or can't tell.

Symptom a. 10 min after column pressure increased, column temperatures go crazy
b. 10 min after column pressure increased, bottoms level decreases
c.
d.
e.

| Working Hypotheses | Initial Evidence | | | | | Diagnostic Actions | | | |
|---|---|---|---|---|---|---|---|---|---|
| | a | b | c | d | e | A | B | C | D |
| 1. tray collapsed stripping section | S | S | | | | ✔ | | | |
| 2. too much bottoms fed to debutanizer | N | S | | | | ✔ | | | |
| 3. too much overheads in feed | N | S | | | | ✔ | | | |
| 4. feed valve FV1 stuck | S | S | | | | | | ✔ | |
| 5. pump F-26 not working | S | S | | | | | ✔ | | |
| 6. check valve on idle pump allows backflow | S | S | | | | | | | ✔ |
| 7. | | | | | | | | | |

Diagnostic actions:

A. readings of instruments on column
B. visit site and listen to pump for cavitation
C. visit site and see location of valve stem on FV-1
D. shut isolation valves on idle pump

**4. Plan**
Select "read instruments" as the first task because it is inexpensive and should help test many of the hypotheses. Many of the key variables are displayed in the control room.

**5. Do it**
Go to the control room, notebook in hand.

**6. Look back**

Comment: This example illustrates the approach one might take in being systematic.

## 2.5 Summary

Research on problem solving and on trouble shooting provide key information about the trouble-shooting process. Based on this research the four key features of the process are:

- the problem-solving process: monitoring, checking and double checking, being organized and systematic and keeping the problem in perspective.
- data handling and critical thinking: data gathering and resolution, based on fundamentals, with valid reasoning and being complete.
- synthesis: having five to seven working hypotheses, being flexible and putting it all together well.
- decision-making: based on criteria, priorities and avoiding bias.

A feedback form is given in Figure 2-3. The Trouble-Shooter's Worksheet was created to aid the process. Its use was illustrated.

## 2.6 Cases to Consider

For each of the five cases given in Section 1.6 complete the Trouble-Shooter's Worksheet 2-4.

**Trouble-Shooter's Worksheet 2-4:**
(© copyright 2003 Donald R. Woods and Thomas E. Marlin)

**1. Engage:** Write down what is said; what you sense, smell, hear. If someone is telling you, then use skilled reflective statements to ensure you accurately obtain the information.

- Emergency priority: Safety? Hazard? Equipment damage?
  shut down □; safe park □. If not, □ then:
- Draw a sketch of the process and mark on values. Provide a description in words of what is going on.
- Manage any panic you might feel by saying "*I want to and I can.* I have a strategy that works. Let's systematically follow it."
- *Monitor:* Have you finished this stage? Can you check? What next?

**2. Define the stated problem:** Systematically classify the *given* information using IS and IS NOT. If the information is not known at the stage check ?□ to remind you to gather this information

| | IS | IS NOT |
|---|---|---|
| WHAT | ________________ | (should be happening but it is not) |
| | ________________ | ________________ |
| WHEN | ?☐ ________________ | ?☐ ________________ |
| WHERE | ?☐ ________________ | ?☐ ________________ |
| WHO | ?☐ ________________ | ?☐ ________________ |

– startup new process ☐ suggest use Basics
– startup after maintenance or change ☐ suggest use Change
– usual operation but changes made in operation but not in equipment ☐ suggest use Basics
– usual operation ☐ suggest use Basics.

- *Monitor:* Have you finished this stage? Can you check? What next?

**3. Explore**: Gather *information to be gathered* ? ☐ in Define stage ☐
*Exercise*? ☐ or a *problem*? ☐.
Strategy: *change* ☐ or *basics* ☐.
Perspectives. *Why? Why? Why?*

________________________________________
*Why?* ↑
________________________________________
*Why?* ↑
________________________________________
*Why?* ↑
________________________________________
*Why?* ↑
________________________________________
*Why?* ↑

Start → ________________________________________

- Prioritize: product quality ☐ ; production rate ☐; profit ☐
- **Goal:** safe-park? ☐: short term now with long term later ☐; long term now ☐

Action to be achieved: **S**pecific terms and **M**easurable: ________________
Attainable? ________________
Reliable? ________________
Timely? ________________
Safe? ________________
$ ________________

- Check consistency of data/symptoms: inter-data consistency?
  OK □ no □
  data consistent with fundamentals?
  OK □ no □
- Likelihood of problem: startup new process □
  maybe mechanical electrical failure
  usual operation □: ambient temp? □ maybe fluids problems;
  high temperature? □ then maybe materials problems
  System? failure of heat exchanger □> rotating equipment □>
  vessels □ > towers □
- Identify key and *What if?*

*What if?* ______________________ then ______________________
*What if?* ______________________ then ______________________
*What if?* ______________________ then ______________________

- List changes made □ and/or trouble-shooting causes based on symptom □.
- Brainstorm root causes:
- Hypotheses: list in Chart; Symptoms: code and list in chart; Analyze with **S** supports; **D** disproves and **N** neutral or can't tell.

Symptom a. ______________________________
b. ______________________________
c. ______________________________
d. ______________________________
e. ______________________________

| Working Hypotheses | Initial Evidence | | | | | Diagnostic Actions | | | |
|---|---|---|---|---|---|---|---|---|---|
| | a | b | c | d | e | A | B | C | D |
| 1 | | | | | | | | | |
| 2 | | | | | | | | | |
| 3 | | | | | | | | | |
| 4 | | | | | | | | | |
| 5 | | | | | | | | | |
| 6 | | | | | | | | | |
| 7 | | | | | | | | | |

Diagnostic actions:

A. ______________________________
B. ______________________________
C. ______________________________
D. ______________________________

# 3
# Rules of Thumb for Trouble Shooting

In Section 3.1 we consider the general rules of thumb for processes and different types of problems, instruments and people and the environment. In Sections 3.2 to 3.10 we consider different types of equipment: transportation, energy exchange, homogeneous phase separations, heterogeneous phase separations, reactions, mixing, size reduction, size enlargement and bins. Sections 3.11 and 3.12 consider "systems" and hazards. Guidelines and trouble-shooting rules of thumb are not available for many pieces of equipment. Often, guidelines for good practice are given since trouble can often results from "poor practice". The style used for presenting information about trouble shooting is as follows. The *"symptom"* is shown in italics in quotes. This is followed by the root causes, separated by a slash, / root cause/ root cause. The causes are listed with the most-likely cause first, next-likely cause second and so on. Some causes are not root causes. Such causes are shown as [cause]*. Those causes are listed in square brackets with an *, for example [corrosion]* might be listed as a cause. But what is the root cause of the corrosion? Such "cause–symptoms" are listed separately with their root causes. For example, *[Corrosion]*:* inadequate stress relief for metals/ wrong metals chosen/ liquid flows at velocities > critical value.

## 3.1
## Overall

Consider general rules of thumb and typical causes, rules of thumb about corrosion, for instrumentation and for people, respectively.

### 3.1.1
### General Rules of Thumb and Typical Causes

Gans et al. (1983) suggests that big failures usually have simple causes, such as a compressor that will not start. On the other hand, small failures (or deviations from the norm) often are caused by complex causes, such as the product does not *quite* meet specifications because of a buildup of contaminants.

*Successful Trouble Shooting for Process Engineers.* Don Woods

ISBN: 3-527-31163-7

The general most likely causes for failure differ depending upon whether this is the

- startup of a new process or
- startup after a shutdown and maintenance or
- fault that develops for an on-going, operating process.

*a) Common Faults for First Time Startup.*

The faults encountered are:

| | |
|---|---|
| 75% Mechanical/electrical failures | leaks, broken agitators, plugged lines, frozen lines, air leaks in seals. |
| 20% Faulty design or poor fabrication | unexpected corrosion, overloaded motors, excessive pressure drop in heat exchangers, flooded towers. |
| 5% Faulty/inadequate initial data | often chosen to be the scapegoat by inexperienced trouble shooters. |

*b) Startup after Maintenance*

Ask questions about what specifically was changed, repaired, or modified.

*c) Trouble for On-going Processes*

For ambient temperature operations, about 80% of the problems experienced are fluid dynamical.

For high-temperature operations, about 70% of the problems experienced are materials failure.

Frequency of failures based on type of equipment:

| | |
|---|---|
| 17% | heat exchangers. |
| 16% | rotating equipment: pumps, compressors, mixers. |
| 14% | vessels. |
| 12% | towers. |
| 10% | piping. |
| 8% | tanks. |
| 8% | reactors. |
| 7% | furnaces. |

Another approach is to consult data for Mean Time Between Failures, MTBF. Some example MTBF, in years, are: reciprocating compressors (nonlubricated), 2.8 years; gas turbines, 3.9; centrifugal pumps, 4; screw compressors, 4; reciprocating compressors (lubricated), 4.1; motors, 11.4; large induction motors, 16. See, for example, H.P. Bloch, "Looking for RCTA databases. Consider Failure statistics", Hydrocarbon Processing, Jan 2002, p 36.

### 3.1.2
### Corrosion as a Cause

Corrosion could cause trouble. Here are some general ideas about corrosion and some details about trouble shooting.

General ideas:

1. The strength of materials depends totally on the environment in which the materials function and not on the handbook values.
2. All engineering solids are reactive chemicals – they corrode.
3. The usual eight forms of material failure are: 1) uniform corrosion: uniform deterioration of the material (32%); 2) stress corrosion: simultaneous presence of stress and corrosive media (24%); 3) pitting: stagnant areas with high halide concentration (16%); 4) intergranular corrosion: most often found in stainless steels in heated areas (14%); 5) erosion: sensitive to high flowrates, local turbulence with particles or entrained gas bubbles. For flowing gas-solids systems the rate of erosion increases linearly with velocity and depends on abrasiveness of particles (9%); 6) crevice corrosion: concentration cells occur in stagnant areas (2%); 7) selective leaching or dealloying: removal of one species from a metallic alloy (1%) and 8) galvanic corrosion: dissimilar metals coupled in the presence of a solution with electrolyte (negligible).
4. Stress corrosion (the second most significant form of corrosion) can start from perfectly smooth surfaces, in dilute environments in material with stresses well below the yield stress.
5. >70% of stress corrosion cracking is related to residual – not applied – stresses.
6. The penetration of stress corrosion cracking as a function of time depends on the alloy composition, structure, pH, environmental species present, stress, electrochemical potential and temperature.

**Trouble shooting:**

*"High concentration of metals (Fe, Cr, Ni, Cu) in solution":* [corrosion]*/ contaminants from upstream processing.

*"Ultrasonic monitoring shows thin walls for pipes, internals or vessels":* faulty ultrasonic instrument/ [corrosion]*/ faulty design. *"Failure of supports, internals, vessels":* [corrosion]*/ faulty design/ unexpected stress or load.

*"Leaks":* [corrosion]*/ faulty installation/ faulty gasket/ faulty alignment.

*[Cavitation in pumps]*:* pump rpm too high/ suction resistance too high/ clogged suction line/ suction pressure too low/ liquid flowrate higher than design/ entrained gas.

*[Corrosion]*:* [corrosive environment]*/ inadequate stress relief for metals/ wrong metals chosen/ liquid flows at velocities > critical velocity for the system; for *amine circuits:* > 1 m/s for carbon steel and > 2.5 m/s for stainless steel/ large step changes in diameter of pipes/ short radii of curvature/ flange or gasket material projects into the pipe/ [cavitation in pumps]*/ improper location of control valves.

*[Corrosive environment]**: temperature too high; *for amine solution*: > 125 °C/ high dissolved oxygen content in liquid/ the liquid concentration differs from design; *for steam*: trace amounts of condensate or condensate level in condensers > expected; *for 316 stainless steel*: trace amount of sodium chloride; *for sulfuric acid*: trace amounts of water diluting concentrated acid; *for amine absorption*: total acid gas loadings > 0.35 mols acid gas/mol MEA, > 0.40 mols acid gas/mol DEA, > 0.45 mols acid gas/mol MDEA; makeup water exceeds specifications; *for amine absorption* units exceeds: 100 ppm TDS, 50 ppm total hardness as calcium ion, 2 ppm chloride, 3 ppm sodium, 3 ppm potassium and 10 ppm dissolved iron; *for sour-water scrubbers*: cyanides present/ pH change/acid carryover from upstream units/ high concentration of halide or electrolyte/ presence of heat stable salts/ bubbles present/ particulates present/ invert soluble precipitates with resulting underlying corrosion/ sequence of alternating oxidation-reduction conditions.

### 3.1.3
### Instruments, Valves and Controllers

Trouble-shooting sensors: Most sensor faults are because of improper selection, incorrect installation or adverse environmental conditions. *"Wrong signal"*: fouled or abraded sensors/ bubbles or solid in fluid/ sensing lines plugged or dry/ electrical interference or grounding/ sensor deformed/ process fluid flow < design or laminar instead of turbulent flow/ contamination via leaky gaskets or O-rings/ wrong materials of construction/ unwanted moisture interferes with measurement or signal/ high connection or wiring resistance/ nozzle flappers plugged or fouled/ incorrect calibration/ orifice plate in backwards/ orifice plate designed incorrectly/ sensor broken/ sensor location faulty/ sensor corroded/ plugged instrument taps: *for sour-water strippers*: water or steam purge of taps malfunctioning or local temperatures < 82 °C at which ammonium polysulfides form. *"Wrong input"*: sensor at wrong location/ insufficient upstream straight pipe for velocity measurement/ feedback linkages shift or have excessive play/ variations in pressure, temperature or composition of the process fluid. *"Fluctuating signal"*: bubbles in the liquid/ flashing because $\Delta p$ across an orifice plate > design or fluid too close to the boiling point causing cavitation.

Trouble-shooting control valves: The signal to the valve should be midrange; otherwise the signal depends whether the valve is fail open or fail close. *"Leaks"*: erosion/ corrosion/ gaskets, packing or bolts at temperatures, pressures and fluids that differ from design. *"Can't control low flowrate"*: miscalibration/ buildup of rust, scale, dirt/[faulty design]*. *"Can't stop flow"*: miscalibration/ damaged seat or plug. *"Excessive flow"*: excessive $\Delta p$. *"Slow response"*: restricted air to actuator/ dirty air filters. *"Noise"*: cavitation/ compressible flow. *"Poor valve action:"* dirt in instrument air/ sticky valve stem/ packing gland too tight/ faulty valve positioner. *"Cycling"*: stiction. *[Faulty design]**: valve stem at design flowrate is not at midrange. *[Stiction]** is the sticking and friction related to valve movement and measured as the difference between the driving values needed to overcome static friction upscale and downscale. Likely cause of small amplitude, continuous cycling: gland too tight/ insufficient driving force/antistiction coating damaged/ faulty valve positioner/ poorly tuned control system/ incorrect valve/ incorrect actuator.

Trouble-shooting transmitters: *"Erratic or fluctuating output":* vibrations/ improper orientation/ loose connections.

Trouble-shooting block valves: Check that the arrow on the valve is in the same direction as the flow through the valve. Test via *"turn and seal"* to check movement and closure. *"Reduced flow":* valve not fully open/ plugged with dirt. *"Poor control by the control system":* block valve on bypass partially open/ block valves upstream or downstream of the control valve partially closed.

Trouble-shooting check valves: *"Noisy":* backpressure too high. *"Reduced flow":* backpressure too high.

Trouble-shooting control systems: *"Oscillation":* feedforward/ poorly tuned/ valve sticks or has excessive hysteresis. *"Returns with offset":* proportional control only. *"Two related variables start to deviate":* lack of ratio controllers/ failure to relate analyses to flows.

### 3.1.4
### Rules of Thumb for People

Nine suggestions are given.

1. Become aware of your own uniqueness and personal style, and how you might differ from the style of others.
2. Honor the seven fundamental rights of individuals, RIGHTS. **R**, to be Respected; **I**, Inform or to have an opinion and express it; **G**, have Goals and needs; **H**, have feelings and express them; **T**, trouble and make mistakes and be forgiven; **S**, select your response to others expectations and claim these rights and honor these in others.
3. Avoid the four behaviors that destroy relationships: Contempt, Criticism, Defensiveness and Withdrawal/ stonewalling.
4. Trust is the glue that holds relationships together. Three elements of trust are benevolence, integrity and competence.
5. Build trust by benevolence through loyalty to others, especially when they are not present and by not doing anything that would embarrass or hurt them. Build trust by integrity by *keeping* commitments to yourself-and others; *clarifying* expectations that you have of yourself-and of others; *showing* personal integrity, and honesty; *apologizing* promptly and sincerely when you know you are wrong; *honoring* the fundamental RIGHTS listed above and avoiding the killers; *listening and understanding* another's perspective; *being truthful;* and *accepting others* "warts and all". Build trust by being competent in your area of expertise.
6. Destroy trust by the reverse of the Builders of trust listed above, and by selectively listening, reading and using material out of context; not accepting the experience of others as being valid; making changes that affect others without consultation; blind-siding by playing the broken record until you've even-

tually worn them out or subtly make changes in the context/issues/wording gradually so that they are unaware of what is happening until it is too late.

7. The 12:1 rule applies to rebuild relationships. 12 positive experiences are needed to overcome 1 negative experience.
8. To improve and grow we need feedback about performance. Give feedback to others to encourage and help them; not for you to get your kicks and put them down. Focus on five strengths for every two areas to improve on.
9. Be skilled at responding assertively. *"When you... I feel .. adjust by..."* .

### 3.1.5
### Trouble-Shooting Teams

Use Worksheet 3-1 after each meeting, set goals and celebrate achievement. Use the framework developed by Francis and Young (1979) for growth; consult Fisher et al. (1995) for more short-term ideas.

#### Trouble-shooting team meetings

These are organized by symptom with possible corrective responses suggested for chair {C} and member {M}.

#### Problems with Purpose and Chairperson

*"No apparent purpose for the meeting"*: {C} don't have a meeting. {M} question the purpose of the meeting. See also Agenda and timing problems.

#### Agenda and Timing Problems

*"No agenda"*: {M}: phone {C} and ask for agenda. Invoke "no agenda, not attendance."/at meeting: "Perhaps the first thing we should do is to create an agenda." / After 5 minutes, "We seem to be lost. Could we draw up an agenda and follow that?"

*"Meeting drags on and on"*: {C} should have circulated an agenda with times for each item and used the 20 minute rule/ {M} "Perhaps we can follow the agenda."/ {M} indicate to {C} ahead of time the amount of time you have available for the meeting and then leave at that time.

*"Get off the track"*: {M} seek direction, purpose, summary of progress. see also Behavior problems: *"Subgroups interrupting and talking"*.

*"Group gets bogged down"*: state problem/ summarize/seek agenda clarification/ invoke 20 min rule.

*"Decisions made just at the end of the meeting"*: state frustration/ suggest tabling/ suggest future corrective way to handle in future. See also Agenda and chairperson problems.

#### Behavior and Participation Problems

*"People come into a meeting cold:"* {C or M} suggest reconvene meeting when all are prepared.

*"Late arrivals":* {C} start meeting on time and continue with the agenda through the disruption of the retardee/ {M} "I realize that not everyone is here but I suggest that we start. It looks like a long agenda to get through."

*"Some people do all the talking and some remain silent":* wrong membership/ encourage quiet ones to contribute/ ask each, in turn, to summarize his/her point of view/ ask a "safe" question of the silent ones/ privately check with the silent ones and reevaluate whether they need to attend/ ask open ended questions/ use nominal group.

*"Sub groups interrupting and talking":* identify problem/ suggest discussing one issue at a time and add subgroup's issues to agenda/ be silent until the side conversation stops. "Thank you." / Interrupt the side conversation.

*"Indecisive members, continual question asker":* ask for their ideas early/ redirect questions he/she asks back to him/her.

#### Conflict or Apparent Conflict

*"Conflict because of differing views:"* restate the importance and value of everyone's opinions, restate the RIGHTS/ attempt to bring conflict into the open/ summarize different views/ focus on different performance or opinions and not personalities/ remind of fundamental RIGHTS.

*"Conflict over facts:"* stop the argument, identify problem as you see it and check that that is a problem/ identify facts we need clarified and probable expert.

*"Conflict over values, goals, criteria, process or norms:"* stop discussion, identify problem as you see it and check that that is a problem/ use problem solving.

*"Resistance to new ideas, we tried that before, it won't work, over my dead body, we don't have the resources":* surface the resistance/ honor the resistance/ invoke consequence of no decision or of repeating what we've always done before/ use consensus building techniques/ reflect on the home turf of the objector and the impact the decision might have on them; explore if this might be brought to the group as an issue to address/ root cause of most resistance is fear of change, apathy, vested interests, not invented here, negativism, overwhelmed by enormity of proposal.

**Worksheet 3-1:** Rating form for teams

Assessment of your team and team meeting Name: ____________________
Date: ____________________

Purpose of team: ____________________________________________ unclear ☐
Purpose of this meeting ____________________________________________ unclear ☐

Agenda for this meeting: detailed, clear and circulated ahead of time ☐
bare minimum circulated ahead of time ☐
none ☐

Three-minute team task to seek consensus about the rating of the Task and Morale:

- **Teamwork: Task** all members clear about and committed to goals; all assume roles willingly; all influence the decisions; know when to disband for individual activity; all provide their unique skills; share information openly; the team is open in seeking input; frank; reflection and building on each other's information; team believe they can do the impossible; all are seen as pulling their fair share of the load.

The degree to which these descriptors describe your team's performance (as substantiated by evidence: meetings, engineering journal, interim report, presentations).

| None of these behaviors | Few of these behaviors but major omissions | | | Most features demonstrated | | All of these behaviors |
|---|---|---|---|---|---|---|
| ☐ | ☐ | ☐ | ☐ | ☐ | ☐ | ☐ |
| 1 | 2 | 3 | 4 | 5 | 6 | 7 |

- **Teamwork: Morale:** Trust high, written communication about any individual difficulties in meeting commitments; cohesive group; pride in membership; high *esprit de corps*; team welcomes conflict and uses methodology to resolve conflicts and disagreements; able to flexibly relieve tension; sense of pride; *we* attitude; mutual respect for the seven fundamental rights of all team members; Absence of contempt, criticism, defensiveness and withdrawal.

The degree to which these descriptors describe your team's performance (as substantiated by evidence: meetings, engineering journal, interim report, presentations).

| None of these behaviors | Few of these behaviors but major omissions | | | Most features demonstrated | | All of these behaviors |
|---|---|---|---|---|---|---|
| ☐ | ☐ | ☐ | ☐ | ☐ | ☐ | ☐ |
| 1 | 2 | 3 | 4 | 5 | 6 | 7 |

Each, in turn, gives a 30-second summary of his/her perception of his/her contribution. This is presented *without discussion.*

**Individual, 30 second reporting of his/her contribution to this meeting:**

________________________________________

________________________________________

________________________________________

________________________________________

Four-minute team task to reach consensus about the five strengths and the two areas for growth.

| Strengths of your team | Areas to work on for growth |
|---|---|
| ________________ | ________________ |
| ________________ | ________________ |
| ________________ | |
| ________________ | |
| ________________ | |

D.R. Woods (2005)

This form should be completed after each meeting and copies used as evidence of growth.

## 3.2 Transportation Problems

**Fundamentals of why fluids move:** Fluids move from high pressure to low pressure, vertically because of gravity force, dragged along by a moving boundary or belt or because of density differences; won't flow out of a sealed vessel or vacuum unless there is a vent break. These are expressed, on the macroscopic level, as Bernoulli's equation. Most trouble-shooting problems encountered are fluid-dynamical problems. Centrifugal pumps are often selected to pump liquids; such pumps operate on their head-capacity curve showing decreasing head with increasing capacity. For pumping liquids, cavitation usually occurs when pumping hot liquids near their boiling temperature or when sucking liquids out of a sump. Whenever plants startup for the first time or after a shutdown, wood, sandwiches, bolts and other crud could have been left unintentionally in the lines. Symptoms and possible causes for specific pieces of equipment are presented as follows: gas-moving equipment for pressure, Section 3.2.1 and for vacuum service, Section 3.2.2. Pumping liquids are considered in Section 3.2.3; pumping solids, Section 3.2.4. Considerations for steam are given in Section 3.2.5.

### 3.2.1
### Gas Moving: Pressure Service

Fans, blowers and centrifugal and reciprocating compressors.

**Fans:** Trouble shooting: *"Noise"*: vortex, flow separation/ loose bearings. *"Discharge pressure low"*: instrument error/ fans in series rotating in the same direction/ operating below the stall point/ density increase. *"Low flowrate"*: instrument error/ flow separation/ pitch angle of blades too shallow/ speed slow/ required system discharge high.

**Blowers:** For rotary lobe: when used for pressure pneumatic conveying install a check valve in the blower discharge. Trouble shooting: *"Discharge pressure high"*: instrument error/ restriction in downstream line/ check valve jammed in closed position/ dirty intake filter. *"Discharge pressure low":* instrument error/ slippage of the drive belts/ relief valve stuck open/ increasing air loss at the rotary valve due to larger clearance opening from wear/ loss of air caused by larger lobe clearance in the blower due to wear/ a leak, such as a ruptured hose, in a vacuum system/ a ruptured bag in the downstream bag house.

**Centrifugal compressors:** Good practice: allow safety margins of design speed 5%, design head 10% and design power 15%. The sonic velocity decreases with an increase in gas molar mass. Trouble shooting: *"Surging"*: insufficient flow/ increased discharge pressure required by the system/ deposit buildup in diffuser. *"Discharge pressure low"*: instrument error/ compressor not up to speed/ excessive inlet temperature/ leak in discharge system. Provide separate anti-surge system for compressors operating in parallel; need careful design of suction piping for double flow compressors.

**Reciprocating-piston compressors**: Good practice: design velocity through valves < 40 m/s. Trouble shooting: major faults: valves and piston rings. *"Knocking"*: frame lubrication inadequate/ head clearance too small/ crosshead clearance too high; *"Vibration"*: pipe support inadequate/ loose flywheel or pulley/ valve LP unloading system defective. *"Discharge pressure high"*: instrument error/ valve LP unloading system defective / required system discharge high. *"Discharge pressure low"*: instrument error/valve LP unloading system defective/ LP valve worn/ system leakage. *"Discharge temperature high"*: instrument error/ LP valve worn/ valve LP unloading system defective/required system discharge pressure high. *"Cooling-water temperature high"*: instrument error/ water flowrate low/fouled area/LP valve worn. *"Valve temperature high"*: instrument error/required system discharge pressure high/ run unloaded too long/LP valve worn. *"Cylinder temperature high"*: instrument error/ required system discharge pressure high/LP valve worn/ wrong speed. *"Flow low"*:

instrument error/LP valve worn/valve LP unloading system defective/dirty suction filter.

### 3.2.2 Gas Moving: Vacuum Service

Liquid piston pump, dry vacuum pump and steam ejectors.

**Liquid piston pump:** Good practice: cool the seals with 0.03 L/s of clean cooling water at pressure at least 35 kPa greater than discharge pressure of pump. Trouble shooting: *"Noisy"*: service liquid level too high/coupling misaligned. *"Capacity low"*: suction leakage/ service liquid temperature too high/ speed too low/ seal water flowrate < design. *"Power excessive"*: service liquid level too high/ coupling misaligned. *"Service liquid temperature high"*: clogged strainer/ partially closed valve/ fouled heat exchanger.

**Dry vacuum pump:** Good practice: size for usual discharge pressure 20–35 kPa gauge to allow for downstream discharge. Vacuum pumps run hot: 50–70 °C. Allow 30-min warmup period before putting on-line. Allow 60 min purge before shutdown. Try not to have the pump discharge into a common header. Multistage pumps tend to run cooler than single stage. Install a check valve on the discharge. If the discharge pressure is > 35 kPa, add a positive displacement blower (designed for 6 $Ndm^3$/ s at design conditions for the vacuum pump) with a bypass that is open for startup.

Trouble shooting: *"Loss of vacuum"*: condensation in the suction line/condensation of species from other units connected to a common exhaust header/ increase in discharge pressure from restriction in downstream processing or pressure blowout in other units connected via common discharge header. *"Excessive corrosion"*: for systems handling acid gas or connected to such systems via common discharge header: warmup period too short/ shutdown purge too short. *"Overheating"*: low cooling-water flow/ fouled cooling system/ inlet gas temperature > 70 °C. *"High amps."*: buildup of polymer caused by operating temperature too high/ polymerizable species gain access via common discharge header.

**Steam ejectors**: Good practice: operability of steam ejectors is very sensitive to the stability in the motive fluid (steam) pressure. Prefer vacuum pumps to steam ejectors. Keep diameter of pipes = diameter of inlet and discharge flanges of ejectors. For distillation columns, as the column overhead mass flowrate increases above design, so will the column overhead pressure and vice versa. Compression ratios per ejector: 6:1 to 15:1. If the inlet gas temperature < 0 °C or below the triple point of water (0.61 Pa) then add steam jacketing to cope with ice formation. Seal for the hot well: submerge > 30 cm. The volume in the hotwell between the pipe and the overflow weir should be 1.5 times the volume in the down spout sealed. Replace any nozzles or diffusers where the area is >7% larger than design.

Trouble shooting: check the last stage first and then move upstream. *"Unstable operation or loss of vacuum"*: steam pressure < 95% or > 120% of design/ steam superheated > 25 °C/ wet steam/ inlet cooling-water temperature hot/ cooling-water flowrate low/ condenser flooded/ heat-exchange surface fouled/ 20–30% higher flow of non-condensibles (light end gases, air leaks or leaks from fired furnaces) / seal lost on barometric condenser/ entrained air in condenser water/ required discharge pressure requirement high/ fluctuating water pressure. *"Water coming out of discharge"*: upstream condenser flooded.

### 3.2.3
### Liquid

The types of pumps include centrifugal, peripheral, reciprocating, rotary, gear and rotary screw.

**Centrifugal:** Good practice: head-capacity curve should not be too flat if pump capacity is controlled by valve positioner. Select pump such that a larger diameter impeller could be installed later. An increase in flowrate causes an increase in required NPSH and a decrease in available NPSH.

Trouble shooting: *"No liquid delivery":* instrument error/not primed/ [cavitation]*/ supply tank empty. *"Liquid flowrate low"*: instrument error/ [cavitation]*/ non-condensibles in liquid/ inlet strainer clogged. *"Intermittent operation"*: [cavitation]*/ not primed/ non-condensibles in liquid. *"Discharge pressure low"*: instrument error/ non-condensibles in liquid/ speed too low/ wrong direction of rotation (or impeller in backwards if double suction). *"Power demand excessive"*: speed too high/ density liquid high/ required system head lower than expected/ viscosity high.

**Peripheral:** Trouble shooting: *"No liquid delivery"*: instrument error/ pump suction problems/ suction valve closed/ impeller plugged. *"Liquid flowrate low"*: instrument error/ speed too low/ incorrect impeller trim/ loose impeller. *"Discharge pressure low"*: instrument error/ speed too low/ incorrect impeller trim/ loose impeller. *"Power demand excessive"*: speed too high/ improper impeller adjustment/ impeller trim error.

**Reciprocating:** Trouble shooting: *"No liquid delivery"*: instrument error/excessive suction lift / [cavitation]*/ non-condensibles in liquid. *"Liquid flowrate low"*: instrument error/ excessive suction lift/ [cavitation]*/ non-condensibles in liquid.

**Rotary**: sometimes NPSH is expressed as Net Inlet Pressure Required, NIPR, (or available NIPA), expressed as kPa absolute (kPa abs). Trouble shooting: *"No flow":* instrument error/ [pump not turning]*/ [pump not primed]*/ relief valve not adjusted correctly or dirt keeping the relief valve open/ wrong direction of rotation/ [cavitation]*/ excessive suction lift. *"Flow < design":* instrument error/ rpm too low/ air leak via bad seals or faulty pipe connections/ [flow going elsewhere]*/ [high slip]*/ suction line clogged/ insufficient liquid supply/ [air or gas in liquid]*. *"Starts but loses prime"*: air leakage/ liquid vaporizing in suction line/ insufficient liquid supply. *"Noisy operation"*: [cavitation]*/ [air or gas in liquid]*/ [mechanical noise related to pump]*/ relief valve chatter/ drive-component noise. *"Power > design":* higher viscous losses than expected/ pressure > design/ fluid viscosity > expected/ fluid "sets up" or solidifies in the line or pump during shut down/ fluid builds up on pump surfaces/ rotating elements bind. *"Short pump service life"*: [corrosion]*/ abrasives present/ speed and pressures > design/ lack of lubrication/ misalignment.

*[Air or gas in liquid]*:* [fluid vaporizes]*/ air bleed missing/ fluid gasifies under operating conditions/ leaks in pumps or piping.

*[Air lock]*:* [fluid vaporizes]*/ air bleed missing/ fluid gasifies under operating conditions.

*[Cavitation]*:* [fluid vaporizes]*

*[Flow goes elsewhere]*:* relief valve faulty or jammed open/ discharge flow diverted to wrong branch line.

*[Fluid vaporizes]*:* [NPSH supplied too small]*/ fluid viscosity > design/ fluid temperature > design/ vapor pressure of fluid too high.

*[High slip]*:* clearance between rotors > specs/ worn pump/ pressure > design.

*[NPSH supplied or NIPA supplied too small]*:* strainer clogged/ temperature too high/ inlet line clogged/ inlet line diameter too small or length too long/ atmospheric pressure < design.

*[Mechanical noise related to pump]*:* wrong assembly/ pump distortion because of wrong piping installation/ pressure > rating/ worn bearings/ worn gears/ loose gears/ twisted shaft/ sheared keys/ worn splines.

*[Pump not primed]*:* valve on inlet line closed/ inlet line clogged/ air leaks/ pump rpm too low/ liquid drains or siphons out during off-periods/ check-valve missing or faulty/ [air lock]*/ worn rotors.

*[Pump not turning]*:* drive motor stopped/ key sheared or missing/ belt drive broken/ pump shaft broken.

**Gear:** Good practice: the higher the viscosity, the lower the rated rpm. On the discharge install a check valve and an expansion chamber or pulsation dampener on the discharge, the latter to reduce noise. For infrequent operation, operating pressure should be 20–30% < rated pressure. For continual operation, operating pressure ≪ rated pressure and rpm < rated rpm. Never allow them to run dry. For startup, idle off-line for about an hour.

Trouble shooting: usually performance does not break down suddenly; instead there is a gradual decrease in performance. Gear pumps are particularly susceptible to cavitation and erosion. *"Low discharge pressure"*: instrument error/ leakage/ low drive power/ faulty relief-valve setting/ [internal leakage]*/ [abrasion]*. *"No liquid delivery"*: instrument error/ suction line clogged/ drive not turning shaft/ check-valve fault. *"Low liquid delivery"*: instrument error/ drive power low/ [internal leakage]*/ [abrasion]*/ [cavitation]*. *"Noisy"*: entrained air in liquid/ liquid doesn't drain from grooves/ misaligned drive and pump shafts/ faulty bearings/ loose mountings/ resonance because mating frequency of gears = natural frequency of gear train / rpm too high/ worn parts/ [cavitation]*. *"Overheating"*: liquid viscosity higher than expected/ liquid feed temperature too low/ faults in drive system such as misaligned drive and pump shafts. *"Shaft won't rotate"*: drive system not working/ material in pump not melted/ temperature too low/ seized pump. *"Significant oscillation in pump suction pressure"*: instrument wrong/ faulty control system/ suction pressure setpoint too low for the process. *"Pump discharge pressure oscillates"*: instrument wrong/ starved feed to pump/ change in viscosity of feed/ damaged pump internals.

*"Inadequate volumetric efficiency":* decrease in viscosity/ pressure in upstream process increase/ worn pump.

*[Abrasion]*:* grit in liquid/[cavitation]*/ pH different from design. *[Cavitation]*:* pump rpm too high/ suction resistance too high/ clogged suction line/ suction pressure too low/ liquid flowrate higher than design/ entrained gas. *[Internal leakage]*:* excessive clearance between the gear and sides or end plates/ [abrasion]*.

**Rotary screw (mono)**: Good practice: stator has one more lobe or screw than the rotor; Reduce rpm for abrasives; starting torque = 4 × initial torque. If particles are present, try to minimize the abrasion by using viscosity > 5000 mPa s. Backflow or "slip" is reduced as viscosity increases. NPSH problems are usually not important except for suction lift, pumping from a vacuum and fluid vapor pressures of > 15 kPa. Trouble shooting: *"No liquid delivery"*: instrument error/ wrong direction of rotation/ insufficient suction lift/ clogged inlet/air leaks on suction/faulty pressure relief valve/ worn pump. *"Rapid wear"*: discharge pressure too high/ pump runs dry/ incorrect materials of construction/ speed too high for abrasives/ viscosity too low for abrasives. *"Noisy"*: insufficient feed flowrate/ air leak in suction/gas in feed liquid/ speed too high/ poor alignment. *"Excessive power"*: rpm too high/ liquid viscosity > design/ operating pressure > design/discharge line plugged/ stator expanded or swollen. *"Failure of the stator"*: bond failure (pH > 10 or local hot spot)/ temperature > design. *"Initially OK but gradual increase in power needed":* swelling of elastomeric stator coating because of chemical attack.

### 3.2.4 Solids

Johanson's[1), 2)] definitions of terms used to characterize solid particles are given in Section 3.7.3. The important terms are AI, RI, HI, FRI, FDI, BDI, CI, RAS and SBI.

**Bucket elevators:** Trouble shooting: major difficulties are unloading and loading: jamming of materials between the buckets and the side of the boot.

**Rotary/star valve:** Trouble shooting: use amps as guide to solids throughput. Keep air velocity high enough to prevent plugging of the air-vent line.

**Pneumatic conveying: dilute phase: for vacuum**: Trouble shooting: air leakage and powder arching in hopper are the major threats. *"No flow or flow < design":* air leaks/ powder arching in feed hopper, see Section 3.10/ low solids flow because of increased air loss in rotary valve/ wrong type of rotary valve used/ insufficient air/ line too long/ vacuum pump problems, see Section 3.2.2 . *"Pressure (vacuum) at suction to blower > design (vacuum < design)":* air leaks/ failure of discharge valve to seal on the receiver. *"Erratic pressure readings":* irregular feed. *"Explosion":* moisture too low/ lines not grounded. *"Does not sound "tinny" when listening with stethoscope":* material accumulated inside pipe at this location.

1) J.R. Johanson, 2002, "Troubleshooting bins, hoppers and feeders," Chem. Eng. Prog. April pp. 24–36.

2) J.R. Johanson, 2000, "Smooth out solids blending problems," Chem. Eng. Prog. April, p. 21.

**Pneumatic conveying: dilute phase: for pressure:** Trouble shooting: use pressure at the outlet of the blower as prime indicator. *"$\Delta p$ across blower > design or 2:1 ratio"*: restriction in downstream conveying line/ check valve jammed closed/ dirty intake filter/ plugged discharge silencer/ increase in feed to the system/ length of pipe > design. *"$\Delta p$ across blower < design"*: slipping v-belts/ air loss at the rotary valve. *"No flow"*: [plugged line]* *"No flow or flow < design"*: overfed fan system/ insufficient air/ insufficient solids/ line too long/ inlet air pressure too low. *"Erratic pressure readings"*: irregular feed. *"Amps on rotary valve < usual"*: solids flow < design/ air loss through the rotary valve/ increased clearances. *"Does not sound "tinny" when listening with stethoscope"*: material accumulated inside pipe at this location. *"Gradual decrease in performance"*: wear on the blower caused by dusty air.

*[Plugged line]**: within the first couple of metres of the beginning of the system: material feed problems/air supply problems. *[Plugged line]** after the first couple of metres: air leak with the plug occurring about 10 m downstream of leak/ erosion of rotary valve causing increase in air leakage.

**Pneumatic conveying: dense phase:** Trouble shooting: *"No flow or flow < design"*: plugged line/ malfunction of line boosters because of stuck check valve/ high humidity. *"Solids fed to conveying line < design"*: ratio of air to fluidize in the blow tank relative to convey is too small/ fault in control system. *"Solids fed to conveying line > design"*: ratio of air to fluidize in the blow tank relative to convey is too large/ fault in control system. *"Solids flow = 0"*: top discharge and the ratio of air to fluidize to convey is too small. *"Solids flow gradually decreases"*: restriction in the discharge pipe/ blinding of the fluidizing membrane.

**Feeder: volumetric for extruder:** Trouble shooting: *"Does not run"*: no power/ jammed. *"Stalls"*: material jam/ current limit set too low. *"Erratic speed control"*: controller poorly tuned/ sensor malfunction/ material jam. *"Feed rate variable"*: particles arching in the hopper/ moisture level too high/ overheated polymer (prematurely fused) feed polymer.

**Feeder: screw conveyor:** Trouble shooting: *"Shear pins on feeder drive break"*: screw diameter < exit hole from bin. *"Motor overload on feeder drive"*: screw conveyor diameter < exit hole from hopper. *"Screw feeder initially OK then motor overloads"*: screw flight spacing in the direction of sold flow decreases markedly/ difference between FDI and BDI < 5% suggests a moderately incompressible solid whose flow is very sensitive to screw flight spacing.

**Feeder from bottom of hopper:** Trouble shooting: *"Feeder motor overloads immediately:"* wrong wiring/foreign material in feeder/ hopper is full and solids give excessive solids pressure because of particle characterization and hopper design / FDI large and large HI. *"Feeder exit flowrate suddenly < expected"*: blockage in hopper outlet/ lumps of particles forming in hopper/ large RI and small HI possibly caused by temperature cycles. *"Feeder exit flowrate gradually < design:"* solids builup in the feeder/ large CI, large AI and RI/ wrong materials of construction in feeder. (Often happens with vibrating feeder.)

**Feeder: belt feeder from the bottom of a hopper:** Trouble shooting: *"Belt feeder initially starts but suddenly stops with motor overload"*: gap between the belt and hopper interface edge is too small/ belt sags between pulleys/ large FDI and small% difference between FDI and BDI.

### 3.2.5
### Steam

*Good practice:* Take steam off the top of the steam header; put condensate into the top of the condensate return header.

## 3.3
## Energy Exchange

The fundamentals for thermal energy exchange are that heat flows from a high temperature to a low temperature. Thermal forms of energy are not always available to do work. Overall energy is conserved; often we write expressions for the mechanical energy balance (on the macroscopic level this is Bernoulli's equation) and the thermal energy balance (on the macroscopic level this is $q = UA\ LMTD$). When trouble shooting heat exchangers, usually the fault is fluid dynamical: liquids don't drain; baffles are placed so that liquid can't go where you expect; vents are missing that prevent us from bleeding off trapped gases. Fluids to watch are water and hydrogen; both have extremes in thermal properties. Thermal expansion will occur when exchangers are brought up to temperature. This may cause a leak at the head-to-tube sheet joint if the difference between the temperature on the tubeside less the temperature of the bolts > 50 °C. For systems involving steam, scrutinize the steam trapping system: ensure that traps are not flooded, that the appropriate trap has been installed, that the bypass is not left open, and that thermodynamic traps are not fed to a common header. Steam should come from a nozzle on the top of the steam main; condensate should be discharged into the top of the condensate header.

More specifically, we list symptoms and possible causes for the following equipment: In this chapter we consider first providing mechanical drives, in Section 3.3.1. Furnaces are considered in Section 3.3.2. Heat exchangers, condensers and reboilers are listed in Section 3.3.3. Sections 3.3.4, 3.3.5 and 3.3.6 consider refrigeration, steam generation and high-temperature heat-transfer fluids, respectively.

### 3.3.1
### Drives

**Engines:** Good practice: use high efficiency motors when replacing or repairing existing installations. Trouble shooting: *"Hammering/knocking"*: loose parts/ seized parts. *"Pre-ignition"*: fuel with unstable hydrocarbons/ incorrect timing. *"Detonation"*: wet fuel/ incorrect timing/ intake air too hot/ glowing carbon on the piston/ leaking valve stem/ worn valve guides. *"Misfiring"*: incorrect timing/ faulty ignition elements/ wrong gap in the spark plugs/ wet fuel/ spark-plug gap coated or filled with carbon or oil. *"Overheat"*: lubrication failure/ inadequate cooling/ poor quality fuel/

fuel to air ratio too lean. *"Sooty exhaust":* incorrect fuel/ air- fuel ratio too rich/ inadequate cooling/ wrong valve adjustment. *"Valve leaking":* inadequate cooling/ valve angle incorrect/ wrong metallurgy. *"Piston blow-by:"* over lubrication. *"Worn bearings":* misaligned crankshaft.

**Electric motor:** Trouble shooting: *"Won't start":* overload trip/ loose connection/ grounded winding/ grounded stator. *"Runs backwards":* reversed phase sequence. *"Excessive noise":* 3 phase machine single phased/ unbalanced load between phases. *"Synchronous motor fails to come up to speed":* faulty power supply or overload trip/ windings grounded. *"Overheat":* unbalanced load between phases/ wrong line voltage/ short circuit in stator winding. DC Motors: *"Won't start":* weak field/ low armature voltage/ open or short circuit in armature or field. *"Runs too slow":* low armature voltage/ overload/ brushes ahead of neutral. *"Runs too fast":* high armature voltage/ weak field/ brushes behind neutral. *"Brushes sparking":* brushes worn/ brushes poorly seated/ incorrect brush pressure/ dirty, rough or eccentric commutator/ brushes off neutral/ short-circuited commutator/ overload/ excessive vibration. *"Brush chatter":* incorrect brush pressure/ high mica/ incorrect brush size. *"Bearings hot":* belt too tight/ misalignment/ shaft bent/ damaged bearings/ wrong type of bearings.

**Steam turbine:** Good practice: consider extracting energy via a steam turbine for any pressure reduction in steam service. Use high-pressure steam for energy; low-pressure steam for heating. Don't operate with wet steam. Trouble shooting: *"Turbine fails to start:"* too many hand valves closed/ nozzles plugged or eroded/ dirt under carbon rings. *"Slow startup":* throttle-valve travel restricted/ steam strainer plugged/ load > rating. *"Insufficient power":* throttle-valve travel restricted/ too many hand valves closed/ oil relay governor set too low. *"Speed increases as load decreases":* throttle-valve travel restricted/ throttle assembly friction/ valve packing friction. *"Governor not operating/ excessive speed variation":* governor droop adjustment needed/ governor lubrication problem/ throttle-valve travel restricted. *"Overspeed trip on load changes":* trip valve set too close to operating speed/ throttle-valve travel restricted/ throttle assembly friction. *"Overspeed trip on normal speed"*: excessive vibration/ dirty trip valve/ trip valve set too close to operating speed. *"Leaking glands":* dirt under carbon rings/ worm or broken carbon rings/ scored shaft.

**Steam turbine used for the generation of electricity:** Trouble shooting: *"Turbine overspeeding":* [load disconnection suddenly]*/ [Trip Throttle Valve stuck]*/ control valve fault// [extraction valve fault]*. *"Bearings damaged":* [turbine overspeeding]*/ [lube oil]*/ excessive vibration/ no lube oil/ bearing temperature too high/ flow of parasitic currents/ [clogging]*/ [electronic pin clogging]*.

*[Clogging]*:* [lube oil]*/ long time without operating.

*[Electronic pin clogging]*:* [lube oil]*/ long time without operating.

*[Extraction valve fault]*:* wear on valve bearing/ loss of hermetic seal.

*[Load disconnection suddenly]*:* operator error/ automatic bus bar protection because of downstream changes in electric system.

*[Lube oil]*:* low pressure/ oil temperature too high/ oil too old/ oxidation/ water contaminates oil.

*[Solenoid valve malfunction]*:* [electronic pin clogging]*/ [clogging]*/ solenoid shorted coil/ faulty control signal/ sensor error.

*[Trip Throttle Valve stuck]*:* [clogging]*/ [solenoid valve malfunction]*

**Gas turbine:** consists of a compressor, combustor and turbine sections. Trouble shooting: *"Combustion noise:"* fouled or clogged combustor/ loose or cracked lining in combustor. *"Vibration":* bearing failure in compressor or turbine/ blade damage in compressor or turbine/ surging compressor/ fouled turbine. *"Exhaust temperature > design":* combustor fouling. *"Exhaust temperature < design":* combustor clogged. *"Thermal efficiency < design":* fouled turbine/ turbine blade damage/ turbine nozzle distortion. *"Mass flow < design":* compressor fouling/ compressor filter clogged/ compressor blades damaged.

### 3.3.2 Thermal Energy: Furnaces

Multi-use including heating, boiling, reactions. Related topics distillation, Section 3.4.2.

For fired furnaces: monitor CO and excess air to reduce rejected energy and improve efficiency, consider the installation of economizers and air preheaters to recover additional heat from the flue gas.

For steam generation: preheat boiler feed water with available low-temperature process streams, maximize the use of heat-transfer surfaces by optimizing soot-blowing frequency and decoking of tubes, flash blowdown to produce low-pressure steam if required.

Trouble shooting: *"Gas temperature > design":* instrument wrong/ insufficient excess air/ process side coking of tubes/ leak of combustible material from process side/ overfiring because of high fuel-gas pressure. *"Gas temperature < design":* instrument fault/ fouling/ too much excess air/ insufficient area/ fuel-gas pressure < design. For convection furnace: *"Exit process gas temperature < design":* excess air/ decrease in flame temperature/ damper has failed closed. *"Pressure inside furnace > design":* instrument wrong/ fouling on the outside of the tubes in the convection section/ exhaust fan failure. *"Faint blue-gray smoke rising from top of furnace":* fouling outside tubes in the convection section/ pressure in furnace > atmospheric. *"Puffing, rhythmic explosions"*: burners short of air for short period causing minor over-firing/ wind action/ start up too fast. *"Tube failure":* localized overheating/ burning acid gases as fuel/ free caustic in water and dryout/ dry out and attack by acid chloride carried over from water demineralization/ breakthrough of acid into water from demineralizer. *"High fuel-gas pressure":* failure of pressure regulator. *"Tube dryout":* tubeside velocity too low. *"Low furnace efficiency":* high combustion air flow/ air leak into the firebox/ high stack temperature/ heat leaks into the system. *"Equipment suddenly begins to underperform":* fouling/ bypass open. *"Temperature-control problems":* missing or damaged insulation/ poor tuning of controller/ furnace not designed for transient state/ unexpected heat of reaction effects/ contaminated fuel/ design error.

### 3.3.3
### Thermal Energy: Fluid Heat Exchangers, Condensers and Boilers

For not truly countercurrent, if the correction factor for the LMTD drops below 0.75 we run the risk of temperature crossover. Provide pressure relief to allow for systems where block valves could isolate trapped fluids. Include impingement baffles at shell inlet nozzles to prevent erosion of tubes and flow-induced vibration. Account for the larger heat exchange that occurs for clean tubes/surfaces; the design was based on reduced heat-transfer coefficients that accounts for ultimate dirty film resistance. Ensure the air is vented. Liquids being heated should leave at the top of the exchanger to prevent the buildup of gases coming out of solution and vice versa for liquids with suspended solids or viscous fluids. Orient baffle windows to facilitate drainage. Slope condensers to remove the condensed phase. Maximum cooling-water temperature is 45 °C. Prefer water or other nonflammable heat-transfer media. For flammable heat transfer fluids, select operating temperature below its atmospheric boiling temperature. If refrigeration is required, prefer less hazardous refrigerant even if this means operating at higher pressures. Use pinch analysis, identify inefficient exchanges and retrofit heat-exchanger networks to maximize heat recovery. Optimize cleaning schedule. Consider "on-line" mechanical cleaning where fouling is a problem. Use turbulence promoters in laminar flow and gas services and where turndown has significantly reduced the heat-transfer coefficient. For air-cooled systems include a trim cooler with water as coolant.

**Shell and tube heat exchangers:** Good practice: to provide lower inventory and intensify, prefer plate exchangers to shell and tube exchangers with the highest surface compactness. Trouble shooting: *"Thermal underperformance on both streams (coolant exit temperature < design; hot exit > design temperature):"* instrument fault/ not enough area/ thermal load reduced via flowrate or change in thermal properties (eg, less hydrogen than design)/ inerts blinding tubes/ [fouling]* more than expected/ tube flooded with condensate (see faulty steam trap, Section 3.5.1) or trap in backwards or insulated inverted bucket steam trap. *"Equipment suddenly begins to underperform":* fouling/ bypass open. *Temperature-control problems":* missing or damaged insulation/ poor tuning of controller/ not designed for transient state/ unexpected heat of reaction effects/ contaminated feeds/ design error/ unexpected heat of solution effects/ changes in properties of the fluids. *"Heat transfer to shell side fluid < design and $\Delta p$ < design":* instrument/ increase in viscosity/ fluid bypasses baffles (baffle cut > 20%, no sealing strips, excessive baffle clearance, shell side nozzles too far from tube bundle)/ stratification/ faulty location of exit nozzles/ faulty baffling/ inlet maldistribution. *"Heat transfer to tubeside < design and uneven (and uneven tube-end erosion at inlet)":* maldistribution to the tubes (axial nozzle entry velocity > tube velocity, for radial nozzle entry velocity > 1.9 tube velocity). *"Heat transfer to one fluid < design and $\Delta p$= design":* instrument fault/ oil contamination of water. *"Thermal overperformance both fluids, and usually $\Delta p$ > design on hot fluid side, perhaps charring of cold stream and freezing of hot stream:"* instrument fault/ cocurrent piped incorrectly as countercurrent/ area too large/ hydrogen concentration in gas stream > design/ clean tubes but design area selected on dirty service. *"Thermal overperformance one* *stream: cold exit temperature > expected":* instrument/ plugged tubes/ inlet velocity < design, fouled screen on pump suction/ pump problems, see Section 3.2.3/ increased heat load. *"Poor control of outlet temperatures (±5 °C)":* poor tuning of control/ instrument fault/ oversized area combined with multipass with local changes in effective MTD with fluid velocity. *"Rapid tube failure or glass or karbate tubes break":* inlet gas velocity too high and directed onto tubes/ gas velocity > 5 m/s causing tube vibration/ surges in cooling water/ surges cause by syphon without vent break. *"Higher Δp when operating at design flows and temperatures":* underdesign/ design for 2-phase stratified flow but slug flow occurs/ gas service but the operating pressure < design.

*"Leaks":* erosion/ corrosion/ vibration/ improper tube finishing/ cavitation/ lack of support for tube bundle/ tube end fatigue. *"Leaks from the gasket at the tube sheet joint":* sudden upsets that cause the ΔT between the flange and the bolt to be > 50 °C. *"Noise/ vibration":* excessive clearance between baffles and tubes/ inlet gas velocity too high and directed onto tubes/ gas velocity > 5 m/s causing tube vibration/ surges in cooling water/ surges cause by syphon without vent break. *"Gradual reduction in heat transfer and increase in Δp":* small tube leaks.

*[Fouling]**: change in pH/ water temperature high and invertly soluble compounds precipitate/ water temperature high and algae and fungi form/ corrosion products/ sublimation/ process condensate freezes/ coolant fouling/ silt deposits/ aggregation and destablization of colloids causing wax and asphaltenes to deposit from hydrocarbons. For more on [fouling] see Section 3.11.

**Shell and tube condensers:** Trouble shooting: *"Condensation duty < design; exit vapor temperature > design, high flowrate of vapor out vent":* instrument fault/ undersized condensers/ change in process gas pressure/ inward leakage of non-condensibles/ change in feed composition/ [fouling]* on the process side/ vapor binding/ vapor pockets/ inert blanketing (usually near the condensate outlet for condensers operated flooded for pressure control)/ condensate flooding, see steam traps, Section 3.5.1/ baffle orientation horizontal not vertical/ excessive entrainment in vapor feed/ baffle window > 45%/ drain line too small/ leakage between the tubesheet and baffles/ bowed tubesheet/ condenser designed for horizontal service installed vertically. *"Condensation duty > design:"* excess condenser area/ clean tubes/ condenser designed for vertical service installed horizontally/ liquid entrainment in feed. *"Condensation duty < design and Δp process > design and excessive flow out vapor vent":* undersized condenser. *"Coolant water temperature > design":* instrument fault/ low coolant flowrate/ high coolant inlet temperature/ cooling tower fault/ excess condenser area. *"Cooling water exit temperature > design and higher steam usage in distillation column reboiler and uneven column operation:"* excess condenser area via overdesign or clean surfaces. *Heat transfer drops off > rate than expected and Δp increases faster than expected":* [fouling]* because of oversized kettle reboiler on distillation column or change in pH or flow regime laminar when design was turbulent or higher level of contamination in fluids or crud carryover from upstream equipment (e.g. silica from catalyst in upstream reactor) or compensation for oversize by reduced coolant flowrate. *"Loss of volatile vapor out vent, high vent-gas temperature, degree of subcooling < design and unusual temperature profile between vapor inlet and condensate outlet:"*

instrument error/ underdesign. *"Loss of volatile vapor out vent, apparent undersized area for condensation of immiscible liquids":* lack of subcooling of condensate/ condenser installed horizontally instead of vertically. *"Fog formation":* high ΔT with noncondensibles present/ high ΔT with wide range of molar mass of the vapor. *"Equipment suddenly begins to underperform":* [fouling]*/ bypass open. *Temperature-control problems":* missing or damaged insulation/ poor tuning of controller/ not designed for transient state/ unexpected heat of reaction effects/ contaminated feeds/ design error/ unexpected heat of solution effects/ changes in properties of the fluids.

*[Fouling]**: change in pH/ water temperature high and invertly soluble compounds precipitate/ water temperature high and algae and fungi form/ corrosion products/ sublimation/ process condensate freezes/ coolant fouling/ silt deposits. For more see Section 3.11.

**Shell and tube reboilers:** Good practice: to provide lower inventory and intensify, prefer thermosyphon reboilers to kettle reboilers. If ΔT > 25 °C, probably the boiling mechanism is film boiling. If ΔT < 25 °C, usually the boiling mechanism is nucleate boiling.

Trouble shooting, general: *"Insufficient boilup":* [fouling on process side]*/ condensate flooding, see steam trap malfunction, Section 3.5.1 including higher pressure in the condensate header/ inadequate heat supply, steam valve closed, superheated steam/ boiling point elevation of the bottoms/ inert blanketing/ film boiling/ increase in pressure for the process side/ feed richer in the higher boiling components/ undersized reboiler/ control system fault/ for distillation, overdesigned condenser. *"Equipment suddenly begins to underperform":* [fouling]*/ bypass open. *Temperature-control problems":* missing or damaged insulation/ poor tuning of controller/ not designed for transient state/ unexpected heat of reaction effects/ contaminated feeds/ design error/ unexpected heat of solution effects/ changes in properties of the fluids.

*"Insufficient boilup and gradual increase in steam pressure to maintain boilup:"* [fouling]*/ inerts in steam. *"Insufficient boilup and gradual decrease in steam pressure to maintain boilup:"* steam blowing, see steam trap malfunction, Section 3.5.1. *"Water contamination":* leak. *"Cycling (30 s–several minutes duration) steam flow, cycling pressure on the process side and, for columns, cycling Δp and cycling level in bottoms":* instrument fault/ condensate in instrument sensing lines/ surging/ [foaming]* in kettle and thermosyphon/ liquid maldistribution/ steam-trap problems, see Section 3.5.1, with orifice Δp across trap < design/ temperature sensor at the feed zone in a distillation column/ collapsed tray in a distillation column. *"Level high in reboiler":* instrument/ inlet or exit pipe nozzle too small/ wrong nozzle orientation/ steam trap fault, see Section 3.5.1/ steam trap is above the reboiler. *"Breathing: puffs of vapor and entrained liquid leave reboiler:"* overdesign/ clean tubes when designed for fouled conditions.

*[Inadequate heat supply]**: wet steam/ too great a Δp across steam valve gives wiredrawing and superheat/ steam valve closed/ control system fault.

**Kettle**: Good practice: rarely underdesigned and should not be used for foams. Trouble shooting: general plus the following symptoms and causes unique to this type of reboiler:

*"Surging"*: poor liquid distribution/ [fouling]*. *"Low boilup rate and gradual increase in steam"*: film instead of nucleate boiling/ too high a $\Delta T$/ clean tubes/ conservative overdesign/ [fouling]*/ flooding with condensate because of steam-trap problems, see Section 3.5.1/ bottom temperature elevation/ increase in column pressure/ feed concentration of light components < design/ not enough heating medium. *"Low boilup rate and decrease in steam pressure"*: steam trap blowing, see Section 3.5.1. *"Low boilup rate, pressure increase in reboiler and surges"*: [foaming]*/ inerts/ leaks/ undersized reboiler/ diameter of vent line too small/ top tubes not covered with liquid/ high liquid level that floods the vapor disengagement space/ inlet feed maldistribution/ inadequate vapor disengagement.

*[Foaming]**: see Section 3.11.

*[Fouling on the process side]**: low liquid level causing vapor-induced fouling/ solids in feed that are trapped by the overflow baffle. For more on fouling see Section 3.11.

**Thermosyphon:** Good practice: vertical thermosyphon reboilers are usually not used for vacuum or extremely high pressure service. Trouble shooting: general plus the following symptoms and causes unique to this type of reboiler:

*"Insufficient boilup"*: [fouling on the process side]*/ insufficient steam flow/ condensate flooding/ low liquid level in distillation column gives low liquid circulation and increased fouling/ high liquid level in distillation column (static head > design) or higher density of feed liquid gives higher boiling temperature and circulation and insufficient vaporization for vertical thermosyphon/ pipe lengths < design/ pipe diameter > design/ process fluid in vertical thermosyphon drops below 30–40% of the tube length. For horizontal thermosyphon, appears to be undersized but the cause is liquid feed maldistribution. *"Surges in boilup"*: process fluid circulation rate too low/ [fouling on the process side]*/ wide boiling range/ overdesign. *"Cycling (30 s–several minutes duration) steam flow, cycling pressure on the process side and, for columns, cycling $\Delta p$ and cycling level in bottoms"*: in addition to general, all natural circulation systems are prone to surging/ feed contains high w/w% of high boilers/ vaporization-induced [fouling]*/ constriction in the vapor line to the distillation column. For horizontal thermosyphon: maldistribution of fluid temperature and liquid.

*[Fouling]**: insufficient static head/ excess friction in the pipes/ on the tubeside the outlet nozzle area < total tube area/ on tubeside the inlet nozzle area < 0.5 total tube area/ rate of vaporization > 25% of circulation rate/ mass rate of vaporization > mass rate of circulation/ natural circulation rate < 3× vaporization rate/ vaporization induced solids. For more see Section 3.11.

**Forced circulation:** operates with sensible heat mode in the tubes. Trouble shooting: general plus the following symptoms and causes unique to this type of reboiler: *"Unstable"*: insufficient NPSH in pump, see Section 3.2.3. *"Insufficient boilup and rapid fouling"*: insufficient circulation/ pump fault, see Section 3.2.3/ plugged circulation lines. *"Insufficient boilup"*: [fouling]*/ circulation rate low/ pump problems, see Section 3.2.3/ no vortex breaker/ excessive circulation and a wide spread in boiling temperatures in bottoms. *"Excessive vapor in flash chamber, unstable distillation column operation and apparent underdesign of overhead condenser"*: overdesign.

**Vertical falling-film evaporator:** see also Absorbers, Section 3.4.8 and Evaporators, Section 3.4.1 and falling-film reactors, Section 3.6.6. Good practice: always check the liquid feed distribution with water before putting on line.

Trouble shooting: *"Boilup < design":* [liquid maldistribution]*.

*[Liquid maldistribution]**: tubes not vertical/ inadequate calming of feed/ variations in weir height.

**Spiral plate exchanger:** Trouble shooting: *"Heat transfer < design":* stratification caused by faulty inlet and exit nozzle location/ baffling/ maldistribution.

**Plate exchanger:** Good practice: put regulating and control valves on the inlet lines, never on the outlet lines, to minimize pressure in the exchanger. Never allow the exchanger to be under a vacuum. Keep temperature < 120 °C; pressure < 2.5 MPa. Trouble shooting: *"Leaking gaskets":* temperature too high/ temperature spike/ pressure too high/ cold fluid stopped but hot fluid continues/ superheated steam/ under vacuum.

**Air cooled:** Good practice: induced draft preferred to forced draft to minimize hot-gas recirculation. Include a water-cooled "trim cooler". Ensure the exit tubes are "flooded" so that the vapor doesn't bypass condenser. If extreme cold conditions are expected, allow for fan to operate in reverse to counteract the overcooling by the natural circulation of cold air. Trouble shooting: *"Insufficient condensation":* instrument fault/ maldistribution along either feed or exit headers/ buildup of non-condensibles in bottom tube rows/insufficient area/ ambient temperature too high/ fan not working/ blades wrong pitch/ baffles stuck/ [fouled tubes]*/ hot-gas recirculation/ tubes not sealed. *"Cycling":* control system/ vent for syphon-break is missing on exit manifold. *"Outlet temperature on tube-side is high":* undersized/ tube [fouling]* on inside or outside/ flow maldistribution on process or air side/ hot air recirculation/ air flowrate too low. *"$\Delta p$ on process side high":* [fouled]* tube side/ increased liquid viscosity/ overcooling/ vapor not condensed. *"Exit air temperature > expected":* low air flowrate/ flow maldistribution on tube side/ ambient air temperature > expected/ unexpected hot air recirculation. *"Exit air temperature < expected":* high air flowrate/ flow maldistribution on tube side/ ambient air temperature < expected. *"Sluggish control":* the use of fan pitch variation as the control variable.

*[Fouling]**: see Section 3.11.

### 3.3.4
### Thermal Energy: Refrigeration

Trouble shooting: use the p-H diagram for the refrigerant as a basis for trouble shooting. *"Compressor discharge pressure < design":* turbine drive problem, power limited/ overloaded centrifugal compressor or valve problem for reciprocating compressor/ wrong composition for the speed/ not enough refrigerant/ compressor fault, see Section 3.2.1. *"Compressor discharge pressure > design:"* fouled condenser/ insufficient air to the cooling tower/ low flowrate of water to the condenser/ air in the refrigerant/ too much refrigerant, level too high. *"Compressor discharge pressure > design and condensing temperature normal":* poor drainage from the condenser/ non condensibles in the refrigerant/ refrigerant letdown valve plugged. *"Suction pressure*

*< design"*: process feedrate < design/ low circulation of refrigerant/ not enough refrigerant in chiller, level too low/ leak causing a loss of refrigerant/ compressor problems, see Section 3.2.1. *"Suction pressure > design":* process coolant load > design/ throttle valve incorrectly adjusted/ low level refrigerant/ compressor problems, see Section 3.2.1. *"Process exit temperature > design, refrigerant temperature from the chiller > design and the approach temperature = design":* chiller pressure > design/ not enough refrigerant/ heavy ends impurities in refrigerant/ level of refrigerant in chiller < design/ expansion valve plugged/ restriction in refrigerant suction line/ unit too small/ process fluid velocity too slow/ too much refrigerant in chiller causing flashing in the compressor suction. *"Process exit temperature > design, refrigerant temperature in chiller = design and approach temperature > design":* fouling on refrigerant side/ fouling on the process side/ chiller unit too small. *"Condenser temperature > design":* [fouled]* condenser

*[Fouling]*:* see Section 3.11.

### 3.3.5 Thermal Energy: Steam Generation

See thermal energy furnaces/ boilers, Section 3.3.2. See Section 3.2.5 for steam distribution

Trouble shooting steam generation: *"Tube failure":* feed water contains impurities/ for forced circulation: water circulation rate too low/ dry spots in tubes/ vibration induced tube failure. *"Wet steam":* rate of steam generation > design causing inadequate demisting in the steam drum. *"Steam production < design":* fuel-gas pressure < design/ soot in flue-gas passages/ thermostat incorrect and burners cut out too soon/ wrong type of fuel-gas burner. *"Stack temperature > design":* not enough air/ overfiring.

### 3.3.6 High-Temperature Heat-Transfer Fluids

Good practice: usually a portion of the liquid is purged and replaced with fresh makeup.

Trouble shooting: *"Rapid cycling of the furnace or heating elements":* [fluid velocity low]*. *"Vapor pressure increased":* [thermal cracking of fluid]*. *"Noisy pump"*: contaminants such as water/ [fluid velocity low]*/ [thermal cracking of fluid]*. *"Pump discharge pressure fluctuates":* contaminants such as water/ [fluid velocity low]*/ [thermal cracking of fluid]*. *"Startup of cold unit takes longer than usual":* [oxidation of fluid]*. *"Heater cannot achieve setpoint":* [oxidation]*/ [thermal cracking]*. *"Poor control":* [control valve plugged]*/ [heater cannot achieve setpoint]*/ control design faulty/ controller not well tuned.

*[Control valve plugged]*:* [thermal cracking]*/ [oxidation]*/ filter plugged/ filter missing/ filter not working.

*[Fluid velocity low]*:* pump problems, see Section 3.2.3/ filter plugged/ controller not well tuned/ wrong location for filter/ crud left in the lines during maintenance.

*[Oxidation]*:* temperature of air in expansion tank > 60 °C/ for higher temperatures in expansion tank, dry inert gas blanket not used in the expansion tank.

*[Thermal cracking]*:* fluid velocity in the furnace or heater < design.

See also trouble-shooting suggestions related to gas-liquid separators, Section 3.5.1, furnaces, Section 3.3.2, and pumps, Section 3.2.3.

## 3.4 Homogeneous Separation

The fundamentals upon which most of these processes are based include: mass is conserved; mass transfers because of bulk movement and diffusion. The rate of mass transfer is proportional to the concentration driving force of the target species and the surface area across which the transfer occurs. Phase equilibrium is a useful starting approximation but usually it is the rate at which the system moves toward equilibrium that is important. Surface phenomena effects, especially foaming and fouling, wetting and dispersed phase stability are issues to consider.

In this section we consider the separation of species contained in a homogeneous phase, such as a liquid or gas. The separation is based on exploiting a fundamental difference that exists between the species. Methods that exploit differences in vapor pressures are evaporation, in Section 3.4.1 and distillation, in Section 3.4.2. Methods exploiting solubility are solution crystallization, Section 3.4.3; absorption, Section 3.4.4, and desorption, Section 3.4.5. Solvent extraction, Section 3.4.6, exploits differences in partition coefficient.

Methods based on exchange equilibrium and molecular geometry include adsorption of species from a gas, Section 3.4.7, and of species from a liquid, Section 3.4.8. Ion exchange, Section 3.4.9, exploits differences in surface activity and exchange equilibrium.

Membrane separations described include reverse osmosis, Section 3.4.10; nanofiltration, Section 3.4.11; and ultra and micro-filtration, Section 3.4.12. Separation of larger sized species are considered "heterogeneous systems" and are considered in Section 3.5.

### 3.4.1 Evaporation

Good practice: keep the pressure drop between the last effect and the inlet to the vacuum device < 3 kPa. Consider vapor recompression for conventional low $\Delta T$ evaporators such as falling film, forced circulation and horizontal tube falling film. Vapor recompression is rarely used on high $\Delta T$ systems such as rising film, calandria and submerged tubes. Trouble shooting: *"Product contamination":* leaking valves/ crud left in storage tanks/ crud left in dead legs in piping/ [corrosion]* products/ unexpected chemical reactions/ sampling fault/ analysis fault/ unexpected solubility effects. *[Corrosion]*:* see Section 3.1.2.

**Vapor recompression evaporators:** *"Evaporation rate < design":* [fouled]* heat-transfer surface/ uneven movement of liquid over heat-transfer surface/ feed property changes/ excessive noncondensibles from leaks or present in feed/ flooded condensate, trap malfunction, Section 3.5.1/ feed temperature < design/ water leakage into the system/ lower compressor suction pressure, see also Section 3.2.1.*"Steam economy low":* instrument fault/ excessive venting especially the first, second and third effects/ vapor exiting through condensate, trap problems, Section 3.5.1 / vapor blowing into product flash tank through the liquor lines/ [foaming]*/ internal afterheaters leaking/ afterheater scaled so that liquor from the colder effect is not correctly preheated for the next effect/ [entrainment]*/ excessive vacuum/ [fouling]*. *"Recovery-boiler efficiency low":* [fouling]*. *"Vibration":* vapor velocity high through the first row of tubes. *"Vacuum problems":* see vacuum, Section 3.2.2.

*[Corrosion]*:* see Section 3.1.2.

*[Entrainment]*:* poor design of deflector/ liquid level above the tubes/ [foaming]*.

*[Foaming]*:* see Section 3.11.

*[Fouling]**: sodium suflate precipitates especially in the first effect/ lignin precipitates especially in the first and second effect/ vapor sulfurization and condensation in third and fourth effects/ velocity too small.

**Falling-film evaporator:** Trouble shooting. *"Evaporation rate < design":* [liquid maldistribution]*/ steam trap malfunction, see Section 3.5.1/ steam flowrate too small/ [foaming]*/ [fouling]*.

*[Foaming]*:* see Section 3.11.

*[Fouling]**: tubular velocity too small: for 5-cm diameter tubes, recommended velocities are: for viscous liquids use 3 m/s; for the finishing effect, 2–2.7 m/s; for the intermediate effects, 1.5–1.8 m/s; for the initial effects, 1.2–1.5 m/s/ pump problems, see Section 3.2.3.

*[Liquid maldistribution]*:* not vertical/ inadequate calming of feed/ variations in weir height.

**Forced-circulation evaporator:** Trouble shooting: usual problems are fouling/scaling and high liquid viscosity.

*[Fouling]**: tubular velocity too small: for 5-cm diameter tubes, recommended velocities are: for viscous liquids use 3 m/s; for the finishing effect, 2–2.7 m/s; for the intermediate effects, 1.5–1.8 m/s; for the initial effects, 1.2–1.5 m/s/ pump problems, see Section 3.2.3. For a more general consideration of fouling see Section 3.11.

*[Liquid maldistribution]*:* not vertical/ inadequate calming of feed/ variations in weir height.

**Multiple-effect evaporator:** Good practice: capacity of one or more effects in series is proportional to (condensing temperature of the steam supplied – temperature of the liquid boiling in the last effect) and the overall heat-transfer coefficient. If foaming occurs, reduce the liquid level in the effect. Trouble shooting: *"Reduced flowrate from last stage to maintain target strength"*: water temperature to contact condenser too high/ insufficient condensing area/ [decreased UA]*/ [foaming]*. *"$\Delta T$ higher than usual before stage "x" and $\Delta T$ lower than usual after stage "x"":* [decreased UA in stage "x"/ [foaming]*. *"Steam usage higher than normal":* steam leak into an effect/ bleed rate too high/ poor trap performance, see 3.5.1. *"Cycling (30 s–several minutes duration) steam flow, cycling pressure on the process side and, for columns, cycling* $\Delta p$ *and cycling level in bottoms":* instrument fault/ condensate in instrument sensing lines/ surging/ [foaming]* in kettle and thermosyphon/ liquid maldistribution/ steam-trap problems, see 3.5.1, with orifice $\Delta p$ across trap < design/ temperature sensor at the feed zone in a distillation column/ collapsed tray in a distillation column/ unsteady vacuum see Section 3.3.2.

*[Decreased UA]**: inadequate condensate removal/ liquid level too low in the effect/ [fouling]*/ inadequate removal of non-condensible gas.

*[Foaming]**: natural occurring surfactants/ pH far from the zpc/ naturally occurring polymers/ solids particles/ corrosion particles/ mechanical foam breaker not rotating/ baffle foam breaker incorrectly designed or damaged/ antifoam ineffective (wrong type or incorrect rate of addition)/ gas velocity too high/ rate of evaporation too fast/ overhead disengaging space insufficient height/ liquid downflow over foam too low, see also Section 3.11.

*[Fouling]**: tubular velocity too small: for 5-cm diameter tubes, recommended velocities are: for viscous liquids use 3 m/s; for the finishing effect, 2–2.7 m/s; for the intermediate effects, 1.5–1.8 m/s; for the initial effects, 1.2–1.5 m/s.

### 3.4.2
### Distillation

Good practice: for trays, add 10% more trays or two trays to improve operability. Weir height: 5 cm with length 75% of the tray diameter to provide a liquid weir overflow rate > 5 and < 20 L/s m of weir into the downcomer. Usually use 15 L/s m. For lower flows use a picket weir. Overall downcomer area should be > 5% total tray area. For foaming liquids increase downcomer area by 50%. The downcomer exit should be at least 1.2 cm below the top edge of the outlet weir. Include four, 6-mm diameter weep holes in each tray for shutdown drainage.

For packing, water test the liquid distributor for good liquid distribution before startup.

*[Surface tension negative]**: If the surface tension of the distillate > surface tension of the bottoms (surface tension negative) prefer the use of trays to packings to minimize potential for liquid film breakup.

*[Surface tension positive]**: If the surface tension of the distillate < surface tension of the bottoms (surface tension positive), the foam above trays might be unexpectedly stable.

Trouble shooting: The relationship between the symptom and the causes partly depends on the control system used. Check the auxiliaries to see if they are at fault: reboilers and condensers, see Section 3.3.3; vacuum, see Section 3.2.2; pumps, see Section 3.2.3. For packed towers, 80% of the causes are liquid maldistribution.

*"$\Delta p$ across the column » design (> 1/2 the column height), reflux flowrate » usual; $\Delta T$ across column < design, overhead composition contains heavies > design; surges in the liquid overhead, bottoms level low or fluctuates, bottoms pressure > design, higher column pressure and higher temperature profile below the flooded portion of the column the temper-*

*ature profile > design and all trays below the flood are dry and bottoms composition off spec"*: [jet flooding]*. *"$\Delta p$ across the column » design, reflux flowrate gradually increasing; $\Delta T$ across column < design, overhead composition contains heavies > design; bottoms level low or fluctuates, bottoms pressure > design, and higher temperature profile below the flooded portion of the column the temperature profile > design and "all trays below the flood are dry""*: [downcomer flooding]*. *"$\Delta p$ across the column > design"*: instrument fault/ high boilup rate/ steam flow to reboiler > design. *"$\Delta p$ across the column < design"*: instrument fault/ [low boilup rate]*, see Section 3.3/ dry trays/ low feedrate/ feed temperature too high.

*"Feed flowrate < design"*: instrument fault/ pump problems, see Section 3.2.3/ filter plugged/ column pressure > design/ feed location higher than design.

*"Temperature of feed > design"*: instrument fault/ preheater fouled/ feed flowrate low/ heating medium temperature < design, see heaters, Section 3.3.3. *"Temperature of bottoms < design"*: instrument fault/ [low boilup]* see Section 3.3.3/ loss of heating medium/ steam trap plugged, see 3.5.1/ feedrate to column > design/ feed concentration of low boilers (overheads) > design/ feed distributor fouled. *"Temperature of bottoms > design"*:instrument fault/ [column pressure > design]*/ high boilup/ overhead condenser vent plugged/ insufficient condensing, see Section 3.3.3. *"Temperature at top > design"*: instrument fault/ bottom temperature > design/ reflux too low/ distillate feed forward too high/ column pressure high/ [flooding]*. *"Temperature at top > design and overhead composition contaminated with too many heavies"*: vapor bypassing caused by excessive vapor velocities (high boilup) or not enough liquid on tray or packing, or downcomers not sealed, or sieve holes corroded larger than design and tray weeps/ reflux too low/ feed contains excessive heavies. *"Temperature at top < design"*: instrument fault/ control temperature too low/ [low boilup]* see Section 3.3.3. *"All temperatures falling simultaneously"*: [low boilup]*. *"All temperatures rising simultaneously:"* pressure rising.

*"Overhead off spec"*: poor tray or packing efficiency/ [maldistribution]*/ not enough trays or packing/ loss of efficiency/ high concentration of non-condensibles/ missing tray/ collapsed tray/ liquid entrainment/ liquid bypass and weeping/ liquid or gas maldistribution. *"Overhead contaminated with heavies and excessive reflux rate and high boilup rate"*: inadequate gas-liquid contact/ insufficient liquid disengagement from vapor/ presence of non-condensibles in feed. *"Overhead and bottoms off spec and decreases across column in both $\Delta T$ and $\Delta p$ "*: [dry trays]*. *"Overhead and bottoms off spec"*: bypass open on reflux control valve. *"Overhead and bottoms off spec, decrease in $\Delta T$ across column and perhaps $\Delta p$ increase and cycling of liquid in the bottoms"*: [damaged tray]*. *"Distillation overhead off spec"*: excessive inerts from upstream/buildup of trace/ purge not sufficient from recycle.

*"Separation performance of column decreases"*: trace amounts of water/ trace amounts of water trapped in column/ [bumping resulting in plate damage]*. *"Level of bottoms > design"*: bottoms pump failure, see Section 3.2.3/ bottoms line plugged. and see implications for reboiler, Section 3.3.3. *"Level in bottoms > design" and [column pressure > design]**/ high boilup/ overhead condenser vent plugged. *"Level in bottoms > design and pressure increase in kettle reboiler and surges"*: [foaming]*, inerts/ leaks in kettle reboiler/ undersized reboiler. See also Section 3.3.3. *"Bottoms off spec"*:

loss of tray efficiency/ contamination of bottoms from pump, (from light oil lubricant in bottoms pump or forced circulation reboiler)/ transient vapor puff from horizontal thermosyphon reboiler, see Section 3.3.3. *"Distillate flow too low":* feedrate low/ feed composition of overhead species low/ [low boilup]*/ reflux too high/ overhead control temperature too low. *"Distillate flow too high:"* feedrate high/ feed composition of overhead species high/ reflux ratio too low.

*"Bottoms and overhead flowrates < design":* [flooding]*/ excessive entrainment/ [foaming]*/ excessive Δp but not flooded/ plugging and fouling/ [maldistribution]*. *"Water hammer in column":* process fluid above the tube sheet of a thermosyphon reboiler. *"Cycling of column temperatures:"* controller fault. *"Product contamination":* leaking valves/ crud left in storage tanks/ crud left in dead legs in piping/ corrosion products/ unexpected chemical reactions/ sampling fault/ analysis fault/ unexpected solubility effects. *"Cycling (30 s–several minutes duration) steam flow, cycling pressure on the process side and, for columns, cycling Δp and cycling level in bottoms":* instrument fault/ condensate in instrument sensing lines/ surging/ [foaming]* in kettle and thermosyphon/ liquid maldistribution/ steam-trap problems, see Section 3.5.1, with orifice Δp across trap < design/ temperature sensor at the feed zone in a distillation column/ collapsed tray in a distillation column.

*[Bumping resulting in plate damage]*:* trace amounts of water.

*[Column pressure > design]*:* [high boilup]*/ overhead condenser vent plugged.

*[Damaged trays]*:* leak of water into high molar mass process fluid/ large slugs of water from leaking condensers or steam reboilers/ startup with level in bottoms > design/ attempt to overcome flooding by pumping out bottoms at high rate/ too rapid a depressurization of column/ unexpected change in phase.

*[Downcomer flooding]**: excessive liquid load/ restrictions/ inward leaking of vapor into downcomer/ wrong feed introduction/ poor design of downcomers on bottom trays/ unsealed downcomers/ [foaming]*.

*[Dry trays}*:* flooded above/ insufficient reflux/ low feedrate/ high boilup / feed temperature too high.

*[Foaming]**: surfactants present/ surface tension positive system/ operating too close to the critical temperature and pressure of the species/ dirt and corrosion solids/ natural occurring surfactants/ pH far from the zpc/ naturally occurring polymers/ solids particles/ corrosion particles/ antifoam ineffective (wrong type or incorrect rate of addition/ gas velocity too high/ vapor velocity too high/ tray spacing too small/ asphaltenes present. A more generic listing of the causes of foaming is given in Section 3.11.

*[Jet flooding]*:* excess loading/ fouled trays/ plugged holes in tray/ restricted transfer area/ poor vapor distribution/ wrong introduction of feed fluid/ [foaming]*/ feed temperature too low/ high boilup/ entrainment of liquid because of excessive vapor velocity through the trays/water in a hydrocarbon column.

*[High boilup]**: see Section 3.3.3.

*[Low boilup]*:* see Section 3.3.3.

*[Maldistribution]*:* weirs not level/ low liquid load/ backmixing/ faulty design.

*[Premature flooding]**: internal damage/[fouling]*/ change in feed composition or temperature/ unexpected entrainment/ [foaming]*/ incorrect design for downco-

mers/ unstable control system/ level control problems/ instrument error/ second liquid phase in the column.

### 3.4.3
### Solution Crystallization

Good practice: to prevent plugging, avoid having natural sumps for suspension type crystallizers. Control the degree of supersaturation. For crystallizers operated with cooling or evaporative crystallization, the supersaturation occurs near the heat exhange surface. For antisolvent or reaction crystallizers, the key control of supersaturation is local (often the mixing). Differentiate among the different types of product or impurities to solve problems: surface contamination, agglomeration traps impurities, inclusions, polymorphism. Check that impurities are soluble for the end point conditions of crystal growth = condition of separation.

Trouble shooting: base approach on mass and energy balances, population or number balances. Follow the population density of number versus size. Must know the type of crystals and the mode of operation.

*"Yield < design"*: initial concentration < design. *"Impure product because of surface contamination"*: poor solid-liquid separation/ poor washing, see Sections 3.5.12, 3.5.13, 3.5.14. *"Impure product because of agglomeration trapped impurities"*: wrong pH/ wrong magma electrolyte concentration/ wrong mixing. *"Impure product because of inclusions"*: supersaturation driving force too large. *"Impure product because of polymorphism"*: change in crystal habit during the crystallization process/ isoelectric point/ mixing problem. *"Crystal habit (shape and aspect ratio) differs from specs"*: wrong temperature during growth/ impurities especially surfactants/ supersaturation level too high. *"Size distribution > design"*: supersaturation too close to metastable limit. *"Filtration rate slow"*: crystals too small/ large size distribution/ fault with filtration, see Sections 3.5.12 and 3.5.13. *"Incrustation, fouling, deposits"*: cold spots/ missing insulation/ low suspension density/ protrusions and rough areas on the process surface/ local supersaturation too high/ cooling surfaces too cold. *"Product contamination"*: leaking valves/ crud left in storage tanks/ crud left in dead legs in piping/ [corrosion]* products/ unexpected chemical reactions/ sampling fault/ analysis fault/ unexpected solubility effects. *[Corrosion]** see Section 3.1.2.

**Vacuum and circulating systems:** Trouble shooting: *"Crystal size too small"*: low suspension density/ high circulation rate/ solids in feed causing nucleation sites/ feed flowrate > design/ excessive turbulence/ local cold spots/ subsurface boiling/ supersaturation too high or too close to the metastable limit. *"Insufficient vacuum"*: see Section 3.2.2/ obstruction in vapor system/ insufficient cooling water to condenser/ temperature of the cooling water to the condensers > design/ air leaks. For steam ejectors: steam pressure < design. For mechanical vacuum pumps: seal water flowrate < design/ rpm < design. *"Liquid level in crystallizer fluctuates wildly"*: check out the vacuum system, see Section 3.2.2/ low steam pressure to the steam ejectors/ fluctuation in the flow of cooling water to the condensers. *"Circulation rate differs from design"*: see pumps, Section 3.2.3. *[Foaming]**: air leaks in pump packing/ air in feed/ air leak in flanges or valve stems. A generic listing of the causes of foaming is given in Section 3.11.

### 3.4.4 Gas Absorption

Good practice: the more selective the absorbent the more difficult it is to regenerate the absorbent. Prefer the use of low holdup internals. Select materials of construction to promote wetting: select critical surface tension of the solid is > the surface tension of the liquid. If the surface tension of the feed liquid > 2 mN/m *larger* than the surface tension of the bottom exit liquid or the absorption of the solute lowers the surface tension (surface tension negative) prefer the use of trays to packings to minimize potential for liquid film breakup. If the surface tension of the feed liquid > 2 mN/m *smaller* than surface tension of the bottom exit liquid (surface tension positive), the foam above trays might be unexpectedly stable; stable films on packing.

Trouble shooting: for multitube cocurrent falling film absorber: *"Concentration of product acid < design, inadequate absorption"*: liquid maldistribution/ gas maldistribution. *"Low heat-transfer coefficient"*: liquid or gas maldistribution. *"Hydraulic instability"*: no vent break on the syphon. *"Product contamination"*: leaking valves/ crud left in storage tanks/ crud left in dead legs in piping/ corrosion products/ unexpected chemical reactions/ sampling fault/ analysis fault/ unexpected solubility effects.

*For amine absorption* of sour gas, Good practice: keep inlet amine solvent temperature at least 5 °C hotter than inlet gas temperature to minimize condensation of volatile hydrocarbons in the inlet gas stream. Trouble shooting: *"Insufficient absorption or off-specification for exit scrubbed gas"*: feed gas concentration off spec/ feed gas temperature or pressure outside operating window: *for amine absorbers*: > 50 °C for $H_2S$ and < 24 °C for $CO_2$/ feed gas pressure has decreased/ [solvent flowrate too low]*; *for glycol dehydration*: 12.5 to 25 L TEG per kg water removed / [solvent incorrect]* / incorrect feed try location/ [column operation faulty]*/ absorber operating conditions differ from design/ [absorber malfunction]*.

*"$\Delta p$ across absorber > design"*: gas flowrate > design/ pressure < design/ [foaming]*/ plugged trays/ plugged demister pads/ collapsed tray or packing. *"$\Delta p$ on column fluctuating"*: [foaming]*. *'Solvent carryover from the top of the column"*: [foaming]*. *"Liquid level in vessels fluctuates"*: [foaming in column]*. *"Change in absorption rate"*: *for amine absorption*: decrease in removal of $H_2S$ and increase in removal of $CO_2$/ [foaming]*. *"Overloaded liquid in downstream gaseous processing equipment"*: [foaming in absorber]*. *"Solvent losses high"*: [physical losses]*/ [entrainment]*/ [solubility]*/ [vaporization]*/ [degradation]*/ [loss elsewhere]*/ *for glycol dehydration* typical losses = 0.015 mL/ $m^3$ gas treated.

*[Amine concentration too high or too low]**: if too high, lack of equilibrium driving force/ if too low, insufficient moles of amine for the feed concentrations.

*[Column operation faulty]**: plugged tray or packing/ poor distribution for packing/liquid flowrate < minimum required for loading/ [gas velocity too high]*/ col-

lapsed trays or packing/ plugged or broken distributors/ [foaming]*/ solvent–stripper overhead temperature too low. see also Section 3.4.2

*[Corrosion]**: see Section 3.1.2.

*[Degradation]**: chemical reaction; for *amine*: reacts with $CO_2$ and $O_2$; forms stable salts: for *glycol*: reacts with $O_2$/ thermal decomposition; for *amine*: surface temperatures > 175 °C ; for *glycol*: surface temperatures > 205 °C.

*[Entrainment: GL]**: demister plugged, missing, collapsed, incorrectly designed/ [flooding]*/ [foaming]*/ inlet liquid line or distributor undersized of plugged/ poor distribution for packing/liquid flowrate < minimum required for loading/ [gas velocity too high]*/ solvent feed temperature > specifications/ [column operation faulty]*/ tray spacing < design. see also GL separators Section 3.5.1

*[Entrainment: L-L]**: fluid velocity too high; example > 10 L/s $m^2$/ liquid distributor orifice velocity > design; *for amine*: for amine > 0.8 m/s; for hydrocarbon > 0.4 m/s/ faulty location of exit nozzles/ interface level wrong location/ faulty control of interface/ no vortex breaker/ exit fluid velocities > design/ insufficient residence time/ [stable emulsion formation]*. see also decanters, Section 3.5.3.

*[Foaming]**: [foam-promoting contaminants]*/ [gas velocity too high]*/ [liquid residence time too low in GL separator]*/ antifoam addition faulty/ faulty mechanical foam breaker/ [liquid environment wrong]*. A generic listing of causes for foaming is given in Section 3.11.

*[Foam promoting contaminants: soluble]**: naturally occurring or synthetic polymers/ naturally occurring or synthetic organics >C10; example lube oils/ naturally occurring or synthetic surfactants; *for amine systems*: the surface active contaminants include condensed hydrocarbons, organic acids, water contaminants, amine-degradation products/ faulty cleaning before startup; surfactants left in vessels.

*[Foam promoting contaminants: solid]**: [corrosion products]*; for *amine systems*: iron sulfides; amine salts formed from organic acids + hydrocarbons/ faulty cleanup before startup; rust left in vessel/ dust/ dirt/ particulates.

*[Flooding]**: see Section 3.4.2.

*[Gas velocity too high]**: vessel diameter too small for gas flow/ column pressure < design/ trays or packing damaged or plugged giving excessive vapor velocity/ temperature too high/ upstream flash separator passing liquids: feed contaminated with excessive volatile species/ stripping gas fed to column too high/ flowmeter error/ design error.

*[Liquid environment wrong]**: pH far from the zpc/ electrolyte concentration too low.

*[Physical losses]**: leak to atmosphere/ purges for sampling/ sampling/ heat exchanger leak/ pump seal flushes/ filter changes/ piping, fitting, valve stems, gaskets, pumps.

*[Solubility losses]**: liquid-liquid systems: system pressure < design/ for amine: concentrations > 40% w/w/ system temperatures too high.

*[Solvent contaminated]**: carryover from upstream equipment; example oil from compressor; brines, corrosion inhibitors, sand, [corrosion products]* / oxygen leaks into storage tank/ inadequate corrosion control, example low pH causing corrosion/ degradation via overheating, ex-hot spots in reboiler tubes or fire tubes/ ineffective

filters/ ineffective cleaning before startup/ *for amine absorbers*: corrosion products/ FeS/ chemicals used to treat well

*[Solvent feed temperature too high]***: fouled exchanger/ undersized heat exchanger/ ambient temperature too high.

*[Solvent flowrate too low]***: flowmeter or sensor error/ absorber pressure > design/ plugged strainer, lines of filters/ low liquid level in pump feed tank/ [cavitation]*/ air-locked pump and see Section 3.2.3 for trouble shooting pumps.

*[Solvent incorrect]***: incorrect concentration of active ingredient: *for amine absorbers*: [amine concentration too high or too low]*; *for glycol dehydration*: solvent concentration TEG < specifications / [solvent stripping inadequate]*/ [solvent feed temperature too high]*/ [solvent contaminated]*.

*[Solvent loss elsewhere]***: upstream units, for example *for glycol dehydration*: glycol dumped with hydrocarbons separated in upstream flash drum/ loss in downstream solvent stripper.

*[Solvent stripping inadequate]***: not enough steam in stripper/ incorrect pressure in stripper/ [foaming]*/ [contaminated solvent]*/ contaminated feed: *for amine strippers*: other sulfur species causing high partial pressure/ leak in the feed preheater contaminating feed with stripped solvent.

*[Vaporization losses]***: system pressure < design/ *for amine*: concentrations > 40% w/w/ system temperatures too high.

### 3.4.5
### Gas Desorption/Stripping

Trouble shooting: *"Solvent or stripped liquid concentration > design"*: boilup rate or steam stripping rate too low/ feed concentration > expected/ feed contamination; *for sour-water stripper*: acid in feed may be chemically bonded with $NH_3$ and prevent adequate stripping of $NH_3$/ [foaming]*/leak in preheater exchanger/ [column malfunction]*.

*"Overhead from stripper < specifications"*: insufficient flowrate of stripping gas/ for *glycol dehydration*: reboiler temperature < 175–200 °C or reboiler too small for required duty or fouling of reboiler tubes/ [foaming]*/ dirty or broken packing or plates/ [fouled or scaled internals]*/ [flooding]*/ top pressure > design/ leak in preheater exchanger/ [feed concentration off specification]*. *"Overhead temperature on stripper > design"*: reflux flowrate too low/ [flooded]*/ [foaming]*/ feed contaminated with light hydrocarbons. *"For sour-water strippers or glycol dehydration: Pressure at reboiler > design"*: instrument error/ top pressure > design/ [$\Delta p$ across column > design]*/ overhead line plugged/ [flooding]*/ for stripper for *glycol dehydration*: slug of hydrocarbon in feed is flash vaporized at reboiler and blows liquid out of stripper. *"For sour-water strippers: odor or $H_2S$ problems at the storage tank"*: 0.6 to 1 m layer of oil on top of water missing/ oil layer exceeds 0.6 to 1 m depth/ faulty inert gas operation. *"Plugging of overhead system"*: top temperature not within the operating window; *for sour-water strippers*: temperature < 82 °C at which ammonium polysulfides form but temperatures too high give excessive water in overhead vapor causing problems for downstream operation / overhead lines not insulated/ insufficient steam tracing

on overhead vapor lines. *"Feed flowrate and composition to the stripper varies":* [instrument error]*/ sampling error/ analysis error/ [faulty separation in flash drum]*/ [foaming in upstream absorber]*/ no intermediate storage tank between the flash drum and the stripper/ storage tank faulty operation or design: *for SWS*: residence time < 3 to 5 days; stratification occurs, bypassing occurs, insufficient mixing in tank/ oil layer on top of water in storage tank exceeds 0.6 to 1 m depth.

*[Column malfunction]*:* [feed concentration off specification]*/ excessive stripping gas or steam velocity/ too much cooling or condensation/ top temperature > design/ insufficient reflux cooling/ packing broken, damaged/ [fouled or scaled internals]*/ [foaming]*/ [flooding]*, see also Section 3.4.2.

*[Feed concentration off specification]*:* [foaming in upstream absorber]*/ for *glycol dehydration:* upstream flash separator passing water; for oil or hydrocarbon in feed to *SWS*": residence time for sour water in flash drum is < 20 min.

*[Flooding]*:* see Section 3.4.2.

*[Foaming]*:* see Section 3.11.

*[Fouling]*:* see Section 3.11

*[Fouled or plugged internals]*: for SWS*: cooling water leak/ pH of feed water too basic/ calcium ion concentration too high causing precipitation when temperatures in stripper exceed 122 °C/ temperature < 82 °C at which ammonium polysulfides form/ overhead lines not insulated.

*[Instrument error]*:* calibration fault/ sensor broken/ sensor location faulty/ sensor corroded/ plugged instrument taps: *for sour-water strippers:* water or steam purge of taps malfunctioning or local temperatures < 82 °C at which ammonium polysulfides form.

*[$\Delta p$ across column > design]** see Section 3.4.2.

### 3.4.6
### Solvent Extraction, SX

Good practice: the dispersed phase should not preferentially wet the materials of construction. If unexpected rapid coalescence occurs, suspect [Marangoni effects]* and change the dispersed phase. Treat the buildup of the "rag" at the interfaces based on the cause: corrosion products or stabilizing particulates, surfactants, or amphoteric precipitates of aluminum or iron. Consider adjusting the pH. Solid particles tend to accumulate at the liquid–liquid interface.

Trouble shooting: *"Poor separation"*: level control fault/ phase velocities too high/ contaminant gives stable dispersion/ smaller drop size than design/ rag formation/ temperature change/ pH change/ decrease in electrolyte concentration. *[Rapid coalescence]**: wrong phase is the continuous phase/ [Marangoni instabilities]*/ pH at the zpc/ high electrolyte concentration in the continuous phase.

*[Marangoni effects]*:* non-equilibrated phases/ local mass transfer leads to local changes in surface tension and stability analysis yields stable interfacial movement.

**For column extractors:** *"Decrease in extraction efficiency":* agitator speed to fast/ excessive backmixing/ flooding.

*[Flooding]**: agitator speed too fast/ feed sparging velocity too high/ drop diameter smaller than design.

**For centrifugal extractors:** for bioprocessing/proteins: Good practice: the partition coefficient sensitive to pH, electrolyte type and concentration.

*"Product contamination":* leaking valves/ crud left in storage tanks/ crud left in dead legs in piping/ corrosion products/ unexpected chemical reactions/ sampling fault/ analysis fault/ unexpected solubility effects.

Other suggestions for trouble-shooting decanters are given in Section 3.5.3. More about stable emulsion formation is given in Section 3.11.

### 3.4.7
### Adsorption: Gas

Trouble shooting: *"Wet gas"*: steam leak/ leaky valves/ inadequate regeneration/ wrong adsorbent/ adsorbent damaged by excessive regeneration temperature/ adsorption cycle too long/ [early breakthrough]*. *"$\Delta p$ high"*: fine particulates in feed/ breakdown of adsorbent/ high gas feedrate. *"Product contamination":* leaking valves/ crud left in storage tanks/ crud left in dead legs in piping/ [corrosion]* products/ unexpected chemical reactions/ sampling fault/ analysis fault/ unexpected solubility effects.

*[Corrosion]*:* see Section 3.1.2. *[Early breakthrough]**: gas short-circuiting bed/ faulty regeneration/ increased concentration in feed/ other contaminants in feed.

### 3.4.8
### Adsorption: Liquid

Good practice: carbon regeneration by multiple hearth furnaces. For edible oils prevent contact with air.

Trouble shooting: *"Early breakthrough"*: liquid short-circuiting bed/ faulty carbon regeneration/ increased concentration in feed/ other contaminants in feed. *"Pressure drop high"*: fine particulates in feed/ breakdown of carbon/ high liquid feedrate.

### 3.4.9
### Ion Exchange

Good practice: use an upstream degasser to remove carbonic acid.

Trouble shooting: usual sources of trouble are change in ions in the feed, the multiport valves improperly seat so that feed or regenerant bypass into the effluent, clogged liquid distributors, clogged underdrains; degradation of the resin and faulty backwash. Organic fouling mainly affects anionic exchangers. *"Throughput capacity < design":* instrument error/ increase in feed concentration/ less resin volume than design/ regenerant concentration < design/ regenerant volume < design, 0.5–3.5 L/s L of resin/ regenerant flowrate < design/ wrong regeneration ion/ contamination of regenerant with high valence ions. *"In the spring, reduced flowrate through the unit* *demineralizing river water":* high concentration of particulates, clay in spring river water.

*"In the summer for a unit demineralizing water, throughput of the cationic exchanger decreeses, exchange capacity for calcium and magnesium decreases but the anionic exchanger is unaffected":* suspect ferric or high-valency cation present in the feed. *"In the summer for a unit demineralizing river water, throughput of the anionic exchanger decreases, exchange capacity decreases but the cationic exchanger is unaffected":* suspect fertilizer runoff with phosphate, carbonic acid and high sulfate anions as contamination. *"Contamination in exit liquid > design":* instrument error/ sampling error/ online too long/ faulty regeneration/ [fouled]*/ [poisoned]*/ high feed concentration of target ions/ feed concentration of high valence co-ions. *"$\Delta p$ > design":* dirt in feed/ water from river in springtime/ instrument error/ temperature/ resin void volume changes/ inlet distribution system blocked/ [resin degradation]* and backwashes into inlet/ backwash rate too high/ underbed blocked.

*[Fouling of the resin]**: iron and high-valence ions/ oil/ mud/ polyelectrolyte/ calcium sulfate precipitate/ silica/ barium sulfate/ carbonic acid/ sulfate or phosphate/ organics/ algae and bacterial fouling.

*[Poison resin]**: cobalticyanide/ polythionate/ ferricyanides/ complex humic acid/ color bodies in sugar juices.

*[Resin degradation]**: ingress of oxidants/ free chlorine in feed/ temperature increase/ [fouled]*/ [poisoned]*/ corrosion products/ [resin fines]*.

*[Resin fines]**: thermal or physical shock/ freeze/thaw.

**WAC**: *"Alkalinity leakage during exhaustion cycle":* inadequate regeneration. *"Hardness leakage during exhaustion cycle":* regeneration fault with calcium sulfate precipitation (if sulfuric acid is the regenerant).

**SAC**: *"Sodium leakage":* inadequate regeneration: wrong concentration, wrong flowrate, wrong length of time. *"Hardness leakage during exhaustion cycle":* regeneration fault with calcium sulfate precipitation (if sulfuric acid is the regenerant).

**WBA**: *"Mineral acid leakage":* under regenerated/ upstream SAC malfunctioning. *"Sodium leakage, high pH and high conductivity":* SAC resins contaminated the bed. *"Silica problems":* series regeneration with SBA with pH falling below the isoelectric point of silica in the resin bed.

**SBA**: *"Increase in silica leakage":* [resin degradation]*. *"Leakage of target ions":* organic fouling. *"Low pH and high conductivity":* organic fouling. *"Increase rinse quantities":* organic fouling. *"Low pH (5.5), increased conductivity, increase silica leakage, increase rinse volumes, loss of throughput capacity:"* organic fouling. *"Product contamination":* leaking valves/ crud left in storage tanks/ crud left in dead legs in piping/ corrosion products/ unexpected chemical reactions/ sampling fault/ analysis fault/ unexpected solubility effects.

### 3.4.10
### Membranes: Reverse Osmosis, RO

Good practice: consider pretreating hydrophobic membranes for aqueous use.

Trouble shooting: "*Permeate flow < design:*" physical fouling (incorrect/incomplete pretreatment/ scaling/ biofouling)/ chemical fouling: (pH shift/ incorrect anti-scalant dosage). "*Permeate quality degradation:*" failure of mechanical seal/ chemical attack of membrane by pH, chlorine or biodegradation / concentration polarization / post-contamination.

### 3.4.11
### Membranes: Nanofiltration

Trouble shooting: "*Permeate flow < design:*" physical fouling (incorrect/incomplete pretreatment/ scaling/ biofouling)/ chemical fouling: (pH shift/ incorrect anti-scalant dosage). "*Permeate quality degradation:*" failure of mechanical seal/ chemical attack of membrane by pH, chlorine or biodegradation / concentration polarization / post-contamination.

### 3.4.12
### Membranes: Ultrafiltration, UF, and Microfiltration

Good practice: for membranes that are not hydrophobic; check the isoelectric or zero point of charge point of the species in solution compared with the charge on the membrane and consider changing the pH of operation so that the surface charges are the same. For hydrophobic membranes treating aqueous feeds, consider pretreating the membrane to make the membrane surfaces hydrophilic.

Trouble shooting: "*Permeate flux < design*:" physical clogging (inadequate pre-screening/ backwash problems/ aeration/ recirculation/ increase in influent solids loading); chemical fouling (change in water quality/ inadequate cleaning). "*Permeate quality < design*:" failure in mechanical seal, breakage of the membrane or hollow fibres / post contamination via regrowth/ degradation of membrane by pH or chlorine.

## 3.5
## Heterogeneous Separations

In heterogeneous phase separation we start with at least two phases. Sections 3.5.1 to 3.5.4 address the separation of gas from liquid, gas from solid, liquid from liquid and gas from liquid, respectively. Liquid-solid separators include drying, Section 3.5.5; screens, Section 3.5.6; settlers, Section 3.5.7; hydrocyclones, Section 3.5.8; thickeners, Section 3.5.9; sedimentation centrifuges, Section 3.5.10; filtering centrifuges, Section 3.5.11; and filters, Section 3.5.12. Trouble shooting screens to separate solids is discussed in Section 3.5.13.

### 3.5.1 Gas–Liquid

Good practice: install a demister.

In this section knockout pots and steam traps are considered.

**Knockout pots:** Trouble shooting: *"Poor separation"*: [foaming]*/ insufficient residence time/ feed and exit nozzles at wrong location/ faulty design. *[Foaming]**: surfactants present/ dirt and corrosion solids/ natural occurring surfactants/ pH far from the zpc/ naturally occurring polymers/ insufficient disengaging space above the liquid/ antifoam ineffective (wrong type or incorrect rate of addition)/residence time insufficient/ designed for a vertical vessel but a horizontal vessel installed/ vapor velocity too high/ mechanical foam breaker not rotating/ baffle foam breaker incorrectly designed or damaged/ asphaltenes present/ liquid downflow velocity through the foam is too low.[3] See also Section 3.11.

**Steam traps:** Good practice: install trap below condensate exit (or with a water seal if the trap is elevated), use a strainer before most traps, use a check valve for bucket traps. Slant pipes to the trap. Use a downstream check valve for each trap discharging to a common header. Pipe diameter should be greater than or equal to the trap inlet pipe diameter. Prefer to install auxiliary trap in parallel instead of a bypass. Do not group trap thermodynamic traps because of their sensitivity to downstream conditions.

Float and thermostatic: usually discharges continuously, low pitched bubbling noise. High pitch noise suggests live steam is blowing.

Balanced thermostatic: leave about 0.6 m of uninsulated pipe upstream of trap. Diagnostics: when bellows placed in boiling water the expansion should be 3 mm.

Inverted bucket: use initial prime to prevent steam blowing. Diagnostics sounds: when it is functioning well: loud initially, then lower pitch bubbling and then silence. Discontinuous discharge. When steam is blowing through the trap, the sound is a steady bubbling if primed with a light load or constant rattling; or continuous high pitched whistling. Diagnostic for loss of prime: close outlet valve for several minutes, then open valve slowly and operation should return to normal. If this fails then check seat and valve.

Thermodynamic: about 6 cycles/minute.

Trouble shooting: the major faults are wrong trap, dirt, steam locking in the trap, group trapping, air binding and water hammer. Too large a trap gives sluggish response and wastes steam. Too small a trap gives poor drainage, backup of condensate. There is a $\Delta T$ across all traps. *"No condensate discharge"*: strainer or line plugged/ steam off/ valves plugged/ no water or steam to the trap/ trap clogged/ wrong trap selected/ worn orifice/ steam pressure too high (inverted bucket)/ orifice enlarged by erosion (bucket trap)/ incorrect $\Delta p$ across the orifice (inverted bucket)/ air vent clogged (inverted bucket or thermostatic air vent on float trap)/ valve seat choked (inverted bucket)/ flabby or elongated bellows (thermostatic)/ superheated

3) Turner, J. et al., 1999, HP June p. 119.

steam caused burst joints or scale (thermostatic). *"Cold trap + no condensate discharge"*: strainer or line plugged/ steam off/ valves plugged/ no water or steam to the trap/ trap clogged. *"Hot trap + no condensate discharge"*: bypass open or leaking/ trap installed at high elevation/ broken syphon/ vacuum in heater coils/ pressure too high (inverted bucket)/ orifice too large (inverted bucket)/ vent hole plugged (inverted bucket)/ defective trap parts (inverted bucket)/ clogged orifice (thermodynamic). *"Live steam blowing, and inlet and exit temperatures are equal":* bypass open or leaking/ worn trap components/ scale in orifice/ valve fails to seat/ trap lost prime (inverted bucket)/ sudden drops in pressure/ [backpressure too high]* (thermodynamic)/ faulty air release (float)/ trap too large (thermodynamic). *"Continuous discharge when it should be discontinuous":* trap too small/ dirt in trap/ high-pressure trap installed incorrectly for low pressure service (bucket trap)/ valve seat clogged with dirt/ excessive water in the steam/ bellow overstressed (thermostatic)/ one trap serves > one unit/ strainer clogged. *"OK when discharging to the atmosphere but not when to a backpressure condensate header":* condensate line diameter too small/ wrong orifice/ interaction with other traps connected to a common header/ condensate line partially plugged/ [backpressure too high]*. *"Slow and uneven heating of upstream equipment":* trap too small/ insufficient air handling capacity/ short-circuiting when units are group trapped. *"Inverted bucket trap loses prime:"* sudden drop in pressure/ faulty seat/ faulty valve. *" Upstream process cycling":* defective float/ multiple sources of condensate to a single trap/ trap flooded from condensate header/ condensate discharged into the bottom of the condensate header/ $\Delta p$ across the orifice is incorrect for the orifice (inverted bucket).

*[Backpressure too high and trap is hot]**: return line too small/ other traps blowing steam / obstruction in return line/ bypass open/ pressure in header too high.

*[Backpressure too high and trap is cold]**: obstruction in return line/ excess vacuum in return line.

## 3.5.2 Gas–Solid

Bag filters and dry cyclones are discussed in this section.

**Bag filters:** Good practice: replace a complete set of bag filters annually. Install a bypass. Limit the number of parallel rows of bags on either side of the walkway to 3–4 rows for 20-cm diameter bags and 2–3 rows for 30-cm diameter bags. For cleaning, use 0.5–0.7 kPa clean, dry air with an air: cloth ratio of 2: 1 for reverse jet and 2.5: 1 for shaking. Trouble shooting: *"Excessive particle emissions":* cleaning too often/ pressure used to clean it too high/ bag breaks/ gas temperature too high and particles crust on movable blowrings and tear bag. *"$\Delta p$ across bags > design":* faulty cleaning/ improper bag tension/ excessive moisture causing blinding/ poor air distribution/ hopper plugged, see Section 3.10.3/ gas velocity > design. *"Short bag life":* excessive cleaning/ high inlet gas velocity/ fines > design/ blinding because of condensation, improper cleaning, excessive dust load or high cake density.

**Dry cyclone**: Trouble shooting: *"Increase in catalyst losses":* [poor separation in cyclone]*. *"Opaque flue gas from the vessel":* [poor separation in cyclone]*. *"Particulate carryover that affects operation of downstream equipment":* [poor separation in cyclone]*. *"Temperature hot spots in upstream reactor":* [maldistribution]*/ local exothermic reactions.

*[Attrition of the particles]*:* local velocities upstream of cyclone > 60 m/s/ particle too fragile.

*[Change in size of particles in the feed]*:* [generation of fines]*/ [coarse particles]*.

*[Coarse particles (diameter > design)]*:* agglomeration of catalyst/ [sintered particles]*/ wrong specifications for catalyst.

*[Dipleg unsealed]*:* solids level does not cover end of dipleg/ $\Delta p$ indicator for catalyst level faulty/ $\Delta p$ indicator for catalyst level OK but bed density incorrect.

*[Generation of fines]*:* [attrition of the catalyst]*/ fines in the new catalyst.

*[Maldistribution]*:* feed distributor poorly designed/ feed distributor plugged.

*[Plugged dipleg]*:* spalled refractory plug/ level of catalyst in bed too high / $\Delta p$ indicator for catalyst level faulty/ $\Delta p$ indicator for catalyst level OK but bed density incorrect. air out periods with a lot of water of steam in vessel.

*[Plugged grid holes]*:* foreign debris entering with fresh catalyst/ faulty grid design.

*[Poor separation in cyclone]*:* [stuck or failed trickle valve]*/ [plugged dipleg]*/ [dipleg unsealed]*/ gas velocity into cyclone too low or too high/ faulty design of cyclone/ solids concentration in feed too high/ cyclone volute plugged/ hole in cyclone body/ pressure surges/ [change in size of particles in feed]*.

*[Sintered particles]*:* high temperature upstream/ [temperature hot spots in the upstream reactor]*.

*[Stuck or failed trickle valve]*:* binding of hinge rings/ angle incorrect/ wrong material/ hinged flapper plate stuck open/ flapper plate missing.

### 3.5.3 Liquid–Liquid

Decanters and hydrocyclones are discussed in this section.

**Decanter:** Good practice: contamination can interfere with the operation. Traditionally this contamination is surfactants, or particulates. The particulates can be corrosion products, amphoteric precipitates of aluminum or iron. Try changing the pH of the water to alter the surface charge on the dispersed drops. The separation capacity of a settler/decanter doubles for every 20 °C increase in temperature. Caution, if, to ease this separation, the temperature is increased, such an increase in temperature will increase the bulk-phase contamination because of the increased cross-contamination by the mutual solubility.

Trouble shooting: *"Entrained droplets in liquid effluent":* sensor error/sampling error (immiscible drops are not being entrained)/ faulty design of separator/ improper cleaning of vessel after shutdown, e.g, rust left in vessel/ pressure fluctuation/ pressure too low causing flashing/[inaccurate sensing of interface]*/ [drop doesn't settle]*/ [drop settles and coalesces but is re-entrained]*/ [drop settles but doesn't coalesce]*/ [stable emulsion formation]*. *"Fluctuation in liquid level":* no vacuum break on syphon line for bottoms/ level sensor error/ poorly tuned controller/ surges in feed.

*[Coalescer pads ineffective]*:* temperature too high/ pH incorrect/ fibers have the same charge as the droplets/ surface tension negative system/ wetting properties of fibers changed/ fibers "weathered" and need to be replaced/ flowrate too slow through fibers/ wrong mix of fibers/ prefiltering ineffective/ surface tension < 1 mN/m for fluoropolymer fibers or < 20 mN/m for usual fibers/ wrong design/ included in decanter but should be separate horizontal coalescer promoter unit/ faulty design. see Section 3.9.2.

*[Density difference decrease]*:* dilution of the dense phase/ reactions that dilute the dense phase; for *sulfuric acid alkylation*: if acid strength < 85% w/w the olefins polymerize with subsequent oxidation of the polymers by sulfuric acid. As a self-perpetuating continuing decrease in acid strength. Alkylate-acid separation is extremely difficult when acid concentration is 40% w/w.

*[Drop doesn't settle]*:* [density difference decrease]*/ [viscosity of the continuous phase increases]*/ [drop size decreases]*/ [residence time for settling too short]*/ [phase inversion or wrong liquid is the continuous phase]*/ pressure too low causing flashing and bubble formation.

*[Drop settles and coalesces but is re-entrained]*:* faulty location of exit nozzles for liquid phases/ distance between exit nozzle and interface is < 0.2 m/ overflow baffle corroded and failure/ interface level at the wrong location/ faulty control of interface/ liquid exit velocities too high/ vortex breaker missing or faulty on underflow line/ no syphon break on underflow line/ liquid exit velocities too high.

*[Drop settles but doesn't coalesce]*:* [phase inversion]*/ pH far from zpc/ surfactants, particulates or polymers present/ electrolyte concentration in the continuous phase < expected/ [coalescer pads ineffective]*/ [drop size decrease]*/ [secondary haze forms]*/ [stable emulsion formation]*/ [interfacial tension too low]*/ [Marangoni effect]*.

*[Drop size decrease]*:* feed distributor plugged/ feed velocity > expected/ feed flows puncture interface/ local turbulence/ distributor orifice velocity > design; *for amine units*: for amine > 0.8 m/s; for hydrocarbon > 0.4 m/s/ [Marangoni effects]*/ upstream pump generates small drops/ [secondary haze forms]*/ poor design of feed distributor.

*[Inaccurate sensing of the interface]*:* instrument fault/ plugged sight glass.

*[Interfacial tension too small]*:* temperature too high/ [surfactants present]* at interface.

*[Marangoni effects]*:* non-equilibrated phases/ local mass transfer leads to local changes in surface tension and stability analysis yields stable interfacial movement.

*[Phase inversion]*:* faulty startup/ walls and internals preferentially wetted by the dispersed phase.

*[Rag buildup]*:* collection of material at the interface: [surfactants present]* / particulates: example, products of [corrosion see Section 3.1.3]*, amphoteric precipitates of aluminum/ naturally occurring or synthetic polymers.

*[Residence time for settling too short]*:* interface height of the continuous phase decreases/ [inaccurate sensing of interface]*/ turbulence in the continuous phase/

flowrate in continuous phase > expected; for example > 3 L/s $m^2$ / sludge settles and reduces effective height of continuous phase/ [phase inversion]*/ inlet conditions faulty.

*[Secondary haze forms]*:* small secondary drops are left behind when larger drop coalesces, need coalescer promoter, see Section 3.9.2.

*[Stable emulsion formation]*:* [surfactants present]* / contamination by particulates: example, products of [corrosion products. see Section 3.1.3]*, amphoteric precipitates of aluminum or iron/ pH far from the zpc/ contamination by polymers/ temperature change/ decrease in electrolyte concentration/ the dispersed phase does not preferentially wet the materials of construction/ coalescence-promoter malfunctioning/ improper cleaning during shutdown/ [rag buildup]*.

*[Surfactants present]*:* formed by reactions/ enter with feed, example oils, hydrocarbons >C10, asphaltenes/ left over from shutdown, example soaps and detergents/ enter with the water, example natural biological species, trace detergents.

*[Viscosity of the continuous phase increases]*:* temperature too low, for *alkylate-acid* separation, temperature < 4.4 °C/ [phase inversion]*/ contamination in the continuous phase/ unexpected reaction in the continuous phase causing viscosity increase.

**Hydrocyclones:** Good practice: control on pressure drop. May be operated as open or flooded underflow. Trouble shooting: *"Incorrect separation":* faulty design/ inlet pressure too low/ wrong $\Delta p$ from feed to overflow/ interfacial tension < 10 mN/m/ wrong volume split/ feed drop size too small.

### 3.5.4 Gas–Liquid–Liquid Separators

Horizontal drum: Good practice: separates gas, oil and water; as for example as an early separation of natural gas upstream of drying or to handle sour water. Typically, a relatively small load of hydrocarbon. Often called a "flash drum". Often follow the flash drum with a storage tank to allow further separation of water and hydrocarbon. Contamination from naturally occurring or synthetic surfactants or polymers, or corrosion products from upstream processing can cause stable foam or emulsion formation.

Trouble shooting: *"Entrained liquid in overhead gas":* sensor error/ [entrainment: GL]*. *"Incomplete separation of oil from water":* faulty design of separator/ residence time of liquid phases too short/ liquid velocity in the decant phases too fast/ [Marangoni instabilities]*/ liquid feed velocity too high/ poor distribution of liquid feeds/ faulty location of exit nozzles for liquid phases/ overflow baffle corroded and failure/ interface level at the wrong location/ faulty control of interface/ no vortex breaker at water and heavy oil exit nozzles/ liquid exit velocities too high/ [emulsification]*/ contaminant gives stable dispersion/ smaller drop size than design/ rag formation/ temperature change/ pH change/ decrease in electrolyte concentration. See Sections 3.5.1 and 3.5.3 for more details.

*[Entrainment: GL]*:* vessel diameter too small for gas flow/ no demister or demister malfunctioning/ vessel pressure < design/ [foaming]*/ inlet liquid line or distributor undersized or plugged.

*[Entrainment: L-L]*:* liquid velocity too high; example > 10 L/s $m^2$/ liquid distributor orifice velocity > design; *for amine*: for amine > 0.8 m/s; for hydrocarbon > 0.4 m/s / faulty location of exit nozzles/ interface level wrong location/ faulty control of interface/ no vortex breaker/ exit fluid velocities > design/ insufficient residence time/ [stable emulsion formation]*.

*[Foaming]*:* see Section 3.11 for generic causes.

*[Marangoni instabilities]*:* non-equilibrated phases/ local mass transfer leads to local changes in surface tension and stability analysis yields stable interfacial movement.

*[Stable emulsion formation]*:* see Section 3.11 for generic causes; Section 3.5.3 for more specific causes. The dispersed phase should not preferentially wet the materials of construction. If unexpected rapid coalescence occurs, suspect Marangoni effects and change the dispersed phase. Treat the buildup of the "rag" at the interfaces based on the cause: corrosion products or stabilizing particulates, surfactants, or amphoteric precipitates of aluminum or iron. Consider adjusting the pH. Solid particles tend to accumulate at the liquid–liquid interface.

### 3.5.5
### Dryer for GS Separation

Trouble shooting: work with an overall mass and energy balance. *For continuous rotary steam-tube dryer. "Product moisture content high"*: insufficient steam flow/ rotational speed too fast/ insufficient area/ particles clump/ moisture content of the feed too high/ faulty design of dryer/ flights damaged/ incorrect angle of inclination/ upstream batch centrifuge gives periodic wet cake. *For fluidized-bed dryer. "Product moisture content high"*: solids buildup on gas sparger in fluidized bed (caused because inlet gas temperature too high). *For spray dryer: "product wet and clumps form inside spray dryer"*/ insufficient gas flow/ inlet gas temperature too low, instrument fault/ feed solids concentration lower than design/ liquid drops larger than design. *For fixed bed-hopper* to dry polymer feed for extruder (hot gas < 120 °C): *"Feed material not dry":* incorrect drying temperature/ solids throughput > design/ instrument error/ input air too moist/ ambient air leaking into drying air circuit/ adsorbent for drying air incorrectly regenerated/ air dryer (gas adsorber) fault. See related unit adsorption: gas, Section 3.4.4.

**Hopper dryer for polymer feed to extruder.** *"Polymer pellets leaving hopper are not dry"*: incorrect drying temperature/solids throughput > design/ instrument error/ input drying air too moist/ ambient air leaking into drying air circuit/ adsorbent for drying air incorrectly regenerated/ air dryer (adsorber) fault see Adsorption: gas, Section 3.4.7.

### 3.5.6
### Screens for Liquid Solid Separation or Dewatering

**Batch Screen pack downstream of extruder:** Good practice: install standby screen pack with diverter valve to bring standby on line when on–line filter blinds. Trouble shooting: *"Solid contamination of product":* mesh size too large/ contamination downstream of screen pack. *"Gels in final product":* gels form in extruder/ $\Delta p$ across screen pack excessive/ screen area too small/ size and type of screen cannot retain gel/ gels form downstream of screen pack/ downstream temperature promotes gel formation. *"$\Delta p$ excessive":* filter media too fine/ screen area too small/ screen temperature too low/ gel formation in extruder. *"Filter media pushes through back support plate in screen pack":* screens on the downstream side of the pack not coarse enough or not rigid enough/ support plate holes too large.

### 3.5.7
### Settlers for LS Separation

**Grit chamber:** Trouble shooting: *"Floating sludge":* sludge decomposing and buoyed to the surface/ infrequent sludge removal/ sludge not removed from the hoppers. *"Excessive sedimentation at the inlet":* fluid velocity too slow. *"Intermittent surging":* intermittent pumping rates/ liquid maldistribution of feed. *"Sludge hard to remove from hopper":* high feed concentration of grit, clay/ low velocity in the sludge withdrawal lines.

### 3.5.8
### Hydrocyclones for LS Separation

Good practice: control on pressure drop.

Trouble shooting: *"Underflow too dilute, underflow appears as smooth inverted cone":* inlet velocity low/ inlet feed pressure low. *"Underflow appears as slow, vertical rope of coarse solids":* underflow opening too small/ feed concentration of solids higher than design. *"No discharge from the underflow":* plugged inlet/ plugged underflow. *"Underflow unsteady and variable inlet pressure":* air-gas in the feed.

### 3.5.9
### Thickener for LS Separation

Good practice: consider the use of flocculants or use a deep cone. Flocculant dosage should be related to feed inlet concentration, see also Section 3.9.3. Include high-pressure water purge lines for both forward and reverse flow at > 1 m/s. Raise and lower the rake once per shift. For startup, pump feed into the empty tank and recycle underflow until the design underflow densities are achieved. Trouble shooting: *"Stalled rake":* uneven central feed distribution/ excessive flocculant causing island formation/ underflow concentration > design/ unpumpable underflow/ trying to maintain the underflow concentration when the feed contains fines > design/ storing too many solids in the thickener/ "sanding out"/ particle shape differs from design. *"Plugged underflow lines":* insufficient fines/ targetting underflow concentration > design/ temperature change/ pump problems, see Section 2.2.3 / suction velocity < 0.6–2.5 m/s. *"Underflow concentration of solids too low":* removal of too much underflow/ flocculation problems that give islands that lead to the feed concentration ratholing directly to underflow. *"Supernatant cloudy"*: feed velocity excessive causing breakup of colloidal flocs/ changes in pH or electrolyte concentration causing floc breakup/ excessive feed turbulence/ excessive vertical drops of feed/ insufficient flocculant added or flocculant feedrate constant instead of proportional to solids concentration in the feed. *"Sanding out"*: too high an underflow concentration/ feed concentration of dense particles > 200 μm is > design/ power failure. *"Torque > expected:"* feed concentration suspended solids > expected.

### 3.5.10 Sedimentation Centrifuges

**Horizontal scroll discharge decanter.** Trouble shooting: *"Centrifuge won't start:"* vibration switch triggered/ no power/ motor or starter failure/ overheated motor/ [broken shear pin]* / lubrication oil flowswitch tripped. *"Centrifuge shuts down"*: blown fuse/ overload relays tripped/ motor overheated/ [broken shear pin]* / lube oil flowswitch tripped. *"Excessive vibration"*: broken isolators/ motor on flexible mounts/ motor bolts loose/ flexible piping not used/ misalignment/ bearing failure or damaged/ loss of plows/, damaged conveyor hub/ solid product buildup in conveyor hub/ conveyor or bowl not balanced/ conveyor flights worn or portion of blade missing/ trunions cracked or broken/ conveyor bowl cracked or broken/ leaking effluent weirs/ plugged solids in the effluent hopper/ not level. *"High moisture in exit solids"*: liquid dams not set alike or set incorrectly/ feed temperature too low/ feedrate too high/ effluent hopper plugged or not vented/ conveyor flights worn. *"High solids in liquid effluent"*: feedrate excessive/ effluent dams set wrong/ no strip installed/ incorrect feed temperature.

*[Shear pins breaks]**: feedrate too high/ solids concentration too high/ foreign material stuck in bowl.

*[Solid and screen bowl shear pins break]**: plugged discharge hopper/ conveyor blades bent or rough/ worn bowl strips/ loose or broken trunion bolts/ bowl inadequately washed/ clearance too large for blade tip to bowl wall/ bowl inside rough/ wrong size shear pin.

### 3.5.11 Filtering Centrifuge

Good practice: monitor pH and temperature.

**Pusher:** Trouble shooting: *"Machine floods"*: feed concentration < design/ feedrate > design/ irregular feedrate/ change in size distribution or particle diameter. *"Unstable cake formation"*: feedrate > design.

### 3.5.12
### Filter for LS Separation

Good practice: Precoat: 0.75 $kg/m^2$ to give a precoat thickness of 1.6 mm. Rate for precoat: concentration between 0.3 and 5% w/w and at a rate of 0.7–1.4 L/s. $m^2$. This should give a $\Delta p = 14$ kPa. For leaf or rotary filters, maintain consistent pressure differential across cake once the cake is formed. Consider adding body feed continuously when filtering gelatinous species.

Trouble shooting: "*Poor clarity*": leak/ cracks in cake/ partially blinded septa/cake washing too fast/ flashing of filtrate/ air in filter of feed liquid/ changes in liquid properties/ incorrect filter aid/ change in temperature of pH/ small diameter particles in feed than design/ process upset. "*Short cycle/ high pressure/ low flow*": flow lines too small/ obstruction in outlet line/ pump sucking air/ pressure differential too low/ wide fluctuations in feedrate/ air trapped in filter/ too high a filtration rate.

### 3.5.13
### Screens for Solid–Solid Separation

**Screen, vibrating:** Good practice: if damp or sticky, predry or use heater above the screen to reduce moisture to < 3%. Avoid resonance frequencies. Usual angle of operation is 12 to 18° ; for wet, inclined vibrating screen to 7 to $11^0$. Capacity decreases if the angle of inclination is too high. Blinding is mainly caused by material that is 1 to 1.5 times the hole size. Feed thickness should not exceed 4× aperture size for 1.6 $Mg/m^3$; and not exceed 2 to 3× aperture size for 0.8 $Mg/m^3$. Trouble shooting: *"Capacity decreases":* angle of inclination too high/ blinding.

## 3.6
## Reactor Problems

Temperature is usually a key variable. Increasing the temperature by 10 °C, doubles the rate of reaction.

Operating temperature should be at least 25 °C less than the maximum temperature for a catalyst.

For **PFTR**: increasing the temperature may break down the catalyst into a powder that causing dusting/ contamination/ plugging problems downstream and increase the pressure drop in the reactor; may cause the catalyst to agglomerate and deactivate with a drop off in conversion and increase in pressure drop; may lead to tube failure; may promote coking. Increasing the velocity of reactants through the reactor decreases the time available for reaction and exit concentration of product should decrease, pressure drop should increase.

Here are typical trouble shooting symptoms and causes for plug flow tubular reactors, stirred tank reactors and reactive extrusion. For PFTR, we consider multi-tube fixed bed catalyst, nonadiabatic, Section 3.6.1; fixed bed, adiabatic, Section 3.6.2; bubble reactors, Section 3.6.3; packed column reactors, Section 3.6.4; trickling bed, Section 3.6.5 and thin film, Section 3.6.6. For STR, we considered batch STR, Section 3.6.7; semibatch, Section 3.6.8; CSTR, Section 3.6.9; fluidized bed, Section 3.6.10. In Section 3.6.11 we consider a mix of CSTR, PFTR with recycle. Finally reactive extrusion is considered in Section 3.6.12.

### 3.6.1
### PFTR: Multitube Fixed-Bed Catalyst, Nonadiabatic

Trouble shooting: *"Pressure surge":* possible shutdown?/[runaway reactor]*. *"$\Delta p$ increases dramatically, top of tubes hot, less conversion than expected":* possible shutdown?/ contamination in feed / [poisoned catalyst]*.

*"Rapid decline in conversion":* unfavorable shift in equilibrium at operating temperature, for exothermic reactions/ [sintering]*/ [agglomeration]*/ poison in new feed . *"Gradual decline in conversion":* sample error/ analysis error/ temperature sensor error/ [catalyst activity lost]*/ [maldistribution]*/ [unacceptable temperature profiles]*/ [inadequate heat transfer]*/ wrong locations of feed, discharge or recycle lines/ faulty design of feed and discharge ports/ wrong internal baffles and internals/ faulty bed-voidage profiles. *"Gradual decline in conversion and axial temperature constant with depth of region increasing with time":* [poisoned catalyst]*. *"Gradual decline in conversion and axial temperatures < usual":* [poisoned catalyst]*. *"Gas exit concentration of reactants high":* sample error/ analysis error/ catalyst selectivity low/ [catalyst activity lost]*. *"Exit concentration of product higher than design":* reactor leaking. *"Change in product distribution":* [maldistribution]* / [poisoned catalyst]*/ feed contaminants/ change in feed/ change in temperature settings.

*"Temperature runaways":* [temperature hot spots]*/ [reactor instability]*. *"Pressure and bed temperature and reactor unsteady":* water in feed/ [maldistribution]*. *"Local high temperature/hot spot with $T > 100\,°C$ above normal":* [maldistribution of gas flow]*/ instrument error/ extraneous feed component that reacts exothermically. *"Local low temperature within the bed":* [maldistribution of gas flow]*/ instrument error/ extraneous feed component that reacts endothermically. *"Exit gas temperature too high":* instrument error/ control-system malfunction/ fouled reactor coolant tubes. *"Temperature varies axially across bed":* [maldistribution]*. *"Soon after startup, temperature of tubewall near top > usual and increasing and perhaps $\Delta p$ increase and less conversion than expected or operating temperatures > usual to obtain expected conversion":* inadequate catalyst regeneration/ contamination in feed; for steam reforming sulfur concentration > specifications/ wrong feed composition; for steam reforming: steam/$CH_4$ < 7 to 10. *"Soon after startup, temperatures over full length of some tubes > usual and perhaps $\Delta p$ > or < usual and may increase with time":* faulty loading of the catalyst/ [maldistribution]*. *"Hot bands or stripes; perhaps $\Delta p$ increase":* low ratio of steam to methane/ [carbon formation; whisker type]*/ wrong feed composition: for steam reforming steam/methane < 7 to 10: 1. *"Hot bands or stripes near top and perhaps over all tube and rapidly increasing $\Delta p$ and conversion < specifications":* [deactivated catalyst by pyrolytic coke formation]*/ feed concentration wrong: for steam reforming high concentration of heavier hydrocarbons/ steam to hydrocarbon ratio low/ [catalyst poisoned]* by sulfur. *"Temperature at inlet high and high $\Delta p$":* [for steam

reforming: steam contaminated with inorganic solids]*. *"Hot bands in top 1/3 of tubes and methane > usual in exit gas and perhaps $\Delta p$ increase":* contamination in feed / [poisoned catalyst]*.

*"$\Delta p$ higher than design":* catalyst degradation/ instrument error/ high gas flow/ sudden coking/ crud left in from construction or revamp. *"$\Delta p$ increasing gradually yet flowrate constant":* [coke formation]*/ [dust or corrosive products from upstream processes]*.

*"Startup after catalyst regeneration, conversions < standard":* [regeneration faulty]*. *"Startup after catalyst replacement, poor selectivity":* bad batch of catalyst/ preconditioning of catalyst faulty/ temperature and pressures incorrectly set/ instrument error for pressure or temperature. *"Startup after catalyst replacement, $\Delta p$ < expected and conversion < standard":* [maldistribution]* and axial variation in temperature/ larger size catalyst. *"Startup after catalyst replacement, conversion < standard and $\Delta p$ increasing":* [maldistribution and axial temperatures different]*/ feed precursors present for polymerization or coking. *"Startup after catalyst replacement, $\Delta p$ for this batch of catalyst > previous batch":* catalyst fines produced during loading/ poor loading. *"Startup after catalyst replacement, conversion < specifications per unit mass of catalyst and more side reactions":* [maldistribution]*/ faulty inlet distributor/ faulty exit distributor. *"Startup after catalyst replacement, poor selectivity":* bad batch of catalyst/ preconditioning of catalyst faulty/ [tube walls not passified]*/ temperature and pressures incorrectly set/ instrument error for pressure or temperature. *"Startup after catalyst replacement, increased side reactions and conversion < specification":* catalyst loading not the same in all tubes.

*[Active species volatized]*:* [regeneration faulty]*/ faulty catalyst design for typical reaction temperature/ [hot spots]*.

*[Agglomeration of packing or catalyst particles]*:* [temperature hot spots]*.

*[Attrition of the catalyst]*:* flowrates > expected/ catalyst too fragile.

*[Carbon buildup]*:* [inadequate regeneration]*/ [excessive carbon formed]*.

*[Catalyst selectivity changes]*:* [poisoned catalyst]*/ feed contaminants/ change in feed/ change in temperature settings.

*[Catalyst activity lost]*:* [carbon buildup]*/[regeneration faulty]*/ [sintered catalyst]*/ excessive regeneration temperature/ [poisoned catalyst]*/ [loss of surface area]*/ [agglomeration]*/ [active species volatized]*.

*[Excessive carbon formed]*:* operating intensity above usual/ feed changes/ temperature hot spots.

*[Dust or corrosive products from upstream processes]*:* in-line filters not working or not installed/ dust in the atmosphere brought in with air/ air filters not working or not installed.

*[Loss of surface area]*:* [sintered catalyst]*/ [carbon buildup]*/ [agglomeration]*.

*[Maldistribution]*:* faulty flow-distributor design/ plugging of flow distributors with fine solids, sticky byproducts or trace polymers/ [sintered catalyst particles]*/ [agglomeration of packing or catalyst particles]*/ fluid feed velocity too high/ faulty loading of catalyst bed/ incorrect flow collector at outlet.

*[Poisoned catalyst]*:* poisons in feed/ flowrate of "counterpoison" insufficient/ poison formed from unwanted reactions.

*[Poisons in feed]*:* depends on reaction/ contamination in feed/ upstream process or equipment upsets/ changes in feed. Poisons for platforming include high sulfur in feed and high feed end point with upstream equipment failure being compressor failure/ water upset/ chloride upset. Poisons for steam reforming: sulfur, arsenic and alkali metals in the hydrocarbon or steam.

*[Reactor instability]*:* control fault/ poor controller tuning/ wrong type of control/ insufficient heat transfer area/ feed temperature exceeds threshold/ coolant temperature exceeds threshold/ coolant flowrate < threshold/ tube diameter too large.

*[Regeneration doesn't remove all carbon from the catalyst]*:* regeneration temperature not hot enough/ regeneration time not long enough/ [maldistribution]*.

*[Regeneration faulty]*:* temperatures too high/ oxygen concentration < standard/ oxygen concentration > standard causing too rapid a burn/ incorrect temperature and time so that coke left on catalyst. [regeneration doesn't remove all carbon from the catalyst]*/ excessive temperature during regeneration.

*[Runaway reactor]*:* feed temperature too high/ [temperature hot spot]*/ cooling water too hot/ feed temperature too high.

*[Sintered catalyst]*:* temperature sensor error/ [temperature hot spots]*/ [maldistribution]*/ temperature in reactor too high/ regeneration temperature too high.

*[Temperature hot spots]*:* bed too deep/ [maldistribution]*/ flowrate < design/ instrument error/ extraneous feed component that reacts exothermically.

*[Tube walls not passified]*:* walls activated unwanted side reactions and faulty passivation treatment/ wrong passivation treatment/ no passivation treatment.

### 3.6.2
### PFTR: Fixed-Bed Catalyst in Vessel: Adiabatic

*Trouble shooting:*[4] gas-catalytic reactions. Temperature and pressure drops across bed are usually key variables. When a hot spot develops, it usually develops at the front end of the bed and gradually moves through the bed. It may take three to four weeks to travel through the full bed. If the hot spot is 100–200 °C above normal, then usually carbon is deposited and the catalyst is irrevocably damaged. Temperature control is critical for exothermic reactions. *"$\Delta p$ rapidly increases":* emergency shutdown? *"Pressure surge":* possible shutdown?/[runaway reactor]*.

*"Rapid decline in conversion":* unfavorable shift in equilibrium at operating temperature, for exothermic reactions/ [sintering]*/ [agglomeration]*/ poison in new feed . *"Gradual decline in conversion":* sample error/ analysis error/ temperature sensor error/ [catalyst activity lost]*/ [maldistribution]*/ [unacceptable temperature profiles]*/ wrong locations of feed, discharge or recycle lines/ faulty design of feed and discharge ports/ wrong internal baffles and internals/ faulty bed-voidage profiles. *"Gradual decline in conversion and axial temperature constant with depth of region increasing with time":* [poisoned catalyst]*. *"Gradual decline in conversion and axial*

4) Based on R.B. Anderson, person communication; H.F. Rase "Fixed bed reactor design and diagnostics", 1990, Wiley and Dutta, S. and R. Gauly, Hydrocarbon Processing, 1999, Sept, 43–50.

temperatures < usual": [poisoned catalyst]*. *"Gas exit concentration of reactants high"*: sample error/ analysis error/ catalyst selectivity low/ [catalyst activity lost]*. *"Exit concentration of product higher than design"*: reactor leaking. *"Change in product distribution":* [maldistribution]* / [poisoned catalyst]*/ feed contaminants/ change in feed/ change in temperature settings.

*"Temperature runaways":* [temperature hot spots]*/ [reactor instability]*. *"Pressure and bed temperature and reactor unsteady"*: water in feed/ [maldistribution]*. *"Local high temperature/hot spot with $T > 100\,°C$ above normal"*: [maldistribution of gas flow]*/ instrument error/ extraneous feed component that reacts exothermically. *"Local low temperature within the bed"*: [maldistribution of gas flow]*/ instrument error/ extraneous feed component that reacts endothermically. *"Exit gas temperature too high"*: instrument error/ control-system malfunction. *"Temperature varies axially across bed":* [maldistribution]*.

*"$\Delta p$ higher than design"*: catalyst degradation/ instrument error/ high gas flow/ sudden coking/ crud left in from construction or revamp. *"$\Delta p$ increasing gradually yet flowrate constant":* [coke formation]*/ [dust or corrosive products from upstream processes]*.

*"Startup after catalyst regeneration, conversions < standard":* [regeneration faulty]*. *"Startup after catalyst replacement, poor selectivity":* bad batch of catalyst/ preconditioning of catalyst faulty/ temperature and pressures incorrectly set/ instrument error for pressure or temperature. *"Startup after catalyst replacement, $\Delta p$ < expected and conversion < standard":* [maldistribution]* and axial variation in temperature/ larger size catalyst. *"Startup after catalyst replacement, conversion < standard and $\Delta p$ increasing"*: [maldistribution and axial temperatures different]*/ feed precursors present for polymerization or coking. *"Startup after catalyst replacement, $\Delta p$ for this batch of catalyst > previous batch":* catalyst fines produced during loading/ poor loading. *"Startup after catalyst replacement, conversion < specifications per unit mass of catalyst and more side reactions":* [maldistribution]*/ faulty inlet distributor/ faulty exit distributor.

*[Active species volatized]*:* [regeneration faulty]*/ faulty catalyst design for typical reaction temperature/ [hot spots]*.

*[Agglomeration of packing or catalyst particles]*:* [temperature hot spots]*.

*[Attrition of the catalyst]**: flowrates > expected/ catalyst too fragile.

*[Carbon buildup]*:* [inadequate regeneration]*/ [excessive carbon formed]*.

*[Catalyst selectivity changes]*:* [poisoned catalyst]*/ feed contaminants/ change in feed/ change in temperature settings.

*[Catalyst activity lost]*:* [carbon buildup]*/[regeneration faulty]*/ [sintered catalyst]*/ excessive regeneration temperature/ [poisoned catalyst]*/ [loss of surface area]*/ [agglomeration]*/ [active species volatized]*.

*[Excessive carbon formed]*:* operating intensity above usual/ feed changes/ temperature hot spots.

*[Dust or corrosive products from upstream processes]*:* in-line filters not working or not installed/ dust in the atmosphere brought in with air/ air filters not working or not installed.

*[Loss of surface area]*:* [sintered catalyst]*/ [carbon buildup]*/ [agglomeration]*.

*[Maldistribution]**: faulty flow-distributor design/ plugging of flow distributors with fine solids, sticky byproducts or trace polymers/ [sintered catalyst particles]*/ [agglomeration of packing or catalyst particles]*/ fluid feed velocity too high/ faulty loading of catalyst bed/ incorrect flow collector at outlet.

*[Poisoned catalyst]**: poisons in feed/ flowrate of "counterpoison" insufficient/ poison formed from unwanted reactions.

*[Poisons in feed]**: depends on reaction/ contamination in feed/ upstream process or equipment upsets/ changes in feed. Poisons for platforming include high sulfur in feed and high feed end point with upstream equipment failure being compressor failure/ water upset/ chloride upset.

*[Reactor instability]**: control fault/ poor controller tuning/ wrong type of control/ feed temperature exceeds threshold.

*[Regeneration doesn't remove all carbon from the catalyst]**: regeneration temperature not hot enough/ regeneration time not long enough/ [maldistribution]*.

*[Regeneration faulty]**: temperatures too high/ oxygen concentration < standard/ oxygen concentration > standard causing too rapid a burn/ incorrect temperature and time so that coke left on catalyst. [regeneration doesn't remove all carbon from the catalyst]*/ excessive temperature during regeneration.

*[Runaway reactor]**: feed temperature too high/ [temperature hot spot]*.

*[Sintered catalyst]**: temperature sensor error/ [temperature hot spots]*/ [maldistribution]*/ temperature in reactor too high/ regeneration temperature too high.

*[Temperature hot spots]**: bed too deep/ [maldistribution]*/ flowrate < design/ instrument error/ extraneous feed component that reacts exothermically.

### 3.6.3
### PFTR: Bubble Reactors, Tray Column Reactors

Wide variety of configurations ranging from tube loop, jet loop, air lift loop and sparger "bubble reactor". Related reactors include CSTR, Section 3.6.9; STR, Section 3.6.7.

Trouble shooting: *Carryover":* [foaming]*.

*[Foaming]**: surfactants present/ dirt and corrosion solids/ natural occurring surfactants/ pH far from the zpc/ naturally occurring polymers/ insufficient disengaging space above the liquid/ antifoam ineffective (wrong type or incorrect rate of addition)/ bubble rate too high/ mechanical foam breaker not rotating/ baffle foam breaker incorrectly designed or damaged/ asphaltenes present/ liquid downflow velocity through the foam is too low. See Section 3.11 for generic causes of [foaming]*.

See Trouble shooting: section STR, Section 3.6.7 for more on trouble shooting aerobic bioreactors.

### 3.6.4
### PFTR: Packed Reactors

These include trickling filters and gas-liquid-solid packed-column bioreactor.

**Trickling filter**: Trouble shooting: *"Plugged: interstitial voids become filled with biological growth"*: packing too small/ packing of variable diameter/ organic to liquid loading > design. *"Ice formation on top filter surface"*: liquid maldistribution/ feed liquid temperature too low/ air temperature too low. *"Odors"*: loss of aerobic conditions/ accumulation of sludge and biological growth/ lack of chlorine in influent/ high organic loadings in feed especially from milk processing and canneries. *[Foaming]**: surfactants present/ dirt and corrosion solids/ natural occurring surfactants/ pH far from the zpc/ naturally occurring polymers/ insufficient disengaging space above the liquid/ antifoam ineffective (wrong type or incorrect rate of addition)/ vapor velocity too high/ mechanical foam breaker not rotating/ baffle foam breaker incorrectly designed or damaged/ asphaltenes present/ liquid downflow velocity through the foam is too low.

**Gas-liquid-solid packed-column bioreactor**: Trouble shooting: *Carryover"*: [foaming]*.

*[Foaming]**: surfactants present/ dirt and corrosion solids/ natural occurring surfactants/ pH far from the zpc/ naturally occurring polymers/ insufficient disengaging space above the liquid/ antifoam ineffective (wrong type or incorrect rate of addition)/ bubble rate too high/ mechanical foam breaker not rotating/ baffle foam breaker incorrectly designed or damaged/ asphaltenes present/ liquid downflow velocity through the foam is too low. See Section 3.11 for generic causes of [foaming]*.

See Trouble shooting: section STR, Section 3.6.7 for more on trouble shooting aerobic bioreactors.

### 3.6.5
### PFTR: Trickle Bed

Good practice: gas-liquid flow cocurrently down through a packed bed of catalyst. Porosity 0.38–0.42. Ensure operation in the correct flow regime. The effectiveness of the solid catalyst and of the gas-liquid mass transfer decreases if solid catalyst is non-wet. For good wetting of the solid keep the surface tension of the solid > surface tension of the liquid. Prevent foaming. The efficiency depends on the skill in initially distributing the gas and the liquid. Use liquid distribution plate similar to design used for packed towers. The liquid distribution plate should have at least 50 holes/$m^2$ of catalyst bed.

Trouble shooting:[5)] For trickle bed reactors with specific applications to hydrotreating.

*"Low conversion":* feed composition change/ wrong catalyst for feed/ sample error/ flowrate error/ feedrate higher but reactor temperature not increased/ temperature profile wrong/ thermocouple fault/ controller fault/ feed bypassing reactor through leak in heat exchanger/ [channeling]*/ [catalyst]*/ [foaming]*/ For *hydrotreating*: [hydrogen starvation]*/ catalyst not presulfided/ [incomplete presulfiding of catalyst]*. *"Sudden loss of activity of catalyst":* heat exchanger leak/ change in feed composition/ For hydrotreating: [hydrogen starvation]*.

*"$\Delta p$ across the catalyst bed > design";* [channeling]*/ cracked hydrocarbon feed stored without effective nitrogen blanket/ solids in feed/ corrosion products from upstream operations/ bypass on feed filter open/ feed distributor fault/ top catalyst support tray has holes that are too small/ bottom catalyst bed support tray holes are too large/ pugged or partially plugged outlet/ crush strength of catalyst exceeded and fines plug bed/ excessive recycle compressor surge causing breakdown of top layer of catalyst.

For hydrotreating: *"Rapid breakthrough of $H_2S$ during catalyst sulfiding":* [channeling]*.

*"Nonuniform bed temperatures across the diameter during sulfiding":* [channeling]*.

*"Color > specifications":* composition change in feed/ catalyst aged..

*[Catalyst]*:* regeneration failed to remove carbon from catalyst/ excessive regeneration temperature > 540 °C causing sintering, > 760 °C molybdenum sublimation, > 820 °C reduction in crush strength and change in alumina/ poisons in feed/ aged catalyst.

*[Channeling]*:* nonuniform catalyst bed density/ low superficial flowrate < 1.4 kg/ s m$^2$/ offset, tilted or faulty feed distributor/ thermal shock to upstream pipes or equipment causes scale to dislodge and buildup on bed/ internal vessel obstructions such as thermowells or supports.

*[Foaming]**: surfactants present/ dirt and corrosion solids/ natural occurring surfactants/ pH far from the zpc/ naturally occurring polymers/ insufficient disengaging space above the liquid/ antifoam ineffective (wrong type or incorrect rate of addition)/residence time insufficient/ designed for a vertical vessel but a horizontal vessel installed/ vapor velocity too high/ mechanical foam breaker not rotating/ baffle foam breaker incorrectly designed or damaged/ asphaltenes present/ liquid downflow velocity through the foam is too low/ operating in the wrong flow regime.

*[Hydrogen starvation]*:* change in feed composition without corresponding change in hydrogen/ leaks/ dissolution of hydrogen in liquid product/ lower concentration of hydrogen in treat gas/ flowrate of treat gas < expected because of recycle compressor fault.

5) Based on Koros, R.M. "Engineering Aspects of Trickle Bed Reactors", pp. 579–630 in "Chemical Reactor Design and Technology" H. de Lasa (Ed), Martinus Nijhoff Publishers, 1986 and M.D. Edgar, D.A. Johnson, J.T. Pistorius and T. Varadi "Trouble Shooting Made easy" HP May 1984 p. 65.

*[Incomplete presulfiding of catalyst]*:* contact with hydrogen at high temperature for too long a time/ maximum temperature of 150–175 °C exceeded/ use of cracked feed/ excessive addition of presulfiding agents.

*[Rapid coking of catalyst]*:* [hydrogen starvation]*/ temperatures too high.

### 3.6.6
### PFTR: Thin Film

Related topics evaporation, Section 3.4.1 for gravity and agitated falling films, absorbers, Section 3.4.8, and shell and tube heat exchangers, Section 3.3.3.See Section 3.3.3 for trouble shooting Vertical falling-film evaporator.

### 3.6.7
### STR: Batch (Backmix)

Trouble shooting: Batch STR used for polymerization and, to a lesser extent, nitration, sulfonation, hydrolysis, neutralization and, to a much lesser extent, dehydrogenation, oxidation and esterification can pose potentially unsafe operation. Key indicators of such potential hazards include *"Sudden increase in pressure"*, *"Unexplained increase in temperature"*, *"Failure of the mixer"*, *"Power failure"*, and *"Loss of cooling water"*. For any of these conditions our first question should be: emergency shut down? Our knowledge of the MSDS information for the species and their interaction with each other and with the environment is critical.

**Aerobic bioreactors:** Trouble shooting: *"inoculation cannot be used for the reactor/ fermenter"*: *[contamination]**. *"product formation is inhibited"*: *[contamination]**. *"target product cannot be separated from contaminating species"*: *[contamination]**. *"fermentation broth cannot be filtered"*: *[contamination]** .*"steam out of air filter yields dark brown liquid"*: media blowback. *"reduction in cell volume, no further product production, no oxygen uptake, no heat production":* [contamination by bacteriophage]* . *"foaming":* air leaks through gaskets, coils, jacket, hatch/ pH shifted away from zpc/ particles present.

*[contamination in the first 24 hours]**: contaminated inoculum/ poor sterilization of tank accessories and content/ unsterile air.

*[contamination comes in after 24 hours]**: air supply/ nutrient recharges/ antifoam feed/ loss of pressure during the run/ lumps in the media/ media blowbacks.

*[contamination]**: *[stock culture contaminated]*/ [raw materials contaminated]*/ [inoculation tank contaminated]*/ [fermenter contaminated]* / [incorrect procedures]*/ [faulty maintenance]*/ [contamination by bacteriophage]**

*[stock culture contaminated]**: foreign microorganisms in culture stock/ contaminated inoculation flash/wrong sterilization procedure/ temperature and pressure instruments wrong/ air left in sterilization chamber/ sterile area contaminated/ cotton plugs contaminated/ *[faulty sterilization]**/ raw material contaminated with spores combined with inadequate germination-sterilization.

*[raw materials contaminated]*:* dry materials not finely ground/ lumps not removed/ insoluble solids not suspended in solution well because of lumping or

inadequate mixing/ lumps too big to be sterilized in the time–temperature available/ mixing inadequate to keep particles suspended/ particles enter air sparger during filling operation/ starches or proteins not prehydrolyzed with enzymes.

*[inoculation tank contaminated]*:* wrong procedures/ tank dirty/ air leak/temperature and pressure instrument fault/ sample line, inoculation fitting dirty/dirty dead spots, debris, corrosion/ media blow back into air filter/ unsterilized air filter/faulty antifoam or pH additive lines.

*[fermenter contaminated]*: [inoculum tank contaminated]** / inoculum line contaminated/ procedure wrong/ tank dirty/ air leak/ leak from the coil or jacket/faulty sensors/ antifoam is not sterile/ dirty gaskets, bottom valve, sample line and valve, vent line valve, vacuum breaker/ nutrient feed tank or line not sterile/ all lines were not up to sterilization temperature/ steam condensate left in lines/ the humidity of the fermenter air upstream of the "sterile filter" is > 90%/ pH and DO probes were not cleaned between runs/ probe holders were not brushed and cleaned with a hypochlorite or formaldehyde solution/ for a previously contaminated vessel the valves and gaskets were not replaced, instrument sensors were not removed and cleaned; high boiling germicide, such as sodium carbonate or sodium phosphate was not used.

*[faulty maintenance]**: braided packing (on agitator shafts for sterile vessels) not receiving enough germicide/ mechanical seals (on agitator shafts for sterile vessels) not lubricated with sterilizing liquid/ instruments faulty/ bolts on flanges not tightened after heat up to 120 °C/ packed bed air filters not packed to correct density of 200–250 kg/m$^3$.

*[faulty sterilization]**: particles too coarse and dry/ particles not wetted/ particles not suspended/raw material contaminated with spores plus inadequate germination-sterilization.

*[contamination by bacteriophage]*:* source usually difficult to trace/ substitute an immune strain/ develop strain resistant to the phage.

*"Carryover":* [foaming]*

*[Foaming]**: surfactants present/ dirt and corrosion solids/ naturally occurring surfactants/ pH far from the zpc/ naturally occurring polymers/ insufficient disengaging space above the liquid/ antifoam ineffective (wrong type or incorrect rate of addition)/ bubble rate too high/ mechanical foam breaker not rotating/ baffle foam breaker incorrectly designed or damaged/ asphaltenes present/ liquid downflow velocity through the foam is too low. See Section 3.11 for generic causes of [foaming]*.

**Anaerobic digesters:** Trouble shooting: *"Sludge temperature fluctuates"*: instrument fault/ fluctuating feedrate. *"Poor heat transfer with the hot water coils, exit water temperature < design: "* sludge solids adhere to heat-transfer surface. *"Temperature constant but production of methane gas < design":* increased accumulation of scum or grit/ excessive acid production with lower pH and volatile acid > 500 mg/L/ organic overload/ toxic metals in feed/ highly acidic feed/ overdigested sludge. *"Foaming":* incomplete digestion/feedrate > design/ inadequate mixing/ temperature too low/ withdrawal of too much product (digested sludge)/ rate of reaction > design/ large quantities of organics in feed/ insufficient reaction volume for the high organic feedrate/ pH shift away from zpc.

### 3.6.8
### STR: Semibatch

Trouble shooting: Semi-batch STR used for polymerization and, to a lesser extent, nitration, sulfonation, hydrolysis, neutralization and, to a much lesser extent, dehydrogenation, oxidation and esterification can pose potentially unsafe operation. Key indicators of such potential hazards include *"Sudden increase in pressure"*, *"Unexplained increase in temperature"*, *"Failure of the mixer"*, *"Power failure"*, and *"Loss of cooling water"*. For any of these conditions our first question should be: emergency shut down? Our knowledge of the MSDS information for the species and their interaction with each other and with the environment is critical.

**Polymerizer:** Trouble shooting: *"Temperature increases suddenly: "* emergency shutdown?/ mixer stopped/ fouling of heat exchanger/ "gel effect" in polymerization reaction/ coagulation and product fouls the walls of the reactor. *"Particle product size < design":* too many nucleation sites/ lower level of oxygen than design/ too much emulsifier/ too much initiator. *"Particle product size > design":* coagulation/ too few initial nucleation sites/ too much oxygen in the feed/ too little emulsifier/ too little initiator/ emulsifier post feed is too late. *"Temperature increases > design":* emergency shutdown?/ coagulation/ emulsifier post feed too late. *"Batch times < design":* too many nucleation sites/ lower level of oxygen than design/ too much emulsifier/ too much initiator. *"Batch times > design":* too few initial nucleation sites/ too much oxygen in the feed/ too little emulsifier/ too little initiator.

**Agitated bubble reactors**: Trouble shooting. Consider increasing the impeller diameter or using a disk turbine to increase mass transfer. Trouble shooting: *"foaming"*: mixer tip speed too high/ linear gas velocity too high/ surfactant contaminants/ decrease in electrolyte concentration in the liquid/ change in pH/ use of turbine impeller/ lack of a gas sparger/ mechanical foam breaker not rotating/ disengagement space not high enough/ mechanical baffle foam breakers faulty/ antifoam ineffective.

*"flooded impeller":* too small a diameter impeller/ speed too slow.

**Gas-liquid-solid bioreactor:** Trouble shooting: *Carryover":* [foaming]*.

*[Foaming]**: surfactants present/ dirt and corrosion solids/ natural occurring surfactants/ pH far from the zpc/ naturally occurring polymers/ insufficient disengaging space above the liquid/ antifoam ineffective (wrong type or incorrect rate of addition)/ bubble rate too high/ mechanical foam breaker not rotating/ baffle foam breaker incorrectly designed or damaged/ asphaltenes present/ liquid downflow velocity through the foam is too low. See Section 3.11 for generic causes of [foaming]*.

See trouble shooting: section STR, Section 3.6.7 for more on trouble shooting bioreactors.

### 3.6.9 CSTR: Mechanical Mixer (Backmix)

Consider complications because of catalyst deposition and erosion. Trouble shooting: CSTR used for polymerization and, to a lesser extent, nitration, sulfonation, hydrolysis, neutralization and, to a much lesser extent, dehydrogenation, oxidation and esterification can pose potentially unsafe operation. Key indicators of such potential hazards include *"Sudden increase in pressure"*, *"Unexplained increase in temperature"*, *"Failure of the mixer"*, *"Power failure"*, and *"Loss of cooling water"*. For any of these conditions our first question should be: emergency shut down? Our knowledge of the MSDS information for the species and their interaction with each other and with the environment is critical. See Semibatch and STR Sections 3.6.8 and 3.6.7 for more.

**Liquid–liquid:** Typically reactor is a CSTR followed by a decanter to separate the phases and recycle the "catalyst" phase to the reactor. Trouble shooting: see decanter: Section 3.5.3a. *"Alkylate is purple":* [stable emulsion formation]*/[density difference decrease]*/ [drops don't settle]*/ [acid runaway]*. *"$\Delta p$ across the alkylate cooler > design":* [stable emulsion formation]*/ [density difference decrease]*/ [drops don't settle]*/ [acid runaway]*/ acid recirculation rate too fast.

*[Acid runaway]*:* excessive contaminants in feed to reactor/ feedrate too fast/ poor contact or mixing between isobutane, olefin and acid/ fresh acid makeup feedrate stopped/ faulty control/ faulty meter/ ratio of acid: hydrocarbon outside range 45–60% v/v/ ratio of isobutane: olefin < 8: 1/ initial reactor temperature too high or > 18 °C/ poor mechanical design for fresh acid addition. *"temperature of the recycled acid is > 1.7 °C hotter than feed entering the reactor":* [acid runaway]*/ alkyl sulfates polymerize in the decanter/ acid recirulation rate too fast.

*[Density difference decrease]*:* dilution of the dense phase/ reactions that dilute the dense phase; for *sulfuric acid alkylation:* if acid strength < 85% w/w the olefins polymerize with subsequent oxidation of the polymers by sulfuric acid. as a self-perpetuating continuing decrease in acid strength. Alkylate-acid separation is extremely difficult when acid concentration is 40% w/w.

*[Drop doesn't settle]*:* [density difference decrease]*/ [viscosity of the continuous phase increases]*/ [drop size decreases]*/ [residence time for settling too short]*/ [phase inversion or wrong liquid is the continuous phase]*/ pressure too low causing flashing and bubble formation.

*[Drop settles and coalesces but is re-entrained]*:* faulty location of exit nozzles for liquid phases/ distance between exit nozzle and interface is < 0.2 m/ overflow baffle corroded and failure/ interface level at the wrong location/ faulty control of interface/ liquid exit velocities too high/ vortex breaker missing or faulty on underflow line/ no syphon break on underflow line/ liquid exit velocities too high.

*[Drop settles but doesn't coalesce]*:* [phase inversion]*/ pH far from zpc/ surfactants, particulates or polymers present/ electrolyte concentration in the continuous phase < expected/ [coalescer pads ineffective]*/ [drop size decrease]*/ [secondary haze forms]*/ [stable emulsion formation]*/ [interfacial tension too low]*/ [Marangoni effect]*.

*[Drop size decrease]*:* feed distributor plugged/ feed velocity > expected/ feed flows puncture interface/ local turbulence/ distributor orifice velocity > design; *for amine units*: for amine > 0.8 m/s; for hydrocarbon > 0.4 m/s/ [Marangoni effects]*/ upstream pump generates small drops/ [secondary haze forms]*/ poor design of feed distributor.

*[Inaccurate sensing of the interface]*:* instrument fault/ plugged sight glass.

*[Interfacial tension too small]*:* temperature too high/ [surfactants present]* at interface.

*[Marangoni effects]*:* non-equilibrated phases/ local mass transfer leads to local changes in surface tension and stability analysis yields stable interfacial movement.

*[Phase inversion]*:* faulty startup/ walls and internals preferentially wetted by the dispersed phase.

*[Rag buildup]*:* collection of material at the interface: [surfactants present]* / particulates: example, products of [corrosion see Section 3.1.2]*, amphoteric precipitates of aluminum/ naturally occurring or synthetic polymers.

*[Residence time for settling too short]*:* interface height of the continuous phase decreases/ [inaccurate sensing of interface]*/ turbulence in the continuous phase/ flowrate in continuous phase > expected; for example > 3 L/s $m^2$ / sludge settles and reduces effective height of continuous phase/ [phase inversion]*/ inlet conditions faulty.

*[Secondary haze forms]*:* small secondary drops are left behind when larger drop coalesces, need coalescer promoter, see Section 3.9.2.

*[Stable emulsion formation]*:* [surfactants present]* / contamination by particulates: example, products of [corrosion products, see Section 3.1.2]*, amphoteric precipitates of aluminum or iron/ pH far from the zpc/ contamination by polymers/ temperature change/ decrease in electrolyte concentration/ the dispersed phase does not preferentially wet the materials of construction/ coalescence–promoter malfunctioning/ improper cleaning during shutdown/ [rag buildup]*.

*[Surfactants present]*:* formed by reactions/ enter with feed, example oils, hydrocarbons >C10, asphaltenes/ left over from shutdown, example soaps and detergents/ enter with the water, example natural biological species, trace detergents.

*[Viscosity of the continuous phase increases]**: temperature too low, for *alkylate-acid* separation, temperature < 4.4 °C/ [phase inversion]*/ contamination in the continuous phase/ unexpected reaction in the continuous phase causing viscosity increase.

### 3.6.10
### STR: Fluidized Bed (Backmix)

Trouble shooting: First we consider fluidized-bed reactors in general, then fluidized combustors or regenerators and then provide specifics for a fluid catalyst cracking unit, FCCU, which consists of a riser or fluidized-bed reactor, cyclone separator, steam stripper, spend catalyst transport, air-oxidizing regenerator, cyclone separator and a regenerated catalyst return.[6)]

**General fluidized-bed reactor:** *"Gradual change in yield":* [carbon buildup]*. *"Poor yield":* [Loss of catalyst activity]*/ [maldistribution]*/ [unacceptable temperature profiles]*/ [inadequate heat transfer]*/ wrong locations of feed, discharge or recycle lines/ faulty design of feed and discharge ports/ [inadequate mixing]*/ [excessive backmixing]*/ wrong internal baffles and internals/ [poor bubbling hydrodynamics]*/ [inadequate solids circulation rates in reactor]*. *"Change in product distribution":* [maldistribution]* / poisoned catalyst/ feed contaminants/ change in feed/ change in temperature settings. *"Temperature hot spots":* [maldistribution]*/ local exothermic reactions. *"Temperature runaways":* temperature hot spots. *"Pressure and bed temperature and reactor unsteady":* water in feed/ reactor grid hole erosion/ [maldistribution]*/ for FCCU: surging regenerator holdup/ unsteady reactor-regenerator differential pressure controller operation/ rough circulation/ incorrect aeration of U-bend/ incorrect aeration of standpipe/ sticky stack slide valves/ sensor control performance for stack slide valve unsatisfactory. *"Particulate carryover that affects operation of downstream equipment":* [poor separation in cyclone]*. *"Shifts in yield distribution":* [Feed contaminated with light hydrocarbons]*/ [sintered catalyst]*/ coarse particles. *"$\Delta p$ increase across the grid":* [plugged grid holes]*/ fluid flow > usual. *"$\Delta p$ across grid < expected":* air flowrate < design/ [eroded grid holes]*/ for FCCU: [Failure of internal seals in regenerator]*. *"Erratic or cycling pressures":* [surging of the catalyst bed]*. *"Catalyst losses increase":* [poor separation in cyclone]*/ insufficient head space above bed/ fluidization velocity too high/ increase in volume of product through unexpected side reactions/ change in feed flowrate/ flowrate instrument error/ velocity through reactor too high/ pressure surges/ attrition of catalyst.

*[Attrition of the catalyst]**: steam flowrate > expected/ air flowrate > expected/ local velocities into the dense phase > 60 m/s/catalyst too fragile.

*[Carbon buildup]**: [inadequate regeneration]*/ [excessive carbon formed]*.

*[Coarse particles (diameter > design)]**: [generation of fines]*/[loss of catalyst fines]*/ [poor separation in cyclone]*/ agglomeration of catalyst/ [sintered catalyst]*/ wrong specifications for catalyst.

*[Eroded grid holes]**: hole velocity too high / materials of construction/ contaminants in fluid. *[Excessive backmixing]**: [maldistribution]*/ [poor bubbling hydrodynamics]*.

6) Based on Luckenbach, E.C. et al. "Encyclopedia Processing and Design", Marcel Dekker 1981 p. 89; Dutta, S. and R. Gualy, "Overhaul process reactors", HP 1999 Sept pp. 43–50, Lieberman, N. P. "Troubleshooting Process Operations", 2nd ed. 1985 Pennwell Books.

*[Excessive carbon formed in cracker]*:* cracker operating intensity above usual; for FCCU excess aromatics in feed / changes in feed/ poor catalyst stripping/ heavier recycle/ leakage of fractionator bottoms into the feed/ [sintered catalyst]*/ [feed contaminated with metals]*/ [feed contaminated with heavy hydrocarbons, especially aromatics]*.

*[Failure of internal seals in regenerator for FCCU]*:* pressure bump during startup/ regenerator pressure too high/ velocity through the grid too low/ low flow of air to the grid/ stresses too high/ erosion/ abnormal conditions with the auxiliary burner on startup.

*[Gas bubbles too big]*:* particles heavier than design/ particles larger than design/ sintered particles/ single fluidized bed too deep.

*[Gas bypassing in fluidized bed]*:* particles heavier than design/ particles larger than design/ agglomerated particles/ single fluidized bed too deep instead of multiple beds in series.

*[Gas velocity too high]*:* [increase in production of light ends in reactor]*.

*[Generation of fines]*:* [attrition of the catalyst]*/ fines in the new catalyst.

*[Inadequate heat transfer]*:* [maldistribution]*/ insufficient heat exchanger area/ design error/ fouled exchanger. see Section 3.3.8.

*[Inadequate mixing]*:* [maldistribution]*/ [poor bubbling hydrodynamics]*. see Section 3.7.1.

*[Inadequate regeneration]*:* [regenerator doesn't remove all carbon from the catalyst]*/ excessive temperature during regeneration/ coarse particles.

*[Loss of catalyst activity]*:* [carbon buildup]*/[inadequate regeneration]*/[sintered catalyst]*/ excessive regeneration temperature/ [poisoned catalyst]*/ [loss of surface area]*.

*[Loss of catalyst fines]*:* insufficient disengaging space above the top of the bed/ agglomeration of catalyst/ [poor separation in cyclone]*/ $\Delta p$ indicator for catalyst level faulty/ $\Delta p$ indicator for catalyst level OK but bed density incorrect.

*[Loss of surface area]*:* [sintered catalyst]*/ [carbon buildup]*.

*[Maldistribution]*:* faulty feed-distributor design/ plugging of fluid distributors with fine solids, sticky byproducts or trace polymers/ [temperature hot spots]*/ [sintered catalyst particles]*/ [poor bubbling hydrodynamics]* / [poor circulation]*.

*[Plugged dipleg]*:* spalled refractory plug/ level of catalyst in bed too high / $\Delta p$ indicator for catalyst level faulty/ $\Delta p$ indicator for catalyst level OK but bed density incorrect. air out periods with a lot of water or steam in vessel.

*[Plugged grid holes]*/* foreign debris entering with fresh catalyst/ faulty grid design/ lumps of coke or refractory in catalyst/ failure of grid hole inserts/ [sintered catalyst]*/ bits of refractory.

*[Poisoned catalyst]*:* poisons in feed/ flowrate of "counterpoison" insufficient/ poison formed from unwanted reactions.

*[Poisons in feed]*:* depends on reaction: for FCCU poisons in the feed include nickel, vanadium and sodium; the counterpoison is a solution of antimony.

*[Poor bubbling hydrodynamics]*:* [Gas bypassing in fluidized bed]*/ [gas bubbles too big]* particles heavier than design/ [particles larger than design]*/ [sintered catalyst]*/ fluid feed velocity too high/ too deep a bed of catalyst/ [maldistribution]*.

*[Poor circulation]*:* coarse particles/ [maldistribution]*.

*[Poor separation in cyclone]*:* [stuck or failed trickle valve]*/ [plugged dipleg]*/ dipleg unsealed/ solids level does not cover end of dipleg/ gas velocity into cyclone too low or too high/ faulty design of cyclone/ solids concentration in feed too high/ cyclone volute plugged/ hole in cyclone body/ $\Delta$p indicator for catalyst level faulty/ $\Delta$p indicator for catalyst level OK but bed density incorrect/ pressure surges.

*"$\Delta p >$ design":* fines in packed beds/ fines in distributors/ fines in exit nozzles/ crud left in from construction or revamp.

*[Reactor instability]*:* control fault/ poor controller tuning/ wrong type of control/ insufficient heat transfer area.

*[Regenerator doesn't remove all carbon from the catalyst]*:* damaged air grid/ insufficient air/ excessive regenerator velocity/ poor spent catalyst initial distribution/ coarse particles.

*[Sintered catalyst]*:* local high temperatures/ [maldistribution]*/ for FCCU [afterburn in regenerator]*/ [Feed contaminated]*/ high temperature in the regenerator/ [temperature hot spots in the reactor]*.

*[Solids conveying lines flow capacity < design]*:* sticky fines buildup in lines/ wrong $\Delta$p across line.

*[Stuck or failed trickle valve]*:* binding of hinge rings/ angle incorrect/ wrong material/ hinged flapper plate stuck open/ flapper plate missing.

*[Surging of the catalyst bed]*:* water in the feed/ [plugged grid holes]*/ faulty grid design/ [grid holes eroded]* /[for FCCU: failure of internal seals in regenerator]*/ for FCCU: [seal failures]*/ hole in the overflow well/ [reactor instability]*/ control fault in $\Delta$p between cracker and regenerator.

*[Unacceptable temperature profiles]*:* fluctuating temperature/ unsteady bed temperatures.

**Specific for a fluidized bed combustion/catalyst regenerator**: *"Increase in catalyst losses":* [poor separation in cyclone]*/ [failure in regenerator plenum]*/ for FCCU [failure of internal seals in regenerator]*.

*[Failure in regenerator plenum]*:* faulty cyclone design/ catalyst feed too high/ regenerator velocity too high/ faulty spray nozzles causing impingement of plenum sprays/ temperatures too high causing failure in plenum.

**Specific for fluid cat cracker unit – including regenerator system:** *"Overloaded wet gas compressor":* for FCCU high hydrogen production/ increase in production of light ends. *"Gas compressor flow reversal":* [poisoned catalyst]*. *"Gas compressor surge":* [poisoned catalyst (that causes production of lower MM species)]*. *"Gas compressor flow reversal":* [poisoned catalyst]*. *"Wet gas compressor surge":* [poisoned catalyst (that causes production of lower MM species)]*.

*"Hydrogen concentration in wet gas increases":* [poisoned catalyst, especially with nickel and vanadium]*/ [feed contaminated with metals, especially nickel and vanadium]*/ [loss of catalyst activity]*/ feed concentration high in hydrogen/ loss of antimony solution addition. *"Increase in the production of light ends":* for FCCU [feed contaminated with metals]*/ feed concentration high in light ends.

*"Erratic or cycling instrument records on holdup, density and the overflow well":* [surging of the catalyst bed]*. *"Opaque flue gas from the regenerator":* [poor separation in

cyclone in regenerator]/ fluidization velocity too high/ increase in volume of product through unexpected side reactions/ change in feed flowrate/ flowrate instrument error. *"Vibration in the preheat system":* [feed contaminated with water]*. *"Δp increase between reactor and fractionator inlet":* [coking in overhead lines]*. *"Δp lower on the regenerator slide valve":* [poisoned catalyst]*. *"Δp between cracker and regenerator incorrect":* fault with the input air blower/ fault with the flue-gas slide valve on the regenerator/ fault with the regenerated catalyst slide valve/ fault with the spent-catalyst slide valve/ fault with the wet gas compressor/ fault downstream of the wet-gas compressor, such as plugged fractionator overhead condensers (with ammonium chloride salts)/ changes in environment air conditions. Regenerator should be about 20 kPa higher than the cracker for Δp across the regenerated catalyst slide valve. *"Δp between cracker and regenerator fluctuating":* fluctuating temperature in cracker/ fluctuating pressure in regenerator/ fluctuating catalyst circulation rate/ fluctuating level in the overflow well/ shift in catalyst between cracker and regenerator/ incorrect aeration of U-bend/ incorrect aeration of standpipe/ sticky stack slide valves/ sensor control performance for stack slide valve unsatisfactory/ moisture in aeration medium/ unsteady control of air/ U-bend vibration. *"Δp across cylcone > expected":* steam flowrate > expected/ air flowrate > expected. *"Pressure fluctuating in regenerator":* incorrect aeration of U-bend/ incorrect aeration of standpipe/ sticky stack slide valves/ sensor control performance for stack slide valve unsatisfactory. *"Plugged pump on the bottoms of the fractionator":* [poor separation in cyclone]*/ velocity through reactor too high/ faulty cyclone design. *"Overflow well level low":* [eroded grid holes]*. *"Overflow well level high":* [plugged grid holes]*. *"Overflow well level fluctuating":* incorrect aeration of U-bend/ incorrect aeration of standpipe/ sticky stack slide valves/ sensor control performance for stack slide valve unsatisfactory/ hole in the overflow well.

*"Catalyst loss from the regenerator increased":* [plugged grid holes]*/ [eroded grid holes]*/ foreign debris entering with fresh catalyst/ faulty grid design/[poor separation in cyclone]*/ steam flowrate > design/ air flowrate > design. *"Catalyst circulation fluctuates":* [Δp between cracker and regenerator fluctuating]*/ fluctuating temperature in cracker/ fluctuating temperature in regenerator/ fluctuating level in the overflow well/ shift in catalyst between cracker and regenerator/ incorrect aeration of U-bend/ incorrect aeration of standpipe/ sensor control performance for air system unsatisfactory/ moisture in aeration medium/ unsteady control of air/ U-bend vibration/ coarse particles/ hole in the overflow well/ incorrect aeration of U-bend/ incorrect aeration of standpipe/ sticky stack slide valves/ sensor control performance for stack slide valve unsatisfactory/ [surging of the catalyst bed]*. *"Catalyst becomes lighter in regenerator gradually":* [afterburn in regenerator]*. *"Catalyst in fractionator bottoms":* [poor separation in cyclone]*/ velocity through reactor too high/ faulty cyclone design. *"Catalyst has salt and pepper appearance after regeneration":* air grid deficiency/ [Failure of internal seals in regenerator]*. *"Reduced rates of spent catalyst withdrawal":* [poor separation in cyclone in regenerator]*.

*"Temperature difference between bed and cyclone inlet in regenerator":* [failure of internal seals in regenerator]*/ [afterburn in regenerator]* and [inadequate regeneration]*. *"Temperatures of bed and cyclone are uneven in the regenerator":* hole in the over-

flow well/ [plugged grid holes]*/ foreign debris entering with the fresh catalyst/ faulty grid design. *"Temperature on regenerator shell or U-bend high":* damaged refractory. *"Temperature increase in the dilute phase relative to the dense phase":* [afterburn in regenerator]*. *"Temperature of the regenerator cannot be lowered":* [low catalyst circulation rate]*. *"Temperatures of regenerator too high > 750 °C":* excessive heat release.

*"Temperature in dilute phase decreases relative to temperature of the dense bed in the regenerator":* [regenerator doesn't remove all carbon from the catalyst]*. *"Feed preheat requirements > usual":* [low catalyst circulation rate]*. *"Unexplained increase in coke":* [poor catalyst stripping]*. *"High bottom sediment and water levels in the slurry oil product":* [poor separation in cyclone in the cracker]*. *"Higher H/C ratio":* [poor catalyst stripping]*. *"Excess oxygen in regenerator high":* [afterburn in regenerator]*/ [plugged grid holes]*/ [eroded grid holes]*/ faulty grid design. *"Ratio of carbon dioxide to carbon monoxide is higher than usual":* [afterburn in regenerator]*. *"Uneven oxygen distribution in the dilute phase":* [Failure of internal seals in regenerator]*. *"Unsteady heat balance":* [surging of the catalyst bed]*. *"Stripping steam flowrate > expected":* flowmeter error/ steam traps faulty/ partially opened valves/ missing restrictive orifice. *"Air flowrate > expected":* flowmeter error/ partially opened valves/ missing restrictive orifice. *"Flow reversal with feed going incorrectly to the regenerator":* [$\Delta p$ across the regenerator slide valve is < design]*.

*[Afterburn in regenerator]*:* for FCCU [failure of internal seals in regenerator]*/ too much excess air/ oxygen recorder reading incorrect/ meter error for feed and recycle flowmeters/ meter error for cyclone flowmeter/ [insufficient carbon production on catalyst during cracking]*/ air flowrate to regenerator too high/ [plugged grid holes]*/ [eroded grid holes]*/ faulty grid design causing localized air-distribution problem.

*[Coking in overhead lines]*:* insulation missing or damaged on transfer line/ extremely cold/ increase in heavies and condensibles in reactor products.

*[Control air flowrate too low]*:* controller for air faulty or poorly tuned.

*[Failure in regenerator plenum]*:* faulty cyclone design/ catalyst feed too high/ regenerator velocity too high/ faulty spray nozzles causing impingement of plenum sprays/ temperatures too high causing failure in plenum.

*[Feed contaminated with metals]*:* abnormal operation in the upstream atmospheric and vacuum units.

*[Feed contaminated with heavy hydrocarbons]*:* leak in heat exchangers/ partly open valves. *[Feed contaminated with light hydrocarbons]*:* leak in heat exchangers/ partly open valves.

*[Feed contaminated with sodium]*:* seawater leak in upstream equipment/ treated boiler feedwater leaks into feed/ upset in upstream caustic unit.

*[Feed contaminated with water]*:* water in feed tanks/ leaks from steam-out connections/ steam leaks in tank heaters/ water not cleaned out of the lines at startup/ moist air not removed from lines at startup.

*[Higher reactor velocities]*:* [feed contaminated with metals]*.

*[Higher regenerator holdup]*:* hole in the overflow well.

*[Increased air requirements in regenerator for the same conversion in the cracker]**: [feed contaminated with heavy hydrocarbons]*/ [inadequate regeneration]*/ [coke on catalyst > usual].

*[Insufficient coke production on catalyst during cracking]**: cracking operation intensity is lower than usual/ higher quality of feed to the cracker than usual for FCCU fewer aromatics in feed.

*[Low catalyst circulation rate]**: partial blockage of the U-bends/ excessive stripping steam/ insufficient aeration/ [control air flowrate too low]*/ differential pressure between cracker and regenerator set incorrectly or fluctuating.

*[Poor catalyst stripping]**: insufficient steam stripping flowrate/ faulty flow controller on steam flow/ faulty design of stripper/ reactor temperature too low/ faulty contacting between steam and catalyst/ circulation rate too high/ coarse particles.

*[$\Delta p$ across the regenerator slide valve is < design]**/ sudden drop in regenerator pressure/ regenerator slide valve sticking partly open/ compressor surge (see Section 3.3.2).

*[Regenerator doesn't remove all carbon from the catalyst]**: [excessive coke formed in cracker]*/ low excess oxygen/ oxygen sensor error/ flowmeter error for air/ [poor air distribution]*/ flowmeter error for feed and recycle/ air flowrate too small.

*[Sodium on catalyst]**: carryover of sodium from upstream units (caustic)/ treated boiler feedwater used in regenerator sprays/ [feed contaminated with sodium]*.

*[Uneven oxygen distribution in the regenerator]**: hole in the overflow well/ [plugged grid holes]*/ foreign debris entering with fresh catalyst/ faulty grid design.

*[Unstable catalyst bed]**: airflow too low/ grid holes eroded/ faulty grid design.

### 3.6.11 Mix of CSTR, PFTR with Recycle

**Conventional activated sludge:** Trouble shooting: *"Increase in sludge volume index"*: high-density inerts in feed and usually caused few operating problems. *"Decrease in sludge volume index and "bulking"*: high concentration of dissolved organics in feed. *"Sludge rises"*: excessive nitration. *"Frothing"*: decrease in aeration suspended solids/ increase in surfactants in feed/ aeration > design/ increase in temperature.

### 3.6.12 Reactive Extrusion

Trouble shooting: *"Inadequate mixing of liquid reactant with polymer"*: liquid flowrate too high/ screw channel under injection not full of polymer/ faulty screw design. *"Residuals in final polymer > design"*: vent temperature too low/ screw speed too low/ polymer feedrate too high/ screw design does not provide enough shear/ vent pressure too high. *"Polymer has crosslinked or degraded"*: screw rpm too high/ degree of fill too low/ feedrate too low/ heat-zone temperatures set too high/ screw-design fault giving excessive shear/ [screw tip pressure too high]*. *"Extruder torque excessive"*: throughput too high/ screw speed too low/ heat-zone temperatures set too low/ faulty screw design. *"Unable to melt material"*: throughput too high/ screw speed too low/ faulty screw design/ material too slippery. *"Residence time too short"*: throughput too high/ [degree of fill too high]* / faulty screw design. *"Residence time too long"*: throughput too low/ [degree of fill too low]* / faulty screw design. *"Gels or crosslinked materials"*: localized initiator concentration too high/ [melt temperature too high]*.

*[Degradation of melt in extruder]**: [RTD too wide]*/ barrel temperature too high/ screw speed too high (causing overheating and shear damage)/ oxygen present/[oxidation]*/ nitrogen purge ineffective/ wrong stabilizer/ wrong screw/ flows not streamlined/ stagnation areas present/ extruder stopped when temperatures > 200 °C/ copolymer not purged with homopolymer before shutdown/[residence time too long]*.

*[Degree of fill too high]*:* feedrate too high/ screw speed too slow.

*[Melt temperature too high]**: screw speed too high/exit barrel zone temperatures too high/ screw tip pressure too high/ degree of fill too low/ [shear intensity too high]*/ heat-zone temperatures set too high/ [screw tip pressure too high]*.

*[RTD too narrow]*:* [degree of fill too high]*

*[Screw tip pressure too high]**: screens plugged/ die or adapter or breaker plates too restrictive and give too much $\Delta p$/ [polymer viscosity too high]*/ temperatures in die assembly too low/ barrel temperature too low/ screw speed too high/ [shear intensity too low]*/ lubricant needed/ flow restriction/ throughput too high/ die land too short/ cold start/ [degradation of melt in extruder]*.

*[Shear intensity too low]**: screw speed too low/ faulty screw design.

For other symptoms see Section 3.9.6.

## 3.7 Mixing Problems

Here we consider mechanical agitation of liquids and liquid-solid systems. Operating information is given for solids blenders, in Section 3.7.3.

### 3.7.1 Mechanical Agitation of Liquid

Good practice: prefer motionless mixers to intensify. For systems where the viscosity increases with time (ex. polymer reactors) prefer turbines to propellers because turbines are power self-limiting. Check shaft wobble to ensure that impeller will not hit vessel walls if turned on in an empty tank. Consider a foot bearing.

Trouble shooting propeller/ impeller mixers: *"Shaft wobble/vibration"*: impeller speed too close to the 1$^{st}$ critical speed/ shaft runout at the impeller and impeller eccentricity too large/ insufficient support.

*"Excessive gear-reducer maintenance"*: excessive load/ high shock loads/ excessive shaft bending.

*"Excessive packing wear"*: insufficient lubrication/ excessive shaft wobble/ shaft is out of round.

*"Failure of the mechanical seal"*: dirty lubricant/ not enough lubricant/ excessive shaft wobble.

### 3.7.2
### Mechanical Mixing of Liquid–Solid

Trouble shooting stirred tanks: *"Solids floating on the surface"*: solids not wetted by liquid/ insufficient vortex.

### 3.7.3
### Solids Blending

Use the Johanson indices to characterize particles: (see also related topic storage bins, Section 3.10)

Arching index [m], AI, = diameter of the circular exit hole from a hopper that will ensure that an arch collapses in a conical bin or circular mixer, values range 0–1.2 m;

Ratholing index [m], RI, = diameter of the circular exit hole from a hopper that will ensure rathole failure and cleanout in a funnel-flow bin or mixer, values range from 0–9 m. (If RI > 3 then likely "lumps".)

Hopper index [degrees], HI = the recommended conical half-angle (measured from the vertical) to ensure flow at the walls. Usually add 3° to account for variability. Values range 14–33° with 304 s/s.

Flow ratio index [kg/s], FRI = maximum solids flowrate expected after deaearation of a powder in a bin. (measures consistency: small FRI for fine, highly compressible particles; Large FRI for particles > 400 μm, incompressible, very permeable.) Values range 0–90 kg/s.

Bin density index [$Mg/m^3$], BDI = bulk specific mass expected in a container full of solids or, in a mixer, when mixer stops and solid is allowed to deaerate. Values 0.3–1.6 $Mg/m^3$.

Feed density index [$Mg/m^3$], FDI = bulk specific mass at the conical hopper or mixer's discharge outlet. Values are 1–60% < BDI.

Chute index [degrees], CI, = recommended chute angle (with the horizontal) at points of solids impact. Values = angle of slide. values = 20–90°. High values suggest particles stick to sides of mixer or bins.

Rough wall angle of slide [degrees], RAS = angle (relative to the horizontal) that causes continual sliding on a solid on an 80-grit sandpaper surface with a pressure of 140 kPa gauge. Values 20–35°. Approximately equal to the angle of repose.

Adhesion angle index [degrees], AAI = difference between angle of slide (with horizontal) after an impact pressure of 7 MPa or (CI–10°) and the angle of slide without impact pressure.

Spring back index [%], SBI = percentage of solids that spring back after consolidation.

Good practice: usually 300 s blend time. We can invest much effort into *mixing* solids, but we must prevent *demixing* after the blends are mixed. The four mechanisms of demixing are 1) sifting, 2) angle of repose, 3) fluidization and 4) air current. Here are the details: 1. demixing via sifting: occurs if the particles are free flowing with mix of particle size with one size > 3× the diameter of the other. 2) demixing via angle of repose: this occurs with moderately free-flowing particles (AI < 0.18 m) with different angles of repose or two different RAS. For particles characterized in this way, the only blender that seems to prevent demixing is the air-pulse blender. 3) Demixing by fluidization: tends to occur if the blend contains > 20% fluidizing fines characterized by AI < 0.18 m; RI < 1.5 m; FRI < 0.76 kg/s plus coarser material with AI < 0.012; RI < 0.6 m and FRI > 7.6 kg/s. This type of demixing tends to occur if the action of the mixer induces air. 4) Demixing by air current: air carriers superfine, easy-flow, non-agglomerating particles into voids. This is a problem if AI < 0.18 m; RI < 1.5 m; FRI < 0.4 kg/s. Again try to avoid mixers whose action induces air.

Trouble shooting for polymer blenders of feedstock for extruder: *"Material does not flow"*: bridging/ see also hoppers, Section 3.10. *"Components do not feed"*: jammed valve or auger/ solids blockage or bridging/ power fault in feeder. *"Inconsistent flowrate"*: bridge or block in blender/ jammed discharge mechanism/ inconsistent feedrates to blender. *"Wrong blend compositions:"* calibration error in feeder.

## 3.8 Size-Decrease Problems

The focus here is on gas liquid systems that include bubble columns, Section 3.8.1, in packed columns, Section 3.8.2, and in agitated tanks.

### 3.8.1 Gas Breakup in Liquid: Bubble Columns

Good practice: electrolytes in the liquid alter the bubble diameter, the holdup, the interfacial area per unit volume in mechanically agitated devices and affects the $k_L$ a for bubble columns.

### 3.8.2 Gas Breakup in Liquid: Packed Columns

Good practice: the critical surface tension of the solid packing should be greater than the surface tension of the liquid to ensure that the liquid film remains intact in a packed contactor.

### 3.8.3
### Gas Breakup in Liquid: Agitated Tanks:

Good practice: consider increasing the impeller diameter or using a disk turbine to increase mass transfer. Trouble shooting: *"Foaming"*: mixer tip speed too high/ linear gas velocity too high/ surfactant contaminants/ decrease in electrolyte concentration in the liquid/ change in pH/ use of turbine impeller/ lack of a gas sparger. *"Flooded impeller":* too small a diameter impeller/ speed too slow.

## 3.9
## Size Enlargement

The fundamentals used to solve troubles for size-increase operations are surface phenomena. Surfaces are attracted to each other by van der Waals forces; surfaces are repelled by the electrochemical double layer or by steric hindrance. Surface energies, contact angles, and wetting are important. Specific symptoms and causes for equipment to coalesce drops in gas are listed in Section 3.9.1; for coalesce liquid drops in a liquid environment, Section 3.9.2, and create solid aggregates or flocs in a liquid environment, Section 3.9.3. Then we consider the creation of larger-size particle clusters by tabletting, Section 3.9.4; and pelleting, Section 3.9.5. The last two processes considered in this section focus on *change in shape* by injection molding and extrusion, Section 3.9.6; and coating, Section 3.9.7.

### 3.9.1
### Size Enlargement: Liquid–Gas: Demisters

Good practice: consider using mesh pads upstream of impact filter beds to reduce the load on the filter bed. Consider installing mesh pads vertically to facilitate drainage and minimize re-entrainment. For non-corrosive and non-fouling, consider installing vane separators downstream of mesh pads to collect larger drops sheared off from the mesh pad. Cannot be used for up to 25% turnup capacity; avoid the use of inertial devices for up to 25% turndown capacity. Trouble shooting: *"Demisters ineffective":* temperature too high/ fibers have the same charge as the droplets/ wetting properties of fibers changed/ fibers "weathered" and need to be replaced/ flowrate too slow through fibers/ wrong mix of fibers/ prefiltering ineffective/ [foaming]*/ wrong design/ re-entrainment. *[Foaming]*:* see Section 3.11.

### 3.9.2
### Size Enlargement: Liquid–Liquid: Coalescers

Good practice: consider decreasing the temperature to decrease the solubility and increase the surface tension. Adjust pH for water flowing through fibrous and mesh beds so that drop and fiber have the opposite surface charge. Promote coalescence in solvent-extraction systems by using surface tension positive configurations. Trouble shooting: *"Coalescer pads ineffective":* temperature too high/ pH incorrect/ fibers have the same charge as the droplets/ surface tension negative system/ wetting properties of fibers changed/ fibers "weathered" and need to be replaced/ flowrate too slow through fibers/ wrong mix of fibers/ prefiltering ineffective/ surface tension < 1 mN/m for fluoropolymer fibers or < 20 mN/m for usual fibers/ wrong design/ included in decanter but should be separate horizontal coalescer promoter unit/ faulty design/ [stable emulsion formed]*. *[Stable emulsion formed]*:* see Section 3.11.

### 3.9.3 Size Enlargement: Solid in Liquid: Coagulation/Flocculation

**Coagulation and flocculation in general**: Trouble shooting: *"Supernatant not clear":* [coagulation doesn't occur]*/ flocculation doesn't occur]*/ [floc doesn't settle out]*/ [floc forms but breaks up]*.

*[Coagulation doesn't occur]*:* wrong dosage of coagulant-flocculant/ wrong counterion/ pH different from expectations/ pH far from zpc/ faulty mixing in the rapid mix/ valence on the counterion too small/ charge on the dispersed particles or drops reversed from expectations.

*[Flocculation doesn't occur]*:* faulty fluid dynamics into the basin/ reel at wrong rpm/ residence time too short/ mixing not tapered/ unexpected turbulence/ too short a residence time between coagulant and subsequent flocculant dosage.

*[Floc doesn't settle out]*:* floc formed is too loose/ settler fault.

*[Floc forms but breaks up]*:* local turbulence > shear strength of floc.

**For water treatment:** Trouble shooting: *"Coagulation-flocculation ineffective, supernatant murky":* pH > 10/ wrong dosage of alum or coagulant / pH < 4/ increase in concentration of particles in feed/ rpm of reels in flocculation basin too slow/ feed temperature < 12 °C/rpm of reels in flocculation basin too fast.

**For latex**: Trouble shooting: *"Exit crumb too small":* brine concentration too high/ temperature too low/ power input too high/ wrong pH. *"Excessive amount of fines in supernatant":* brine concentration too high/ wrong pH/ temperature too low. *"Strength of the resulting crumb < specifications":* pH too high and brine concentration too high.

### 3.9.4 Size Enlargement: Solids: Tabletting

Trouble shooting: *"Product tablet weight > design":* sample error/ lab error/excessive fines.

### 3.9.5 Size Enlargement: Solids: Pelleting

Trouble shooting: For *strand* pelletizer for polymer resin. *"Pellet diameter too small":* hole too small for the desired throughput/ extruder output too low. *"Pellet diameter too large":* output too high/ speed too low/ feedroll speed too low/ output too high for die-size. *"Pellet too short or too long":* mismatch ratio of feedroll speed versus rotor teeth speed. *"Strands dropping":* feedroll pressure too small/ throughput too low/die-plate has too many holes. *"Pellet cuts are angled":* feed not perpendicular to strands/ strands overlapping. *"Pellet oval shaped":* feedroll pressure too high/ inadequate cooling before cutting. *"Pellet has tails":* incorrect clearance between rotor and cutters. For *water-ring* pelletizer for polymer resin: *"Pellet diameter too small":* hole too small for the desired throughput/ throughput too large. *"Pellet diameter too large":* output too low. *"Pellet too short or too long":* mismatch throughput versus cutter speed. *"Blocked holes":* nonuniform pressure on the die face/ throughput too low/die-plate has too many holes. *"Pellet oval shaped":* cutter speed too high/ inadequate cooling. *"Pellet has tails":* incorrect clearance between die and cutters/ worn cutter blades.

### 3.9.6 Solids: Modify Size and Shape: Injection Molding and Extruders

Consider first injection molding machines, then extruders.

#### 3.9.6.1 Injection molding machines

Good practice: resin should be dried $< 0.02\%$. Do not use resin that has been out of the dryer for $> 20$ min. Cold molds are difficult to fill and require higher injection pressures. Hot molds, generally, give better finish and less molded-in stress. Melt temperature is very sensitive to very small changes in rpm or backpressure despite sensor or controller set point. Measure with hand-held pyrometer or laser sensor. Need slower fill rate for sprue-gated parts to prevent blush, splash or jetting. If the walls are $> 5$ mm, then slow fill helps reduce sinks and voids.

Backpressures of 0.35 to 0.7 MPa help ensure homogeneous melt and consistent shot size. As backpressure increases, melt temperature increases. Holding or back-pressures that are 0.4 to 0.8 of injection pressures are typical. To purge a machine, acrylic is recommended.

***Trouble shooting:*** injection molding: Basically the cause can be with the material, the machine, the operator, the operating conditions, the mold or the part design. To check on the material, try material from another supplier; to check the machine, use same material, conditions, mold on another machine; if the trouble is random, then it is probably the machine; try a different operator on the machine; trouble appears in same location in the product, then flow conditions and look for problems from the front of the piston to the gate. The symptom-cause information is presented as issues related to appearance (color, surface finish and transparency), strength and shape defects and operation and symptoms.

- **Appearance** (Color, surface finish, transparency)

**Color**: *"Discoloration" (typically appears before burn marks appear; location appears at the weld line or where air is trapped in the mold)*: [contamination in heating cylinder]*/ sensor error/ control error/ [degradation, mechanical]*/ [degradation, thermal]*/ [melt too hot]*/ [melt not homogeneous]*/ overall cycle too long/ [contamination in hopper and feed zones]*/ incorrect cooling of ram and feed zone/ [venting in mold insufficient]*/ [residence time too long]*/ cooling time too short/ dryer residence time too long / excessive clearance between screw and barrel / clamp pressure too high/ injection forward time too long/ gate too small/ runner-sprue-nozzle too small. *"Black specs inside transparent product":* faulty cleanout of machine from previous molding operations/ failure to purge when not running for extended times/ nozzle too hot/ barrel temperature in the feed area is too low combined with high screw speed or high backpressure/ sensor error/ sensor located too far from heater bands/ hangup in nozzle tip/ nozzle adapter and end-cap. *"Brown streaks/burning":* wet feed/ [melt too hot]*/ [shear heating in the nozzle]*/ [degradation, mechanical]*/ loose nozzle/ wrong nozzle/ dead spots in hot manifold/ mold should be cold runner system/ gate or runner too small/ [contamination]*/ injection speed too fast/ booster time too long/ injection pressure too high/ mold design lacks vents at burn location/ gate size too small or at wrong location/ plunger has insufficient tolerance to allow air to escape back around the plunger/ poor part design/[venting of mold insufficient]*/ [residence time too long]*. *"Brown streaks at the weld lines or at the end of flow paths, black or charred marks":* [air trapped in mold]*. *"Brown streaks at the same location":* nozzle loose, wrong, too hot/ [shear heating (at gate, runner, cavity restrictions)]*. *"Brown streaks dispersed throughout":* material fault at the hopper: wet material. *"Weld burns":* [melt too hot]*/ injection speed too fast/ [mold too cold]*/ injection hold time too long/ injection pressure too high/ faulty nozzle heating bands/ [air trapped in mold]*.

**Surface finish**: *"Sink marks" (difficult to remove by changing processing conditions):* [cooling insufficient before removal from mold]*/ [short shot]*/ [melt too hot, causing excessive shrinkage]*/ [solidification at mold wall too slow]*/ wrong location for gate/ holding pressure too low/ injection speed to fast/ backpressure too high/ gates too small/ faulty runners/ booster time too low/ fault with nozzle, sprue or runners/ [mold temperature non-uniform]*/ moist feed/ [thick sections continue to shrink after the melt path is frozen]*/ hold time too short/ holding pressure too low/ [backflow from mold]*/ lubricant insufficient/ volatiles in feed/ [solidification at mold wall delayed]*/ [viscosity too high]*/ cooling-water temperature too low /excessive cushion in front of ram/ size of nozzle, sprue and runner too small/ [air trapped]*. *"Fine ridges running perpendicular to the flow front":* [melt too cold]*/ [short shot]*. *"Flow lines" see also "jetting":* feed moist/ injection pressure too low/ [melt too cold]*/ screw not rotating during injection/ injection speed too slow/ backpressure too low/ nozzle orifice too small/ [mold too cold]*/ gates too small/ [venting of mold insufficient]*/ feedrate too small/injection rate too low/ injection hold time too low/ booster time too low/ relocate gates/ faults with nozzle, runner, sprue or gate/ clamp pressure too high/ unequal filling rates between cavities/ core position incorrect/ gates too small/ mold design fault with non-uniform thickness of sections or excessive heavy bosses or ribs. *"Low gloss, dull or rough surface":* moist feed/ injection pressure too low/ [mold too cold]*/ [melt too cold]*/ injection rate too slow/ relocate gates/ mold cooling time too short/ surface of sprue, runner, or cavity rough/ [contamination]*/ [venting of mold insufficient]*/ screw rpm too low/ injection made without screw rotation/ injection speed too fast/ backpressure too low/ nozzle orifice too small/ increase or decrease mold temperature/ gate size too small/ [melt too hot]*/ diameter or depth of cold slug is too small/ wrong location of gate/ wrong location for water channels/ particle size not uniform/ too many fines in feed/ wrong type of lubricant. *"Streaks on part":* stock temperature too high or too low/ screw rpm too high/ nozzle or shutoff valve no tight/ injection speed too fast/ backpressure too low/ cooling and mold-open time too short. *"Splay marks: coarse lines or lumps":* [degradation of melt thermally]*/ injection rate too fast/ increase or decrease the mold temperature/ screw decompression too long/ overall cycle too long/ [contamination, fluid]*/ [shot size too large]*/ [drooling]*/ screw decompression is missing from the molding cycle/ gates too small/ fault in the hot-runner system/ nozzle orifice too small/ sprue and runner size too small/ gate not perpendicular to runner. *"Splay marks: fine lines":* wet feed/ residual non-aqueous volatiles in feed. *"Blush at the gate, dull spot in the part at the gate":* moist feed/ [melt fracture at the gate]*/ [mold too cold]*/ injection pressure too low/ [melt too hot]*/ injection speed too fast/ injection hold time too short/ nozzle diameter too small/ gate land area too large/ diameter of sprue, runner and /or nozzle too small/ depth or diameter of cold slug too small/ wrong location of gate/ [venting of mold insufficient]*. *"Silver streaks":* moist feed/ [nozzle or cylinder too hot]*/ plasticizing capacity, in kg/s, of machine is exceeded/ variation in temperature of feed in hopper/ plastic temperature is too high/ injection pressure too high/ air trapped between granules in the cold end of the machine/ [mold too cold]*/ injection speed too fast/ lack of or excessive external lubrication/ feed is mixture of course and fine particles as with reground/ rear cylinder temperature too high/ [venting of mold insufficient]*/ gates not balanced or at wrong location/ insufficient addition of zinc stearate when using reground/ [air trapped in melt]*/ [degradation of melt, thermally]*/ [air trapped in mold]*/ [cold slugs at the nozzle or hot tip]*/ [contamination]*/ faulty mold design with too many sharp corners or edges. *"Drag marks":* rough surface of mold/ injection pressure too high/ injection hold time too long. *"Worn tracks on part":* [melt too cold]*/ [nozzle too cold]*/ screw rpm too low/ injection speed too fast/ backpressure too low/ nozzle orifice too small/ gates too small/ cold slug well too small. *"Jetting" dull spots and disturbances that look like a jet:* moist feed/ [melt too cold]* / [mold too cold]*/ injection rate too fast/ nozzle diameter too small/ depth or diameter of cold slug too small/ diameter of sprue, and runner too small/ wrong location of gate, incorrectly at a thick section / [nozzle too cold]*/ gate too small/ gate land length too long. *"Wave marks":* feedrate too small/ injection pressure too low/ [melt too hot]*/ [mold too hot]*/ clamp pressure too low/ injection hold time too short/ cycle time too short/ wrong location of water channels/ stock temperature either too hot or too cold/ injection speed too fast or too slow/ nozzle diameter too small/ moist feed. *"Flashing": the flow of material into unwanted areas; if at the end of the flow paths then its cause is usually [shot size too big]; flash in the runner system may indicate continued holding pressure after the gates freeze off:* [melt too hot]*/ injection pressure too high/ injection hold time too long/ injection speed too fast/ clamp pressure too low/ [mold too hot]*/ rework the mold design/ vents too deep/ damaged mold/ misaligned platen/ wet feed/ [shot size too big]*/ [feedrate into mold too high]* / erratic feed/ [design of part faulty]*/ erratic cycle time. *"Weld lines, knit lines":* injection rate too small/ injection pressure too low/ injection hold time too short/ [mold too cold]*/ [melt too cold]*/ vent missing in location of weld/ overflow well missing next to the weld area/ wrong gate location/ too much filler.

**Transparency:** *"Cloudiness or haze for clear plastics":* [contamination]*/ moist feed / [melt too cold]*/ faulty adjustment of barrel temperature profile/ injection pressure too low/ backpressure too low/ [mold too cold]*. *"Bubbles in clear plastics":* moist feed/ [melt too hot]*/ injection pressure too low/ injection rate too fast/ injection hold time too short/ booster time too low/ [mold too cold]*/ mold cooling time too short/ [cooled too fast]*.

- **Strength or Shape Defects**

*"Voids":* [short shot]*/ [mold: external surfaces solidify and shrinkage occurs internally]*/ [thick sections continue to shrink after the melt path is frozen]*/ injection rate too fast/ [melt too hot]*/ booster time too short/ molding cooling time too short/ [cooled too fast]*/ feed moist/ insufficient blowing agent. *"Blisters":* feed moist/ injection pressure too low/ backpressure too low. *"Lamination, peeling":* moist feed/ [mold too cold]*/ [melt too cold]*/ injection speed too fast/ nozzle diameter too small/ gate land area too large/ depth or diameter of the cold slug is too small/ diameter of sprue, runners and or nozzle too small/ [contamination]*/ backpressure too low/ injection made without screw rotation/ screw rpm too low/ [nozzle too cold]* / injection rate too low. *"Warpage, part distortion"(usually caused by non-uniform shrinkage as the molded part cools from ejection temperature to room temperature):* incorrect differential in mold temperatures to account for geometry or mold design/ incorrect handling after ejection/ injection hold time too short and stopped before gate freezes/ cooling time too short/ injection pressure too high or too low/ [mold too cold]*/ shrink fixtures and jigs to promote uniform cooling are missing/ wrong gate locations and too few/ gates too small/ faulty part design/ uneven cooling system on molds/ injection pressure too high/ [melt too cold]*/ holding pressure and time too long/ screw not rotating with injection done/ injection speed too fast/ backpressure too low/ mold temperature either too hot or too low/ time for cooling and mold-open are too short. *"Weld weak":* [mold too cold]*/ injection speed too slow/ [melt too cold]*/ injection pressure too low/ nozzle opening too small/ gate land area too large/ sprue, runner or gate size too small. *"Brittle":* feed moist/ [mold too cold]*/ injection rate too fast/ [melt too cold]*/ injection pressure too high/ gate diameter too small/ nozzle orifice too small/ not enough gates or gates at wrong location/ injection pressure too low/ gate land area too large/ sprue, runner or gate size too small/ gates too small/ cold slug well too small/ holding pressure and time too long/ screw rpm too low/ screw should rotate during injection/ injection speed too fast/ backpressure too low/ mold temperature either too high or too low/ sensor error/ [degradation material]*/ [stress high in part]*/ faulty mold design with notches causing local stress. *"Dimensional variation":* faulty feedrate/ [melt too cold]*/ injection pressure too low/ [mold too cold]*/ injection hold time too short/ injection speed too fast/ cycle time too short/ nozzle diameter too small/ faulty gate location/ gate land area too large/ diameter of sprue, runner and or nozzle too small/ incorrect location of water channels/ [control of machine faulty]*/ [mold con-

ditions wrong]*/ poor part design/ moist feed/ irregular particle size/ batch to batch variation in feed. *"Cracking":* [mold too cold]*/ wrong mold design/ ejection pins poorly located and give unbalanced push/ [shot size too large]*. *"Low heat distortion temperature":* [variation in section thickness]*/ [mold too cold]*/ mismatch between cylinder and mold temperatures/ feedrate too high/ pressure too high/ plunger dwell too long/ excessive temperature variation between front and back of mold/ freezing in the gate because gate orifice too large.

- **Operation**

*"Sticking in cavity":* [mold too hot]*/ [melt too hot]*/ injection pressure too high/ injection hold time too long/ injection hold time too short/ gate land area too large/ diameter of sprue, runner or gate too small/ mold surface is rough/ injection speed too high/ faulty mold design/ incorrect radius of nozzle and sprue bushing/ mold release not used/ air was not provided for ejection/ hold pressure too high/ feed not adjusted to provide a constant cushion/ cooling time too long or too short/ cavity or core temperatures do not have the < 7 °C differential between mold halves/ nozzle too hot/ mold has undercuts and insufficient drafts. *"Sticking parts":* [mold too hot]*/ injection pressure too high/ rough surface on mold/ holding pressure too high/ wet feed/ cooling time too short/ faulty design of ejector/ highly polished or chrome-plated mold surfaces. *"Sticking in sprue bushing":* injection pressure too high/ injection hold time too long/ booster time too long/ cooling time too short/ mold is too hot at the sprue bushing/ nozzle pulled back from mold/ [nozzle too cold]*/ incorrect seat between the sprue and mold/ nozzle orifice is not 7.5 mm smaller in diameter than OD of sprue/ rough surface on sprue/ sprue puller ineffective/ sprue does not have sufficient draft angle for easy release/ screw decompression too low or missing. *"Runner breaks":* holding pressure and time too long/ [mold too hot]*/ sprue, runners and gates are rough/ incorrect radius in the nozzle and sprue bushing/ time for cooling and open time too short. *"Discoloration of sprue":* [melt too hot]*/ nozzle or shutoff valve not tightened/ injection speed too high/ nozzle orifice diameter too small/ [mold too cold]*/ cold slug well too small/ gate diameter too small. *"Drooling at nozzle"*: shutoff valve dirty or clogged/ injection too soon/ wrong nozzle pressure/ poor radius of nozzle and sprue bushing/ [nozzle too hot]*. *"Screw does not return"*: screw rpm too low/ backpressure too high/ wet feed/ hopper out of feed/ obstruction/ temperature in the rear zone too high. *"Ejection of part poor":* rough mold walls/ [shot size too large]*/ knockout system inadequate/ insufficient taper. *"Cycle erratic":* operator/ [pressure erratic]*/ [feedrate erratic]*/ [cylinder temperature cycles]*. *"Cycle too long":* [cooling cycle too long]*/ [heating cycle too long]*/ [operator issues]*/ material should be more heat resistant.

- **Symptoms**

*[Air trapped in melt]*:* screw decompression/ backpressure too low.

*[Air trapped in mold]*:* [venting of mold insufficient]*/ gate diameter too small/ [mold too hot]*.

*[Backflow from the mold]*:* suckback/ faulty non-return valve.

*[Backpressure too high]*:* injection rate too fast.

*[Barrel too hot]**: melt temperature > 271 °C/ sensor error/ faulty barrel heater control system/ worn or incorrectly fitted screw and barrel configuration.

*[Contamination]**: dirty machine/ dirty hopper/ moist feed/ too many volatiles in feed/ [degradation]*/ lubricant or oil on mold/ incorrect mold lubricant/ feed contaminated during material handling/ faulty raw material from supplier/ poor shutdown procedures.

*[Contamination, fluid]**: water or oil leaking into mold cavity.

*[Control of machine faulty]**: incorrect screw stop action/ inconsistent screw speed/ malfunction of non-return valve/ worn non-return valve/ uneven control of backpressure/ faulty temperature sensor/ heater band faulty/ control system fault or poorly tuned/ machine has inadequate plasticizing capacity/ inconsistent control of cycle.

*[Cooling cycle too long]**: [melt too hot]*/ [mold too hot]*/ inadequate cooling in local heavy sections.

*[Cooling insufficient before removal from mold]**: faulty mold design especially for rib design/ injection speed too slow/ injection hold time too short/ injection pressure too low/ melt too hot/ mold too hot/ [venting of mold insufficient]*/ sprue and runners too small diameter/ gate too small/ gate land length too long/ gate not close to thicker areas/ core missing from heavy section.

*[Cooled too fast]**: [mold too cold]*/ mold cooling time too short.

*[Cylinder overheated]**: nozzle too hot/ cylinder temperature too high.

*[Cylinder temperature cycles]**: controller fault/ sensor error/ incorrect line voltage/ power factor problems/ heater bands faulty/ variation in feed temperature.

*[Degradation, mechanical]**: barrel temperature in the feed area is too low combined with high screw speed or high backpressure/ short transition section in screw/ radius between the screw root and the flighter is too small/ small tolerance between the plunger and the wall/ fine material trapped between the plunger and the wall/ excessive reground/ rear cylinder temperature too low/ plunger off-center.

*[Degradation in the extruder of melt thermally]**: temperature sensor error/ [melt too hot]*/ temperature controller fault/ improperly designed or defective non-return valve.

*[Design of part faulty]**: incorrect mold dimensions/ unequalized filling rate in cavity/ mold not sealing because of flash between surfaces/ [venting of mold insufficient]*/ vents too large/ gate land area too large/ runner, sprue and gate dimensions incorrect.

*[Drooling, introduces solid material into part giving defects]**: wet feed/ [melt too hot]*/ suckback pressure too low/ injection pressure too high/ injection forward time too long/ injection boost time too long/ shutoff valve dirty or clogged/ injection too soon/ poor radius of nozzle and sprue bushing/ [nozzle too hot]*.

*[Feed rate erratic]**: feeding mechanism/ bridging in hopper/ hopper design fault.

*[Feed rate into mold too high]**: injection feedrate too fast/ feed setting too high/ sensor error.

*[Flow of polymer into the cavity uneven during high velocity flow into an open area]**: injection rate too fast/ faulty gate location/ gate too small.

*[Granules not melted]*:* plastic temperature too low/ cycle too short for cylinder capacity/ nozzle diameter too large.

*[Heating cycle too long]*:* insufficient heating capacity.

*[Injection too slow]*:* screw rpm too high/ backpressure too high/ injection speed too slow/ injection pressure too low/ injection forward time too short/ booster time too short/ cycle too short.

*[Insufficient plastic in mold]*:* thick sections, bosses, ribs/ not enough feed/ injection pressure too low/ plunger forward time too short/ unbalanced gates/ piece ejected too hot/ variation in mold open time/ no cushion in front of injection ram with volumetric feed.

*[Melt not homogeneous]*:* backpressure too low.

*[Melt too cold]*:* sensor error/ control system error/ lack temperature confirmation via hand-held pyrometer or laser sensor/ cylinder too cold/ screw rpm too low/ backpressure too low/ insufficient plasticizing capacity of machine/ [nozzle too cold]*/ heating band fault/ excessive flow length in mold.

*[Melt too cold at the nozzle or hot tip]*:* nozzle too cold/ temperature sensor error/ too few heater bands/ heater bands too far from nozzle tip/ hot tip heat source too far from orifice or faulty/ sharp corners near the gate.

*[Melt too hot]*:* sensor error/ control system error/ lack temperature confirmation via hand-held pyrometer or laser sensor/ cylinder too hot/ screw rpm too high/backpressure too high/ [mold too hot]*/ [nozzle too hot]*/ injection rate too slow/ gate too large/ gate land too short/ resin too hot/ holding pressure and time too long/ moist feed/ cooling and mold-open time too long/ [residence time too long]*.

*[Melt too hot, localized overheating]*:* [barrel too hot]*/ faulty barrel heater control system/ [nozzle too hot]*/ sensor error/ faulty or incorrectly designed check valve/ worn or incorrectly fitted screw and barrel configuration.

*[Melt fracture at the gate]*:* [melt too cold]*/ temperature sensor error/ injection rate too fast/ gate too small/ sharp edge in gate area/ cold slug well in the runner too small.

*[Mold conditions wrong]*:* [mold temperature non-uniform]*/ injection pressure low/ injection forward time too short/ injection boost time too short/ [barrel too hot]*/ [nozzle too hot]*/ inconsistent control of cycle.

*[Mold temperature non-uniform or erratic]*:* poorly designed water or coolant lines/ [venting of mold insufficient]*/ coolant supply fault.

*[Mold: external surfaces solidify and shrinkage occurs internally]*:* [mold too cold]*/ mold includes sections that are "too thick"/ [melt too cold]*.

*[Mold too cold]*:* cooling lines in wrong location/ coolant too cold/ coolant flowrate too high/ sensor error.

*[Mold too hot]*:* cooling lines in wrong location/ coolant too hot/ coolant flowrate too low/ sensor error.

*[Non-uniform shrinkage as the molded part cools from ejection temperature to room temperature]*:* wrong packing times/ wrong packing pressures/ wrong gate location/ cooling system fault/ temperature sensor fault/ need separate temperature adjustment for mold halves.

*[Nozzle too hot]*:* sensor error/ control error/ temperature setpoint at nozzle too hot/ localized heater bands on the nozzle instead of being spread along the nozzle.

*[Operator issues]*:* slow setup of mold/ need to trim "flashing"/ poor monitoring of cycle times/ excessive machine dead time.

*[Premature gate freeze-off]*:* gate size too small.

*[Pressure too low]*:* injection pressure too low/ loss of injection pressure during the cycle/ feed control set too high causing lower injection pressure.

*[Pressure too high]*:* injection pressure too high/ injection time too long/ boost time too long.

*[Pressure erratic]*:* sensor error/ control system tuning fault/ leaks in the hydraulics.

*[Residence time too long]*:* machine provides shot size that is too large/ dead spots in hot manifold/ temperature too high/ poorly designed manifold system/ [contamination]*.

*[Resin feedrate too low]*:* no material in the hopper/ hopper throat partially blocked/ feed control set too low/ faulty control of feed system/ bridging in the hopper/ faulty hopper design.

*[Shear heating of melt]*:* injection rate too fast/ injection pressure too high/ gates too small/ nozzle orifice too small $< 0.8$ of sprue bushing/ nozzle dirty/ sharp corners/ injection rate too fast/ shutoff nozzle used instead of a general purpose nozzle/ improperly designed or defective non-return valve.

*[Short shot]*:* [resin feedrate too low]*/ injection pressure too low/ [mold too cold]*/ injection speed too low/ [melt too cold]*/ injection hold time too short/ cycle time too short/ diameter of gate, sprue, and runner too small/ nozzle orifice too small/ gate land length too long/ incorrect gate location/ [venting of mold insufficient]*/ [nozzle too cold]*/ nozzle dirty/ shutoff valve dirty/ inject with screw not-rotating/ machine undersized for the shot required/ cycling from wet to dry resin/ excessive flow length in mold/ excessive feed buildup in cylinder/ [mold temperature non-uniform]*/ [air trapped in mold]*/ not enough external lubricant/ poor balance of plastic flow into multiple cavity mold/ holding pressure too low.

*[Shot size too large]*:* resin feedrate too high/ injection pressure too high/ machine shot size much larger than mold requirement.

*[Shrinkage excessive]*:* [melt too hot]*.

*[Solidification at the mold wall delayed]*:* [mold too hot]*.

*[Stress high in part]*:* [mold too cold]*/ [melt too cold]*/ injection pressure too high/ faulty post-mold conditioning/ faulty mold design.

*[Thick sections continue to shrink after the melt path is frozen]*:* faulty mold design, with too much variation in part cross section/ [premature gate freeze-off]*.

*[Venting of mold insufficient]*:* injection rate too fast/ booster time too long/ injection pressure too high/ vents plugged/ not enough vents/ clamp pressure too high/ wrong location of gates relative to vents/ [melt too hot]*/ [mold too hot]*.

*[Viscosity of melt too high]*:* [melt too cold]*/ wrong resin.

#### 3.9.6.2 Extruders for Polymers

Good practice: for startup, the barrel heaters are critical because screw is not rotating. Major concerns about cold start. Rear-barrel temperature usually remains important because it affects the "bite" or rate of solids conveyed in the feed. Barrel temp. must be set appropriately for polymer. Head and die temperatures = desired melt temp (except where want gloss, flow distribution or pressure control).

Screw speed is changed by reducing the motor speed by one of three options: 1) 10 to 20 in two stages: either pair of gears or pulley but second stage is always gears with the screw set in the middle of the last big "bull" gear. For very slow-moving extruders (e.g. twins for rigid PVC) there are usually three stages of reduction to get to < 30 rpm. Most extruder drives are constant torque with max power only available at top screw speed with the reduction ratio sometimes mismatched to the job.

For maximum solids conveying "Stick to the barrel and slip on the screw". Most plastics normally slip on the root of the screw as long as the feed temperature < melt temperature with those that are most likely to stick being highly plasticized PVC, amorphous PET and certain polyolefin copolymers. For amorphous PET covert the feed to less-sticky semicrystalline form by heating to high temperature for at least an hour in an agitated hopper.

Particles must stick to the barrel; trouble occurs with a "slippery feed" such as HDPE and fluoroplastics.

Material is the biggest cost (usually 80%) so reuse as much trim and scrap as possible and keep close thickness tolerances so as not to have excessive thickness.

Shear rate is important because this affects viscosity. All common plastics are "shear thinning" eg. PVC flow is 10× faster if double the push but LLDPE flow increases 3 to 4 times for double the push.

**Single screw:** typically operated 100% filled. Usually flood feed.

**Twin screw:** typically operate 20–100% filled. Cannot be flood fed if running at high speeds.

**Twin screw with vent:** melt seal is about 1 L/D upstream of the vent; feed screw section under the vent operate < 0.5 full. During startup increase the vacuum gradually. Use low degree of fill.

**Coating wire and cable:** preheat the wire to about 120 (for HDPE) to 175 °C (cellular PE) to minimize shrinkage.

*Trouble shooting:* Extruders: the approach usually is to 1) adjust the temperature profile, 2) check the hardware such as the thermocouples, controllers, speed, 3) alter the processing conditions or 4) change the resin or the screw and barrel design. The symptom-cause information is presented as issues related to production, off-spec thickness or shape, off-spec strength, off-spec surface features, and usual symptoms.

- **Production:** *"Throughput < design":* [low bulk density of feed]*/ wrong screw design/ worn screw or barrel elements/ screw rpm too low/ wrong temperature set points/ caking on the feed screw/ caking on the feed port. *"Slow and steady reduction in throughput":* buildup of contaminants on screen pack. *"Torque > design":* feedrate too high/ [degree of fill too high]*/ screw speed too low/ heat zone set points too low/faulty screw design.*"Machine stalls above a certain speed or with certain materials":* constant-torque drive (magnetic clutch)/ AC-DC drive system with constant-drive and constant torque combination. *"Feed from hopper not feeding smoothly":* material too light and fluffy for gravity feed/ material damp/ bridging/ screw channels in the feed zone are not deep enough/ too much external lubricant. *"Drive amps > design"*: polymer viscosity too high/ screw pumping too high/ screw speed too high/ barrel temperatures set too low. *"Amps high for melt pump drive or pump won't rotate":* shear pin/ degraded polymer caught in gears. *"Variation in drive amps":* [solids conveying instabilities]*/% regrind too high/ feed bulk density wrong. *"Cycling motor amps":* [surging]*. *"Extruder noisy"*: loss of feed/ foreign or metal contaminant in feed/ bent screw/ bent barrel/ 1/2 heater burned out. *"Local temperature fluctuations with cycles < 5 min":* instrument circuitry fault/ inconsistent melt/ poor heater contact/ thermocouples poorly seated/ sensor error/ poor sensor location/heating element fault/ controller fault/ [solid conveying instabilities]*. *"Barrel temperatures differ from the set temperatures"*: controller fault/ burnt-out heater/ blue screw syndrome where the rear end bites off more than the front end can pump. *"Real wall temperature > the set point"*: the rear end bites off more than the front end can pump. *"Variations in melt pressure"*: drift or cycle time variation > 1 min: [low feeding efficiency]*/ low friction characteristics/ [low bulk density of feed]*/[melting too soon]*/ adequate early barrel pressure but [melting unstable]*/first barrel heating too high/screw tip pressure too low. *"Screw tip pressure too low":* no resin in feed hopper/ bridging in feed hopper/ temperature too high in extruder entrance zone/ polymer wrapped around screw. Related topic *[Screw tip pressure too high]**. *"Unstable melt pressure"*: screw speed too high/ degree of fill too low/ screw design gives mixing of melt inadequate or low shear/ cycling control on heat zones/ feeder problems. *"Unstable pumping"*: for *vented* extruders: first stage [surge]*/ poor screw balance between stages. *"Material flow out vent"*: for *vented* extruders: poor vent-diverter design/ first stage pumping rate too fast/ screw tip pressure too high/ flowrate > design/ temperature set points for last barrels too low or heaters faulty/ screw design gives localized pressure under the vent/vacuum too high/ if adding liquids, then poor mixing.

- Product thickness or shape does not meet specifications. *"Small size variation":* variation in drive speed/ wrong screw design/ variation in puller. *"Large size variation":* [surging]*. Here are more specific details: *"Variation in thickness in transverse direction and always in the same place":* see "Variation in local temperature". *"Variation in thickness in transverse direction and floating across or around the product":* [mixing of melt inadequate]*/ die temperature setting wrong/ dirty die/[screw tip pressure too low]*/backpressure on extruder < 20 MPa/ wrong design or die or screw/ [degradation of melt in extruder]*/ for blown film: air ring not centered or level/ thermocouple error/ bubble subjected to hot or cold air/ polymer feed has > 6 °C variation in melt temperature. *"Waviness or ridges around the circumference":* [surging]*/ nonuniform water cascade/ uneven take-off speed/ vibration in the take-off equipment. *"Variation in thickness in the direction of extrusion":* [surging]*/ puller slip or incorrect control of tension/ drawdown too much or too fast/ poor alignment/variation in take-up reels/ erratic variation in feed materials/ hot-lips controller cycling/ untuned controller/ faulty controller/ temperature variation in die/ variation in motor load/ variation in melt pressure/ damaged orifices in die or feedblock/ holes in die too large/ incorrect barrel temperature profile/ faulty adjustment of die/ faulty screw design/ plugged screen pack/ temperature sensor fault in barrel/ hopper bridging/ throughput too high/ gels/ for *blown film*: inconsistent nip roll speed control/ frost line too low/ polymer melt temperatures too low/ bubble-cooling control fault/ variation in air flow from blower/ cooling air nonuniform/ gap opening too large/ cooling air flowrate too low. *"Periodic variation in thickness in direction of extrusion":* spinneret temperature too low/ orifice wrong diameter/ wrong drawdown ratio. *"Cyclical variation in thickness in the direction of extrusion":* [draw resistance instability]*.

*"**Filament** breaks":* [surging]*/ some die holes blocked/ temperature variation in head or die/ melt temperature too low or too hot/ drawdown too great/ holes in die too large/ moisture/ [contamination]*/ [melt too hot]*/ gap between bath and die too large. *"Wrong filament shape: correct cross section but too large":* too little pull/ draw distance from die to take-off is too short/ take-off speed too slow/ die-land length too short/ melt temperature too low. *"Wrong filament shape: distorted cross section":* unequal die temperatures/ die incorrect shape. *"Wrong filament shape: size OK but warped":* cooling too intense / linear take-off speed too fast. *"Filament oval cross section":* filaments too hot while passing over rolls/ rolls too hot/ die holes oval/ temperature gradients in die/ tension too high in take-up rolls.

*"Holes in blown film or coating":* moisture in resin/ die lip gap too large/ air gap too small/ vacuum too high; for coating: moisture/ substrate too rough/ coating thickness too thin/ contamination/ decomposition/ compound temperature too high/ see also *"Gels"*.

- Product does not meet strength specifications. *"Product strength < specs for all samples":* faults with the feed/ [degradation of melt in extruder]*. *"Product strength < specs for some samples":* [contamination]*/faults with the feed/ [degradation of melt in extruder]*". *Pipe strength < specs":* melt temperature too low/ throughput too fast/land length too short/ air gap too short/ excessive drawdown at cold temperatures/ too much scrap in feed/ moisture in resin/ dirty metal surfaces/ material sticking on extruder parts/ short die-land length/ high internal angular discontinuities into the die-land section/ linear extrusion speeds excessive/ uneven water coolant cascade/ misaligned sleeve/[mixing of melt inadequate]*. *"Stiffness < design":* for *cast or sheet*: chill roll temperature too low/ low resin density. *For coatings*: *"Poor adhesion":* a variety of apparently contrary causes related to polymer viscosity [polymer viscosity too high]*, degradation especially [oxidation]*, tackiness, temperature:

melt temperature too low or high/ air gap too small/ chill roll temperature too low or hot/ line speed too fast/ poor match between coating and substrate/ substrate problems. *"Low tenacity":* ratio of roll speeds too small/ [degradation of melt in extruder]*/ wrong resin/ nicks in die. *"For wire and cable: covering separates from wire/adhesion":* wire not preheated hot enough/ melt temperature too low/ dirty or moist wire/ [degradation of melt in extruder]*/ cooled too fast/ air-cooling gap too short/ air trapped between wire and coating [trapped air]*. *"Low modulus of elasticity":* melt temperature too low/ air gap distance too short. *"Interfacial instability for coextruded film":* excessive shear stress at the die gap > 0.06 MPa/ throughput too high/ die gap too narrow, melt temperature too low/ polymer viscosity too high/ the relative velocities where polymer flows combine differ by > 4: 1.

- **Appearance:** gloss, fisheyes, shark skin, pits, holes, clarity. Some are surface effects, such as shark skin, regular, ridged, surface deformity with ridges perpendicular to extrusion direction. Others are defects of the whole body of extrudate, caused by [melt fracture]*. These include spiral, bamboo, regular ripple. *"Rough surface or dullness":* [contamination]*/ moisture/ linear speed too fast or screw speed too fast/ die holes too small/ die temperature too low/ die land too short/ [mixing of melt inadequate]*/ [melt fracture]*/ no vents used/ hopper vacuum inadequate/ [screw tip pressure too low]*/ discontinuity in the melt flowlines / low melt temperature/ dirty metal surfaces/ material sticking on extruder parts/ uneven water coolant cascade/ misaligned sleeve/faulty screw design/ screw too hot/ extrudate too hot in the coolant bath causing boiling. *"Pits on surface"*: [contamination]*/ moisture/ water sprays onto extrudate just after exiting the die/ water bath too hot. *"Fisheyes in film"*: moisture/ damp polymer/ too many volatiles in polymer. *"Gels":* [contamination]*/[degradation of melt in extruder]*/ [shear intensity too low]*/ [screw tip pressure too low]*/ number and density of the screen pack too low/ moisture too high/ screw speed too low/ incompatible blend/ [residence time too long]*/ lack of streamlines in extruder/ incorrect startup procedures/ [melting inadequate]*/ [melt too hot]*/ for reactive: localized initiator concentration too high. *"Shark skin"*: [melt fracture]*/ die temperature too low at the land end/ linear extrusion speed too high/ throughput too high/ viscosity of polymer too high/ MWD of polymer too narrow/ lubricant additive missing/ [shear intensity too high]*/ die gap too small.

*"Polymer buildup on die"*: melt temperature too low/ throughput too high/ die gap too small/ wrong screw design/ low level of antioxidants.

*"Porous or bubbles in product"*: poor melt quality at vent/ plugged vent opening/ insufficient vent vacuum/ excessive volatiles in feed/ screw speed too high/ vacuum vent needed. *"Spotted, warped or pocked surface"*: [mixing of melt inadequate]*/ moisture/ roll too cold/ contamination/ screen size too large/ dirty die/ [trapped air]*/ dirt on rolls/ drafty air/ wrong tension/ boiling on extrudate in cooling bath. *"Lines on the product"*: surface scratches on tip or die/ local buildup/ [die swell too high]*/ throughput too high/ polymer adhesion on channels, tip or die/ incorrect contact in

the quench tank/ melt temperature too low/ throughput too fast/land length too short. *"Indented pock marks on pipe after water cooling"*: coolant water spray velocity too high. *"Raised pock marks on pipe"*: water drops on surface in the air-drying zone. *"Discolored material"*: temperatures too high/ wrong formulation/ discontinuities inside extruder.

- **Symptoms**

*[Contamination]**: contaminated feed/ contaminated additives/ dirty die/ polymer on die lips.

*[Degradation of melt in extruder]**: [RTD too wide]*/ barrel temperature too high/ screw speed too high (causing overheating and shear damage)/ oxygen present/[oxidation]*/ nitrogen purge ineffective/ wrong stabilizer/ wrong screw/ flows not streamlined/ stagnation areas present/ extruder stopped when temperatures > 200 °C/ copolymer not purged with homopolymer before shutdown/[residence time too long]*.

*[Degree of fill too high]**: feedrate too high/ screw speed too slow.

*[Die swell too high]**: tip too short/abrupt change in flow near tip or die/ melt temperatures too low in die assembly.

*[Draw resistance instability]**: for blown film, fiber spinning, blow molding: draw ratio too high.

*[Extrusion instabilities]**: screw speed too high/ screw temperature too high/ barrel temperature at delivery end too high/ channel depth too high in the metering section/ the length of the compression section too short/ read barrel end temperatures too low/ diehead pressure is too low.

*[Feedrate too high]**: screw speed too fast/ feed from hopper too fast.

*[Gels]**: [contamination]*/[degradation of melt in extruder]*/ [shear intensity too low]*/[screw tip pressure too low]*/ number and density of the screen pack too low/ moisture too high/ screw speed too low/ incompatible blend/ [residence time too long]*/ lack of streamlines in extruder/ incorrect startup procedures/ [melting inadequate]*/ [melt too hot]*/ for reactive: localized initiator concentration too high.

*[Low bulk density of feed]**:% regrind too high/ grind too coarse.

*[Low feeding efficiency]**: low friction characteristics.

*["Melt fracture" where the critical shear stress of polymer (about 0.1 to 0.4 MPa) exceeded in the die; excessive shear stress at the wall > 0.1 MPa]**: exit speed at the die is too fast/ melt too cold/ throughput excessive/ die land too short/ die opening too small/ entrance to die not sufficiently streamlined/ screw speed too high/ MM and melt viscosity too high/ cross-sectional area in exit flow channel too small/ external lubricant additive missing.

*[Melt too hot]**: screw speed too high/exit barrel zone temperatures too high/ degree of fill too low/ [shear intensity too high]*/ heat-zone temperatures set too high/ [screw tip pressure too high]*.

*[Melting inadequate]**: barrel or die temperature too low/ screw speed too fast/ screw design gives insufficient mixing/ [shear intensity too low]*/ feedrate too high/ material too "slippery"/[degree of fill too high]*/ additional component has too low a melting point/[residence time too short]*.

*[Melting too soon]**: wrong bulk density of feed.

*[Melting unstable]**: especially for screws with high compression ratio and short compression length: insufficient melt capacity/ too large a channel depth in the metering section/ temperature in the metering end of the screw too high/ wrong screw design.

*[Mixing of melt inadequate]**: [screw tip pressure too low]*/ [feedrate too high]*/ screw speed too high and [residence time too short]*/ screw speed to low and [shear intensity too low]*/ [degree of fill too high]*/ faulty screw design for mixing/ temperature set points incorrect/ instrument error in temperature sensors/ temperatures too high/ loading excessive for one component/ no static mixer included/ for reactive extrusion: liquid flowrate too high/ screw channel under injection not full of polymer.

*[Oxidation]**: temperature too high/ screw speed too low/ [residence time too long]*/ oxygen present/ nitrogen purge ineffective/ antioxident stabilizer ineffective/ [trapped.air]*/ hopper vacuum inadequate.

*[Polymer viscosity too high]**: temperature too low/ wrong blend/ [shear intensity too low]*.

*[Residence time too short]**: screw speed too high/ too much feed/ [degree of fill too high]*/ poor screw design.

*[Residence time distribution, RTD, too wide]**: [degree of fill too low]*/ feedrate too small/ screws speed too fast.

*[Screw tip pressure too high]**: screens plugged/ die or adapter or breaker plates too restrictive and give too much $\Delta p$/ [polymer viscosity too high]*/ temperatures in die assembly too low/ barrel temperature too low/ screw speed too high/ [shear intensity too low]*/ lubricant needed/ flow restriction/ throughput too high/ die land too short/ cold start/ [degradation of melt in extruder]*.

*[Shear intensity too low]**: screw speed too low/ faulty screw design.

*[Solids conveying instability]**: feed hopper fault/ internal deformation of the solid bed in the screw channel/ insufficient friction against the barrel surface.

*[Surging]**: *30–90 s*: feed particles are not sufficiently softened (usually at the beginning of the second compression zone) / too rapid compression screws/ [low feeding efficiency]*/ low friction characteristics/ low bulk density of feed/[melting too soon]*/ adequate early barrel pressure but [melting unstable]*/ first barrel heating too high/ screw speed too fast/ faulty screw design/ additional compound slippery/ bridging in resin feed hopper/ feed-zone temperature too high/ [screw tip pressure too low]*/ compound temperature too high/screw too high/ nonuniform take-off speed/ take-off speed too high/ throughput too fast/ controller fault/ feed resin not mixed well/ melt temperature too low.

*[Trapped air in extruder]**: unvented extruder/ wrong screw design/ pressure too low/ rear-barrel temperature too high/ screw speed too high/ vacuum too low in feed hopper/ powder feed instead of pellets.

### 3.9.7
### Coating

Trouble shooting: *"Poor adhesion"*: a variety of apparently contrary causes related to polymer viscosity, degradation, oxidation, tackiness, temperature: melt temperature too low or high/ air gap too small/ chill roll temperature too low or hot/ line speed too fast/ poor match between coating and substrate. *"Rough wavy surface (apple-sauce)"*: wrong resin/ temperature too low or high. *"Edge tear"*: draw ratio too high/ die end temperature too low/ temperature too low or high. *"Oxidation"*: temperature too high/ screw speed too low/ flows not streamlined/ extruder stopped when temperatures > 200 °C/ copolymer not purged with homopolymer before shutdown. "Pinholes in coating": substrate too rough/ coating thickness too thin. *"Surging"*: bridging in resin feed hopper/ feed-zone temperature too high/ wrong screw design. *"Voids"*: moisture/ leaks in resin handling system/ inadequate drying and storage/ [thermal degradation]*/ [gels]*. *"Die lines"*: nicks in die/ dirty lips/ particles in die.

*"Pin holes and breaks"*: coating too thin/ contamination/ decomposition/ compound temperature too high/ moisture.

*"Web tears"*: compound temperature too low/ too much drawdown/ die lip opening too large. *"Poor adhesion"*: compound temperature too low/ substrate problems. *"Excessive neck-in"*: die-to-roll gap too large/ material temperature too high/ die-lip opening too large/ throughput too low/ use resin with lower Melt Index/die-land length too long.

See Symptoms for Section 3.9.6 for the following [Contamination]*; [Degradation of melt in extruder]*; [Screw tip pressure too low]*; and [Shear intensity too low]*.

*[Gels]*:* [contamination]*/[degradation of melt in extruder]*/ [shear intensity too low]*/[screw tip pressure too low]*/ number and density of the screen pack too low/ moisture too high/ screw speed too low/ incompatible blend/ [residence time too long]*/ lack of streamlines in extruder/ incorrect startup procedures/ [melting inadequate]*/ [melt too hot]*/ for reactive: localized initiator concentration too high.

*[Surging]**: screw speed too high/ take-off speed too high/ backpressure too low/ compound temperature too high.

*[Thermal degradation/crosslinking]**: polymer temperature too high/ screw speed too low.

## 3.10
## Vessels, Bins, Hoppers and Storage Tanks

**Bins and hoppers:** In general cohesive strength of powders increases with consolidation pressure. Trouble shooting: *"No flow":* [arching]*/ [rat holing]*. *"Erratic flow":* obstructions alternating between arching and ratholing. *"Flooding or flushing" when a rathole collapses it entrains air, becomes fluidized and the material floods through the outlet uncontrollably: "* fine powders such as pigments, additives and precipitates that tend to rathole/ insufficient residence time in hopper for deaeration. *"Flow-rate limitation":* fine particles where movement of the interstitial air causes an adverse $\Delta p$.

*"Limited live capacity":* [ratholing]*. *"Product degradation":* [ratholing]*. *"Incomplete or non-uniform processing":* [ratholing]*.

*[Arching]**: particle diameter large compared to outlet/ cohesive particles probably caused by moisture or compaction.

*[Ratholing]**: cohesive particles probably caused by increased moisture or by compaction (fine powders < 100 μm such as pigments, additives and precipitates).

## 3.11 "Systems" Thinking

Processes are complex systems in which the performance of one piece of equipment interacts with and influences another. Typically when we trouble shoot we tend to focus on individual pieces of equipment and forget that perhaps something away up stream or even downstream could be causing the local difficulty.

To some extent we are already familiar with *systems* and system interactions when we consider a distillation column. This is not one piece of equipment. Rather it is a complex collection of a column with a series of trays linked by vapor and liquid flow with condenser, reboiler, pumps, controllers all interacting. Thus, troubles on the condenser are reflected in the performance of the reboiler and everything in between. We are skilled at being able to make those connections for a distillation "system". However, when we have a crystallizer, screen, centrifuge, dryer combination we may not be able to easily visualize how the performance of one can dramatically affect the others.

In "systems thinking" the focus can be on *how* performance from one unit is transmitted to units elsewhere in the system. Since equipment is linked by pipes and conveyors carrying fluids and solids, the interaction is through the temperature, flowrates, pressures, compositions and cycling in the streams. Another perspective to "systems thinking" is startup and how the system can move from cold, air-filled conditions to the on-line temperatures, pressures and compositions. A third perspective is how the *environment system* affects the *process system.*

A useful approach it to reflect on the principle that anything that is created inside the system must go somewhere; where does it go and how does it get out of the system? For example, mass is conserved.

1. When you start a plant, all the equipment is filled with air; where does the air go?
2. If a dissolved trace metal gets into the system, where does it go?
3. If a particle breaks down where do the particles go?
4. If recycle is used then any trapped species will build up unless there is a bleed or purge.

In trouble shooting "systems", some common issues include: solvent losses somewhere in the system; fouling; foaming or stable emulsion formation that cause equipment malfunction and carryover; corrosion; recycle causing a buildup of species that may not be removed from the system without adequate bleeds or blowdown. Although many of these are considered for specific pieces of equipment, we include a generic consideration of some of these here. In this listing, the concept or symptom is shown in parentheses and italics, for example, *"Foaming"*, followed by possible causes separated by /. If the cause is not a root cause, then it is represented in square bracket plus an *, [foam-promoting systems]*. These intermediate causes are then listed alphabetically .

- *"Fouling":* velocity too slow/ [particulate fouling]* for example, rust, corrosion products from upstream, scale from upstream units, oil, grease, mud or silt/ [precipitation fouling]* for example sodium sulfate, calcium sulfate, lignin / [biological fouling]* species present such as algae and fungi/ [chemical reaction fouling]*, example coke formation and polymerization fouling/ [flocculation fouling]* or destabilization of colloids, for example asphaltenes or waxes from hydrocarbons/ corrosion products for this unit, see Section 3.1.2/ [solidification fouling]* or incrustation such as the freezing on a solid layer on the surface or crystallization/ [condensation fouling]* such as vaporization of sulfur.

*[Biological fouling]*:* temperature, pH and nutrients promote growth of algae and fungi/ biomaterials present.

*[Chemical reaction fouling]*:* polymerizable species in the feed/ high temperature causing cracking/ high wall temperatures/ stagnant regions near the wall or velocity too slow $< 1$ m/s/ reactant droplets preferentially wet the solid surface/ addition of "fouling suppressant" insufficient, for PVC polymerization oxalic acid or its salt or ammonium or alkali metal borate/ pH change.

*[Condensation fouling]*:* wall temperature too low/ contamination in the vapor.

*[Flocculation fouling]*:* pH at the zpc/ low concentration of electrolyte/ increase in humic acid concentration in water in the fall and spring/ colloids present.

*[Particulate fouling]*:* filter not working or not present/ contaminant in feed/ upset upstream/ erosion/ increase in silt and clays in water in the spring.

*[Precipitation fouling* giving scale or sludge*]*:* soluble species present in feed/ temperature high for invertly soluble/ temperature too low for incrustation or crystal formation.

*[Solidification fouling]*:* wall temperature too low/ missing insulation/ cold spots on wall/ sublimation.

- *"Foaming":* [foam-promoting systems]*/ [foam-promoting contaminants]*/ [gas velocity too high]*/ [liquid residence time too low in GL separator]*/ antifoam addition faulty/ faulty mechanical foam breaker/ [liquid environment wrong]*.

*[Foam promoting contaminants: soluble]*:* naturally occurring or synthetic polymers/ naturally occurring or synthetic organics >C10; example lube oils, asphaltenes/ naturally occurring or synthetic surfactants; *for amine systems*: the surface active contaminants include condensed hydrocarbons, organic acids, water contaminants, amine degradation products/ faulty cleaning before startup; surfactants left in vessels.

*[Foam promoting contaminants: solid]*:* [corrosion products, see Section 3.1.2]*; for *amine systems*: iron sulfides/ faulty cleanup before startup; rust left in vessel/ dust/ particulates.

*[Foam promoting systems]*:* those that foam naturally: methyl ethyl ketone, aerobic fermentation, textile dyeing foam more readily than amine and glycol absorption systems and latex strippping > amine, glycol and Sulfolane strippers > slightly foam promoting: fluorine systems such as freon, $BF_3$/ systems operating close to the critical temperature and pressure/ surface tension positive system/ [Marangoni effects]*.

*[Gas velocity too high]*:* temperature too high/design error/ [foaming]*/ vessel diameter too small for gas flow/ column pressure < design/ trays or packing damaged or plugged giving excessive vapor velocity/ temperature too high/ upstream flash separator passing liquids: feed contaminated with excessive volatile species/ stripping gas fed to column too high/ input stripping gas flowmeter error/ design error.

*[Inaccurate sensing of the interface]*:* instrument fault/ plugged sight glass.

*[Liquid environment wrong]*:* pH far from the zpc/ electrolyte concentration too low.

*[Liquid residence time too low in gas liquid separator]*:* interface height decreases/ [inaccurate sensing of interface]*/ turbulence in the liquid phase/ flowrate > expected/ sludge settles and reduces effective height of phase/ inlet conditions faulty.

- *"Corrosion"*: see Section 3.1.2.
- *"Stable emulsion formation":* contamination by naturally occurring or synthetic surfactants: example, lubricating oils/ contamination by particulates: example, products of [corrosion, see Section 3.1.2]*, amphoteric precipitates of aluminum or iron/ pH far from the zpc/ contamination by polymers/ temperature change/ decrease in electrolyte concentration/ the dispersed phase does not preferentially wet the materials of construction/ coalescence–promoter malfunctioning/ improper cleaning during shutdown/ [rag buildup]*.

*[Marangoni effects]*:* non-equilibrated phases/ local mass transfer leads to local changes in surface tension and hence stable interfacial movement

*[Rag buildup]*:* collection of material at the interface: naturally occurring or synthetic surfactants: example, lubricating oils/ particulates: example, products of [corrosion, see Section 3.1.2]*, amphoteric precipitates of aluminum/ naturally occurring or synthetic polymers.

- Other possible causes of trouble for "systems" include:

*"Bumping resulting in plate damage":* trace amounts of water.

*"Catalyst contamination"*: trace amounts of water.

*"Conversion less than expected"*: temperature spike causing catalyst decomposition/ temperature sensor reading low/ [catalyst contamination]*.

*[Cycling]**: two vessels in series on level control/batch processes in sequence but out of synchronization/ no intermediate storage.

*"Distillation overhead off spec"*: excessive inerts from upstream/buildup of trace/ purge not sufficient from recycle.

*[High $\Delta p$ across bed of catalyst/particles/resin]**: temperature spike/ temperature sensor reading low.

*"Leaks"*: temperature spike/ pressure spike/ temperature sensor reading low/ glands not tight enough around rotating shafts/ valve stem leaks/ thermal expansion of the different metals and parts not correctly accounted for, example, flanges and gaskets on high-temperature heat exchangers.

*"Particle agglomeration in pneumatic conveying"*: increased humidity in air/ trace contaminants in air.

*"Separation performance of column decreases":* trace amounts of water/ trace amounts of water trapped in column/ *[bumping resulting in plate damage]**.

*"Temperature runaway in reactor":* operator error: overcharge reactor/ trace water/ [poor temperature control]*.

## 3.12
## Health, Fire and Stability

In general, we should all be aware of the hazards before we encounter trouble on the plant. Our response should be instantaneous in identifying the degree to which a trouble on the plant or process poses a hazard.

### 3.12.1
### Individual Species

Many indicators are available to guide us about the hazard posed by individual species. The hazards are usually considered to be health hazard (toxic or causing death; causing genetic defects; causing long term disability; causing specific illnesses such as asthma, cancer; impacting on the environment); flammable hazard and explosive hazard.

A simple guideline is the NFPA ratings. These are a scale 0 to 4 with the higher number indicating the extreme in hazard (see Woods, 1994, Data for Process Design and Engineering Practice, Prentice Hall). MSDS documentation is available for most chemicals. Some sources include http://www.msdssearch.net or http://www.msdsxchange.com although these direct you to manufacturer's MSDS information. The quality of the documentation varies from company to company. Also consult http://toxnet.nlm.nih.gov.

#### Health

Vinyl species, benzene ring compounds. Carcinogenic to humans: asbestos, benzene, benzidine, $\beta$-naphthylamine, vinyl chloride. Proven carcinogenic in animal trials: acrylonitrile, butadience, N-nitrosamines

#### Flammability

Species with low Ignition temperatures, say in the range 85–100 °C. Recall that the ignition temperature for combustion in air could be lower if in the presence of pure oxygen or chlorine. For example, for toluene the ignition temperature in air of 535 °C is reduced in chlorine to 175 °C.

#### Stability/explosiveness

Guidelines from structure: azide, perchlorates, nitro compounds, peroxides, vinyl species.

For reactivity or potential for explosion or thermal runaway: if heat of decomposition > 0.2–0.3 MJ/kg; those > 0.5 MJ/kg may be explosive and those > 0.8 MJ/kg may be detonable.

#### Hazardous reactants and types of reactions

Rosenmund reduction, oxidation of nitrous acids; oxidation of low molar mass peracids; alkylation of alkali acetylides, alkylation via Arndt–Eistert; alkylation of diazoalkane and aldehyde; alkylation of diazoalkane; condensation of carbon disulfide with aminoacetamide; esterification of carboxylic acid and diazomethane; esterification of acetylene and carboxylic acid-vinyl ester; reactions involving concentrated peroxides and peracids.

#### Heats of reaction

*Extremely exothermic*: direct oxidation of hydrocarbons with air; chlorinations, polymerizations of polyethylene without diluent. > 3 MJ/kg (or > 150 MJ/kmol: examples: combustion –900 MJ/kmol; hydrogenation of nitroaromatics, –560 MJ/kmol; nitro-decomposition, –400 MJ/kmol; diazo-decomposition, –140 MJ/kmol).

*Strongly exothermic*: nitrations, polymerization of propylene, styrene butadience. 1.2–3 MJ/kg (nitration, –130 MJ/lmol; amination –120 MJ/kmol; sulfonation or neutralization with sulfuric acid, –105 MJ/kmol; epoxidation, –96 MJ/kmol; diazotization).

*Moderately exothermic*: condensations or polymerization reactions of species with molar mass 60–200, 0.6–1.2 MJ/kg.

Exothermic heat of reaction gives an adiabatic temperature rise > 100–200 °C.

Impact sensitivity of < 60 J for solids and < 10 J for liquids.

For powders: explosive usually if diameter is < 200 μm with highest probability of dust explosion if diameter < 65 μm. Note upper and lower concentration for explosive mixture.

### 3.12.2
### Combinations

With water, with air with other chemicals:

Chemicals that react violently with water include: acetyl chloride, aluminum alkyds, aluminum alkyl halides, aluminum chloride, calcium hydride, calcium oxide, ethyl aluminum dichloride, fluorine, lithium hydride, phosphorous pentoxide, phosphorous trichloride, potassium, silicon tetrachloride, sodium, sulfur trioxide, sulfuric acid, thionyl chloride, titanium tetrachloride, zinc alkyls. Moderately: acetic anhydride, activated alumina, aluminum phosphide, calcium, calcium carbide, calcium phosphide, lithium, activated molecular sieves, potassium hydroxide, activated silica, sodium hydroxide, sodium peroxide.

# 4
# Trouble Shooting in Action: Examples

In this chapter we meet five engineers as they trouble shoot the five cases introduced in Chapter 1. While you are reading these cases focus on the trouble-shooting process used. Compare the approaches taken here with the approaches you took in addressing the problems posed at the end of Chapter 1. The engineers have a range of practical experience. In **Case #3**, Michelle has three years experience. In **Case #4**, Pierre has 12 years experience. For **Case #5**, Dave has relatively no practical experience. For **Case #6**, Saadia has 10 years experience. For **Case #7**, Frank has 25 years of experience. These cases have been carefully selected to provide a range of approaches, degrees of difficulty and to illustrate the variety of trouble shooting approaches taken.

Each of the five scripts consists of about three parts with each part concluding with a few questions for you to consider. This reflective break was introduced to give you a chance to reflect on how you would have handled the case, and to decide what you should do next. I recommend that as you read each script, that you *play the game*. At the end of each case an assessment is given of the problem-solving processes used by each of the trouble shooters. Their scores range from F to A+; the richness is in the detail of the feedback. Not everyone will follow Lieberman's (1985), Gans's (1983), Kister's (1979) or my style in trouble shooting. The key is to identify your style and develop confidence in using it.

The cases do not have to be addressed in any order. Select what you would prefer. Enjoy!

## 4.1
## Case #3: The Case of the Cycling Column

Michelle[1)] graduated three years ago. She is on her journey to become an excellent trouble shooter but she lacks much practical experience. "Cycling column? Called in as chief trouble shooter!" exclaimed Michelle. Well, I'll do my best, she thought. Michelle recalled the Trouble-Shooter's Worksheet. She carefully noted the key infor-

1) In all these cases, the names of the engineers have been changed.

*Successful Trouble Shooting for Process Engineers.* Don Woods

ISBN: 3-527-31163-7

mation: startup after maintenance, iC4, and cycling level in the bottom of a distillation column. From the diagram, it looks like a thermosyphon reboiler and an inverted bucket steam trap. Well, *I want to and I can!* I have an organized strategy to apply.

*What are the strengths and weaknesses of Michelle's approach so far?*
*What would you have done differently?*
*Would you head out to the plant directly or are some quick checks to be done from your files first?*

Let's check, thought Michelle. This isn't an emergency. Until I learn more I cannot think of a "safe-park" condition I should impose. The key information is that these symptoms are occurring just after shutdown so I really need to find out what was changed and what was worked on. She cautioned that she should change her thinking from "cycling level in the bottom of the column" "to *apparent* cycling in the level". Michelle quickly scanned the P&IDs for the iC4 unit and confirmed that the diagrams showed a thermosyphon reboiler and an inverted bucket trap with a bypass. She checked in Chapter 3 for the cycling and reboilers and then checked steam traps:

**Reboilers in general:**
*"Cycling (30 s – several minutes duration) steam flow, cycling pressure on the process side and, for columns, cycling $\Delta p$ and cycling level in bottoms":* instrument fault/ controller fault/ condensate in instrument sensing lines/ surging/ foaming in kettle and thermosyphon/ liquid maldistribution/ steam-trap problems, see Section 3.5, with orifice $\Delta p$ across trap $<$ design/ temperature sensor at the feed zone in a distillation column/ collapsed tray in a distillation column.

For thermosyphon:

*"Cycling (30 s – several minutes duration) steam flow, cycling pressure on the process side and, for columns, cycling $\Delta p$ and cycling level in bottoms":* in addition to general, all natural circulation systems are prone to surging/ feed contains high w/w% of high boilers/ vaporization-induced fouling/ constriction in the vapor line to the distillation column. For horizontal thermosyphon: maldistribution of fluid temperature and liquid.

Section 3.5 steam traps:

- *Good practice:* Install a demister.

**Steam traps:** install trap below condensate exit (or with a water seal if the trap is elevated), use a strainer before most traps, use a check valve for bucket traps. Slant pipes to the trap. Use a downstream check valve for each trap discharging to a common header. Pipe diameter $^{3}$ trap inlet pipe diameter. Prefer to install auxiliary trap in parallel instead of a bypass. Do not group trap thermodynamic traps because of their sensitivity to downstream conditions.

Float and thermostatic: usually discharges continuously, low pitched bubbling noise. High-pitch noise suggests live steam is blowing.

Balanced thermostatic: leave about 0.6 m of uninsulated pipe upstream of trap.

Inverted bucket: use initial prime to prevent steam blowing.

Thermodynamic: about 6 cycles/minute

Trouble shooting: the major faults are wrong trap, dirt, steam locking in the trap, group trapping, air binding and water hammer. Too large a trap gives sluggish response and wastes steam. Too small a trap gives poor drainage, backup of condensate. *"Cold trap + no condensate discharge"*; steam pressure too high/ no water or steam to the trap/ plugged line or strainer/ orifice enlarged by erosion (bucket trap)/ incorrect $\Delta p$ across the orifice (inverted bucket)/ bucket vent clogged (inverted bucket)/ current operating pressure > design/ trap clogged. *"Hot trap + no condensate discharge"*: bypass open or leaking/ trap installed at high elevation/ broken syphon/ vacuum in heater coils. *"Live steam blowing"*: bypass open or leaking/ worn trap components/ scale in orifice/ valve fails to seat/ trap lost prime (inverted bucket)/ sudden drops in pressure/ backpressure too high (thermodynamic). *"Continuous discharge when it should be discontinuous"*: trap too small/ dirt in trap/ high-pressure trap installed incorrectly for low pressure service (bucket trap)/ valve seat clogged with dirt/ excessive water in the steam/ bellow overstressed (thermostatic). *"OK when discharging to the atmosphere but not when to a backpressure condensate header"*: condensate line diameter too small/ wrong orifice/ interaction with other traps connected to a common header/ condensate line partially plugged/ [backpressure too high]*. *"Slow and uneven heating of upstream equipment"*: trap too small/ insufficient air handling capacity/ short-circuiting when units are group trapped.

*[Backpressure too high]**: return line too small/ other traps blowing steam/ obstruction in return line/ excess vacuum in return line.

Michelle grabbed her notebook and headed out for the unit. She mentally went over some hypotheses.

*What are the strengths and weaknesses of Michelle's approach so far?*
*What would you have done differently?*
*What is the problem?*
*What are your hypotheses?*

Wait a minute. Here I am thinking of hypotheses when I should slow down. First let's do an IS and IS NOT, define the problem and focus on what was done during the shutdown. She joined a group of operators at the base of the column. You could hear the cycling in the steam control valve. Her temptation was to immediately put the control system on manual and see what happens but she methodically gathered the following information:

During the shutdown the condensate from this inverted bucket trap, that previously discharged to atmosphere, was repiped to discharge into a 200 kPa g condensate header. The condensate discharged into the top of the header. Steam to the reboiler was "saturated" at 1.7 MPa g. For this unit the only other maintenance was instrument checks and visual inspection of the trays. The visual inspection involved opening the access holes. No faults were found in the instruments and no changes were made to the settings of the controllers. Michelle then wrote the following in her notebook:

| | |
|---|---|
| What IS: | cycling in the level in the sight glass. IS NOT: changes in feedrate or composition; no other apparent cycling upstream or in the feed. |
| Where IS: | on this unit. |
| When IS: | used to work OK before the shutdown. |
| Who IS: | this is the first shift. |

What is the problem? Problem SMARTS$

Specific: stop the level in the sight glass from cycling. Measure via level in glass
Attainable? Should be. Depends on the root cause.
Reliable? Depends on the root cause.
Timely? Several simple tests should be able to resolve this.
Safe? Doesn't seem to be an issue here.
$ yes, we are losing money every minute.

She confirmed that the steam entered the top of the reboiler. However, she thought "all natural circulation systems are subject to surging".

She quickly listed some hypotheses:

- collapsed tray.
- change in downstream pressure affecting bucket trap.
- instruments wrong; liquid level is not cycling.
- control system.
- restriction in the vapor line.
- change in concentration of high boilers in the feed.
- temperature sensor incorrectly located in the feed zone.

She discarded "instrument wrong" because she could see the level rising and falling with her own eyes. It may lag or lead the actual level in the column but "something is cycling in there." Collapsed tray is an unlikely possibility because no one went down through the column checking each plate. They "visually inspected" by looking in the access holes. She reflected on the type of tests she could do and prioritized them. 1. Checking the control system was relatively inexpensive and fast. 2. Opening the bypass on the trap and/or changing the condensate to discharge to atmosphere temporarily are other options to check out while she rechecks the specs and sizing used on the bucket trap.

1) When the control system was put on manual the level gradually increased. "Let's think about this", she said, "steam enters the thermosyphon reboiler, condenses and boils a given amount of bottoms. This causes a pressure differential and fresh bottoms cycles into the tubes. However, if the trap does not remove the condensate then the condensate builds up, decreases the area, decreases the heat transferred and hence the boilup. The level increases because feed continues to the column but the boil up rate is decreased." When the set point was increased manually, the valve stem on the steam moves up, the liquid level appears in the sight glass and continues to drop

but shortly thereafter the level appears in the glass and is rising. I conclude that the control system is not at fault.

2) Open the bypass and manually try to change the bypass setting to "level out" the level. Frustrating as this is to try to adjust, this seems to provide some level of control of the cycling.

Michelle concludes that the plant could operate by having the bypass partly open or by unhooking the condensate from the main and discharging to atmosphere as it had operated before. She elects to do the former while she returns to her office to check on the sizing and selection of the trap.

*What are the strengths and weaknesses of Michelle's approach so far?*
*What would you have done differently?*
*Did she address the symptoms or the root cause?*
*What tests and questions would you have posed?*
*What corrective action would you have taken?*

When Michelle consulted the design file, she found that the trap had been selected to handle a design condensate flowrate of 0.3 kg/s for an inlet pressure of 1.7 MPa and atmospheric discharge. Over the years the production rate had increased so that, from her energy balance calculations, the actual condensate flowrate was about 0.6 kg/s. However, throughout these changes the trap had not been changed and was now operating at its maximum condensate capacity. When the differential pressure across the trap was reduced, the condensate handling capacity was reduced by about 10% so that the trap was now undersized. To correct this Michelle selected a new trap, or a different orifice in the existing trap, to be installed to handle 1.2 kg/s condensate under the steam pressure of 1.7 MPa and the downstream pressure of 200 kPa. This followed the general guideline of selecting inverted bucket traps based on double the normal capacity. She also recommended that a check valve and strainer be installed upstream of the trap.

*Discussion*

Overall degree of difficulty of this problem is 4/10, relatively easy. Michelle took half a day to solve this problem. Let's reflect on Michelle's approach following the guidelines in the feedback form. Overall, Michelle recognized her inexperience and tried to follow the Worksheet and to draw on the suggestions from Chapter 3. Michelle treated this as a "*problem*" and not an *exercise* right from the beginning.

**Problem Solving**

She was systematic, organized, used the IS and IS NOT approach and wrote things down. Not much verbal monitoring was apparent and limited checking and double checking. She became a little too hasty in the hypothesis checking and prioritization stages. She discarded some hypotheses without explicitly noting this. This worked OK for her in this case but she should be developing good habits. Overall a rating of B.

**Data Handling**

Actions were based on fundamentals and she astutely enriched her experience by checking files and trouble shooting symptoms/cases (as given, for example, in Chapter 3). Her reasoning is OK for what she described. However, it is difficult to assess because she mentally discarded hypotheses and didn't systematically check the evidence with the hypotheses. This worked for her this time because there were only two pieces of evidence: changes were made to the condensate system in the turnaround and the level in the sight glass cycles. Overall rating B.

**Synthesis**

She listed a variety of hypotheses. Considered many viewpoints initially. Narrowed into the two hypotheses that were easy to check. Overall rating B+.

**Decision Making**

She seemed to continually try to prioritize. No obvious bias was apparent. She should have discussed her decision to "operate on bypass" with the operators because they have to *buy in* to this change. The criteria she used were not apparent. Overall rating C.

*Strengths:* systematic, based on fundamentals, wrote things down, variety of hypotheses, used resources.

*Areas to work on:* more explicitly aware of the process used, think about the operators.

## 4.2 Case #4: Platformer Fires

Pierre is frustrated. Another fire. "That hydrogen just leaks out of every crevice. And the high temperatures we are talking about, 500 °C. Heh, how can we hope to ratchet the flange tight at room temperature and hope that it will remain tight when we heat it up!" He pulled out the book on flange design and starting doing calculations of the relative thermal expansion of metals for 1 1/4 Cr–$^1/_2$ Mo. And we hydrostatically tested the tube bundle based on the pressure differential of 2–2.7 MPa to test for leaks between the tube sheet and the shell. We were careful to use the differential, instead of the absolute pressure of 4.8 MPa.

*How would you characterize Pierre's approach so far?*
*What are the strengths and weaknesses of his approach so far?*
*What would you have done differently?*
*What is the problem?*
*What are your hypotheses?*

Pierre tossed the pencil down on his desk. Let's go back to basics. I have a gas at high pressure and temperature that is leaking through a crevice, igniting and flaming. We have used all our best mechanical engineering brains to design the flange to prevent leaking. Different gaskets, different tightening approaches. These were designed so that the temperature differential on the bolts was negligible even though the tubesheets were very thick. They've done all the thermal expansion stuff already. All the bolts were supposedly torqued the same. I've got to think out of the box! Let's pull out the Trouble-Shooter's Worksheet I used to use.

Is it a safety hazard? Yes siree! It is a fire hazard. Hydrogen + oxygen + spark. What's going on is pretty straightforward, except that it's not clear whether the naphtha feed is on the tube or shell side. OK, that completes **Engage**. **Define the stated problem** is next.

| | |
|---|---|
| What IS and IS NOT: | fires are on the effluent exchanger. IS NOT elsewhere nor on other units. |
| When IS and IS NOT: | ever since we started up |
| Where IS and IS NOT: | IS around the flanges on the shell and tube. IS NOT on the exchanger |

This is a fundamentals problem. OK. I think I have written down the facts so far.

**Explore**. This is a problem! But, I need to bring out the files and clarify information. From the files, Pierre found that:

a) the naphtha is on the tube side and at the higher pressure; the hydrogen-rich gas is on the shell side.
b) the platformer temperature is kept < 540 °C to prevent thermal degradation of the catalyst. The hydrogen-rich gas is 500 °C and used to preheat the naphtha.
c) from the diagrams of the effluent exchanger, the most likely hydrogen leak seems to be between the flange from the shell, the head and the bonnet.

OK, Let's put this into perspective via the *Why? Why?* The problem as I see it is to "prevent hydrogen-rich gas from leaking out the flange" so I'll put that at the start.

Perspectives. *Why? Why? Why?*

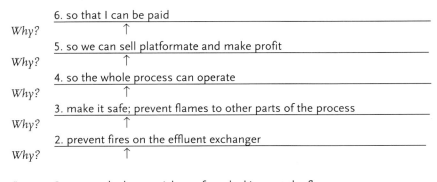

6. so that I can be paid

*Why?* ↑

5. so we can sell platformate and make profit

*Why?* ↑

4. so the whole process can operate

*Why?* ↑

3. make it safe; prevent flames to other parts of the process

*Why?* ↑

2. prevent fires on the effluent exchanger

*Why?* ↑

Start → 1. prevent hydrogen-rich gas from leaking out the flange

That was pretty useful. Maybe I have been working on the wrong problem! Maybe I should just let the hydrogen leak out and focus on how to prevent fires on the effluent exchanger.

*What are the strengths and weaknesses of Pierre's approach so far?*
*What would you have done differently?*
*What would you do next?*

How to prevent fires? A fire is hydrogen + oxygen + spark; I said that right at the start. OK, if I have the hydrogen, then I can remove the oxygen or the spark or both. Let's brainstorm to remove the oxygen: nitrogen blanket, steam blanket,... I'm going to stop there. I'll just install a circular sparge ring whose circumference is about 5 cm beyond the flange, drill holes on the inside and sparge low-pressure steam at the flange. The heat from the steam will tend to minimize the temperature differential between the inside and the outside of the flange and the steam should displace the oxygen. Viola, no fire. Perhaps in the future they will develop Platforming processes that operate at lower temperatures and pressures, said Pierre wishfully.

*Discussion*

The overall degree of difficulty for this problem is about 3/10, this is relatively easy. However, it wasn't easy at the start for our frustrated Pierre. His frustration shows through in his approach. Fortunately, he got out of the box by using the *Why? Why? Why?* technique. Consider now some feedback for Pierre.

**Problem Solving**

Pierre started with more frustration and lack of confidence. He forgot that he had an organized strategy that had been helped him in the past. He used very little monitoring, checking and double checking and even when he started to use the Worksheet he used it rather superficially. He did use the IS and IS NOT effectively and tried the *Why? Why? Why?* Fortunately, he used the latter effectively. Overall rating C–.

**Data Handling**

He gathered information from the files, and tried to reformulate the problem based on fundamentals. However, he really didn't get to hypothesis generation. His reasoning was OK based on what he told us explicitly. He didn't formally draw on past experience. Overall rating D.

**Synthesis**

He failed to explicitly list hypotheses. Considered very few viewpoints. Overall rating D to F.

**Decision Making**

He seemed to place the blame on the mechanical engineers and the inability to design a flange that would keep the gas from escaping. He focused initially on preventing the gas from leaking. No apparent criteria were given for the decisions. Overall rating D.

*Strengths:* got an answer, could think outside the box, was somewhat systematic once he started to use the Worksheet, used resources, based on fundamentals.

*Areas to work on:* self-confidence, be more systematic.

## 4.3
## Case #5: The Sulfuric Acid Pump

Dave is a new graduate engineer. Dave notes that the evidence points to cavitation. However, being relatively new on the trouble shooting scene he starts to write down a description of the process. Sulfuric acid is stored below the level of the pump. The pump has to "lift" the acid up to the intake. The storage tank is open to the atmosphere. Ok, let me check. Have I understood this correctly? Dave runs his pen around the diagram and checks that he has included all that he thinks is pertinent about the situation. He underlines the evidence and inserts "?" when he has something he is concerned about: the vertical dimensions, two hour cycle of operation, receives "acid"? via return lines from all over the plant; dilute acid (?density and corrosiveness); when the site gauge reads 0.7 m left in the tank, the pump makes a "crackling" noise that the operator says "sounds like cavitation". Concern about erosion of the impeller (? Why, because of acid? Because of cavitation?). OK, "*I want to and I can*!" says Dave as he takes out his trusty Trouble-Shooter's Worksheet. "Ok, what's my next step? I think I am finished with **Engage**. But wait a minute. What about Safety? Hazard? Safe-hold operation? Other than the fact that this is acid, I don't see this as an emergency priority so I can proceed with **Define the stated problem.**

He quickly wrote down the IS and IS NOT information. Dave noted that he probably had enough information here to "solve the problem" without going out on the site. He recalled the cardinal rule of trouble shooting "Go out on the site and see it!" Dave acknowledged this but felt he should do some detailed calculations before going out. His focus would be on the "Basics" around NPSH and suction lift since this trouble occurred from startup.

*What were the strengths and weaknesses of Dave's approach so far?*
*What would you have done?*
*Would you go directly to the plant or would you do calculations first?*
*What calculations would Dave make?*
*Is this really an NPSH problem?*
*Is this a suction-lift problem?*

Before Dave started into the calculations he looked out and saw the rain beating down. It was a good time to stay in the office! He thought. Then, he remembered that when it's raining the atmospheric pressure is usually low. Since this process is open to the atmosphere, he noted that he should include a possible lowering of the atmospheric pressure as a potential cause of the lack of NPSH. Oh, Oh! I did it again, I said lack of NPSH instead of saying *apparent* lack of NPSH. By this time Dave realized that he had skipped into the **Explore** stage without doing his usual checking and double checking. He perused what he had done, said a few positive "yeps" and then started focusing on the SMARTS$. His goal is to stop the "crackling" noise when the level drops to 0.7 m. In his mind Dave realized that he had changed the problem from a "crackling" problem to a "cavitation" problem to a lack of NPSH problem. Hmmmm.

Dave decided to continue his analysis of the NPSH and suction lift. He hauled out some texts and posed a number of *What if?* questions: *what if* the acid return liquid is more dense? Less dense? *What if* the atmospheric pressure really drops? *What if* the vent to atmosphere becomes plugged? *What if* the site gauge is reading wrong? *What if* gas or air is dissolved in the acid? Dave stopped his ruminations and consulted his files on cavitation:

> *[cavitation]**: design fault/ liquid too hot/ non-condensibles in liquid/ vortex entraining gas/ decrease in density of the liquid/ blockage or excessive $\Delta p$ on suction/ suction velocity too high/ increase in rpm without increase in NPSH/ increase flowrate increases NPSH demand. For suction-lift situations: suction lift too high/ low atmospheric pressure (for open systems)/ air leakage into suction line.

Dave reflected on this information and decided that the key hypothesis to check was "design error". He realized he should list a variety of hypotheses about the root cause of "cavitation" but he wanted to do the simple checks first. If the design estimations looked OK, then he would list a range of hypotheses. However, the design calculations seemed too fundamental to ignore. He would: 1) estimate the *NSPH supplied*; 2) check the files and see the *NPSH required* from the vendor's information and 3) compare. He systematically did each in turn.

1) Estimate the *NPSH supplied*. The design files showed that the typical "dilute acid" had a density of 1.2 $Mg/m^3$ (whereas 98% acid had a density of 1.84 $Mg/m^3$) and at the temperature of operation the vapor pressure of water is about 5 kPa; that of dilute sulfuric acid is about 3 kPa. It was raining and the atmospheric pressure was low, he estimated it to be 90 kPa abs. He plugged values into the equation for *NPSH supplied* for an atmospheric pressure of 90 kPa, and an estimated friction loss of 0.7 m= (Atmospheric pressure–vapor pressure of fluid at operating temperature) converted to head–head of lift–head loss in friction.

   7.4 m–3.6 m–0.7 m= 3.1 m of acid.

   He rechecked his calculations. OK.
2) From the vendor files he found that the *NPSH required* was 1.4 m "of water at 21 °C" for this pump at 1800 rpm and the design flowrate of 15 L/s. Since the *NPSH required* is for a pressure drop in the horizontal plane, *NPSH required* is relatively independent of the density and temperature of the fluid (although he remembered that some corrections have been published for the *NPSH required* for hydrocarbons that show the *NPSH required* as being about 0.6 times the *NPSH required* for water. However, for conservative purposes use *NPSH required* for cold water.) OK, so *NPSH required* = 1.4 m.
3) Compare. Recommendations about the relationship between the *NPSH supplied* and required vary. Some suggest that *NPSH supplied* should be 20% higher than that required; or 0.5 m of water higher than that required or the ratio of *NPSH supplied/NPSH required* > 1.4. For this problem *NPSH supplied* is 160% higher; larger by 3.1–1.4 = 1.7 m and the ratio is 2.6. Let's check.

Dave went over his calculations. I'd better return to my earlier *What ifs? What if* the acid density is 1.4 and not 1.2? Dave quickly rechecked his values and came to about 2.1 m acid. This still meets the requirements. *What if* the pump flowrate is 20 L/s? From the vendor's data Dave noted that the NPSH required increased to about 2.5 m water. OK, thought Dave; it doesn't look as though it's a design error. I'll now go back and formally list some hypotheses consistent with a "crackling noise" that an operator interprets as being cavitation. Let's see, and Dave checked over the list for "cavitation" and selected

- liquid too hot
- non-condensibles in liquid
- air leakage into suction line
- vortex entraining gas
- decrease in density of the liquid
- blockage or excessive $\Delta p$ on suction
- suction velocity too high
- increase in rpm

Wait a minute! I'm depending solely on a list from some book. Let me think from basics about this.

*What were the strengths and weaknesses of Dave's approach so far?*
*How well is Dave following the Trouble-Shooter's Worksheet?*
*What would you have done?*
*Do the calculations prove that there should be sufficient NPSH supplied?*
*What are your hypotheses?*
*What would you do next?*

Dave put on his raincoat and headed out to the plant. By now he had zeroed in on the hypothesis that a vortex was forming when the level reached 0.7 m. After he talked with the operator and looked around he decided to test his hypothesis by reducing the flowrate from the pump. This should lower the velocity in the suction line, and lower the friction loss in the suction line. Under these conditions, the liquid level should drop lower in the tank before it starts cavitating. His discussions with the operator and his view of the layout confirmed what he had thought. He realized, however, that if he had come out to the plant earlier he could have hooked up the utility air line to the vent, increased the pressure in the vessel and checked to see if it was an *NPSH required* being insufficient quite easily.

When the operator reduced the flowrate the liquid level dropped to 0.58 m before cavitation. Ahah, cavitation is caused by vortex formation.

Dave shifted gears now to think about how to correct this temporarily and in the long run. The key goal is to prevent cavitation (and the resulting erosion of the working parts of the pump) and to lengthen the pump cycle. To prevent the vortex formation he could float, temporarily, a wooden egg-carton construction on the surface to serve as a vortex breaker. Although Dave thought of floating a double layer of ping pong balls on the surface he realized that if a vortex did form and the balls were sucked into the pump there would be real trouble! This wooden floating vortex breaker should allow pumpage to lower levels in the tank. Another option would be to reduce the flowrate as the level drops. This would mean a signal would have to be transmitted to a location near where the pump exit flowrate could be controlled. At the next shutdown, a well-designed vortex breaker could be attached over the exit pipe; or a control valve could be installed on the pump discharge that gradually closed as the level in the tank dropped.

Dave reflected on how he handled this, one of his first real trouble-shooting problems. He made notes about what he would do the same and what he would do differently on his next trouble-shooting assignment.

*Discussion*

Overall degree of difficulty of this problem is 5/10, relatively easy. Dave took a day to solve this problem. Overall, Dave tried to follow the Worksheet and to draw on the suggestions from Chapter 3. This was a "*problem*" and not an *exercise* for Dave, and he handled it that way.

**Problem Solving**

Dave tried to be systematic, organized, used the IS and IS NOT approach and wrote things down. He included much verbal monitoring and checking and double checking. Overall a rating of A.

**Data Handling**

Actions were based on fundamentals and he used books, files and trouble shooting symptoms/causes (as given, for example, in Chapter 3). He should have visited the site earlier. He systematically checked the evidence with the single hypothesis. Overall rating B–.

**Synthesis**

He focused early on the process operator's judgment about the symptom. He used the resources in Chapter 3 to create a list of hypotheses but then prioritized these and checked first the design error. Many hypotheses should be kept active concurrently. Overall rating C.

**Decision Making**

He prioritized. No obvious bias was apparent. Overall rating B.

*Strengths:* use of resources, monitoring, checking and double checking, being systematic, based on fundamentals.

*Areas to work on:* visit the site, keep more hypotheses active.

## 4.4
## Case #6: The Case of the Utility Dryer

Saadia perused the trouble-shooting case that appeared on her desk. Over the past ten years she had very successfully solved a range of problems with dryers, but usually the adsorbers were hooked in parallel with one unit on-line and the other being regenerated. For this case, the units are in series with the regeneration occurring before the adsorption. "Very interesting," she murmured, "but I'm going to treat this case as a *problem* and not an *exercise*." She retrieved the Trouble-Shooter's Worksheet and checked off the items. This is not an emergency; except that we may need dry instrument air from a skid mount on these cold days. Let's be sure that I understand this particular process. She traced the path of the feed air on the diagram and noted that all traps discharged below ground and were inaccessible for sampling. "Fortunately, there seems to be enough sampling ports," she noted as she checked off S1 to S4. "Often the problems in the past have been leakage of steam from the regenerator into the air, leaky valves, inadequate regeneration, wrong adsorbent or adsorbent damaged by excessive regeneration temperature, or adsorption cycle too long or change in concentration in feed." "Hold on a minute," Saadia said to herself, "you said you were going to treat this as a *problem* and not relive your past achievements! Make sure you understand this flow diagram and then move to define the stated problem." Saadia looked again at the flow diagram and traced the flows for Bed B being regenerated, then cooled. She wrote down succinctly: P1 reads 550 kPa ? Cycle set on V1 for 2 h? Then 1 h ? And valves V2 and V3 for 3 h? TRC 1 = 175 °C? Flowrate about 1/2 design? Proportioning valve actually shut? Is the air really wet leaving our unit? Is this the first time this is reported? Did the vendor give us a performance check on this unit? Current weather? OK, I think I have finished with the **Engage** part. Now to **Define the stated problem**.

| | |
|---|---|
| What IS and IS NOT: | instrument air lines freezing; claims of "wet" air leaving the drying bed. |
| When IS and IS NOT: | on colder nights; not reported at other times |
| Where IS and IS NOT: | on instrument lines. No reports for air supplied elsewhere in the plant. |
| Who IS and IS NOT: | plant operators claim following vendor's instructions. |

"I'll focus on the basics. I think I'm ready to really get into this problem. Let's **Explore**" said Saadia.

*What were the strengths and weaknesses of Saadia's approach so far?*
*What would you have done?*
*What would you do next?*

Before I define the real problem, I want to gather three forms of baseline data: 1) I'll pull the vendor's specs on this unit; 2) I'll do a simple spot-check of performance data, with some samples and 3) do some simple rule-of thumb checks.

- Vendor's specs. The design conditions guaranteed by vendor's file said:
  - Adsorbent: 5000 kg of activated alumina H-151 per bed:
  - Drying cycle: 3 hours
  - Regeneration cycle: 2 hours
  - Cooling cycle: 1 hour

- Dry air product, standard cubic decimetres per second, $Ndm^3/s$, FRI 1: 7300
- Minimum air flow for regeneration, $Ndm^3/s$: 2360
- Proportional valve: part closed but closed enough to ensure minimum air flow for regeneration.
- Inlet air temperature, °C, T1: 32.2
- Inlet air pressure, kPa g, P1: 550
- Moisture content of inlet air, saturation temperature at the design pressure, °C: 32.2
- Final moisture content in "dry air", °C, S4:–42.7 at pressure
- Pressure drop, kPa, P1–P4 = 13.8
- Steam pressure to the heater, MPa g, P2: 1.0 min–5.5 max.

- Simple spot-Test I. Readings should be made of all the instruments 1 hour into the cycle. Use an open cup dew point apparatus attached to the sample lines S1 (inlet dew point), S3 (after the separator) and S4 (dry air dew point). I realize this is the dew point at atmospheric pressure and I have to convert it to pressure conditions if I want to compare. The results were:
  - Pressure, Inlet air, P1, kPa: 689
  - Temperature, Inlet air, T1, °C: 22.2
  - Dew point, Inlet (atmos) S1, °C:–6.7
  - Dew point, Inlet (press), S1, need to calculate.
  - Air flow to regeneration, all because pressure P3 is full on to close the valve.
  - Pressure, Steam P2: 5.2 MPa
  - Temperature, Steam T2: 277 °C
  - Temperature, TRC 3 exit gas from heater: set 177 °C
  - Temperature after regenerator, T4, °C: 104
  - Temperature after cooler, T5, °C: 24
  - Dew point, (atmos) exit of Separator, S3, °C: unable to measure because it fogged up immediately at ambient temperature and pressure.
  - Pressure, Effluent air, P4, kPa g: 662
  - Air flow, FR 1, $Ndm^3/s$: 4000
  - Exit (atmos) Dew point, S4, °C:–48.3
  - $\Delta p$, kPa, calculated P1–P4, 27

- Simple checks: "Lots of numbers, but I want to select the key ones," said Saadia, astutely.
  - Anything unexpected with the feed, the steam, the regenerator, the separator? No, temperatures and conditions pretty consistent with design conditions from the vendor: except flowrate = about 1/2 design.
  - Dew-point conversion between atmospheric pressure and pressure conditions of about 762 kPa abs is a) determine the partial pressure, pp, of the water vapor at the dew point, atmospheric; multiply by the ratio of the process pressure absolute to atmospheric to obtain the partial pressure of water vapor under the process conditions; from humidity tables/charts

find the dew point comparable to this partial pressure of water. Alternatively, I could just express everything in ppm water. Since Saadia realized she would be doing this conversion often she created a simple program to speed her calculations. For the test run, the Exit (press) Dew Point, corresponding to sample location S4 = –29.5 °C or a moisture content of about 50 ppm or more than 3× greater than expected of 15 ppm.

- According to vendor's specs of 32.2 °C saturation at 650 kPa abs, the incoming moisture content is 7375 ppm. For the test conditions, the incoming moisture content was 3430 ppm. Hence, the exhaust specs of 15 ppm should be easy to achieve if everything is working as expected.
- Interesting: at sample location S3, exit from the separator, since today's temperature is 21 °C the moisture content is > 24,000 ppm or 2.4% v/v moisture. "I'll call this a surprise because I have no other data as to what I should expect here. This suggests a carryover of mist from the separator."
- Surprises: exit amount of moisture in the air is = 3 times greater than the target value of 15 ppm and pressure drop = double expected. Before she proceeded further Saadia took several minutes to recheck her calculations. Let's see, Dew point (atmospheric) of –48.3 °C corresponds to a Dew point (at pressure conditions) of about –30 °C (which is consistent with the dew point under pressure conditions not meeting the specs of –42.7 °C).

Saadia realized that this simple test showed that the problem was the exit moisture content (from the sample at S4) in the exhaust, "dried" air is about 3 times higher than expected. This is consistent with "freezing" lines when the outside temperatures are cold. The other symptom is the pressure drop is double the expected value even when the flow rate is 1/2 the design value.

"It's time for hypotheses! ... and tests" declared Saadia, whereupon she added following list to her chart:

Symptom
a. exit air "wet": 3× higher than specifications
b. pressure drop double expected value even when the flowrate 1/2 design
c. ______
d. ______
e. ______

| Working Hypotheses | Initial Evidence | | | | | Diagnostic Actions | | | |
|---|---|---|---|---|---|---|---|---|---|
| | a | b | c | d | e | A | B | C | D |
| 1. Steam leak | S | N | | | | | | | |
| 2. Excessive moisture carryover from the separator | S | N | | | | | | | |
| 3. Valve S2 leaking | S | N | | | | | | | |
| 4. Adsorbent lacking adsorption capacity. | S | N | | | | | | | |
| 5. Absorber on-line too long: breakthrough | S | | | | | | | | |
| 6. Not enough regeneration time | S | | | | | | | | |
| 7. Condenser not cooling sufficient | S | | | | | | | | |
| 8. Instruments wrong, pressure | N | S | | | | | | | |
| 9. Absorbent broken down | ? | S | | | | | | | |
| 10. Temperature TRC, T3 wrong | S | S | | | | | | | |

For the Diagnostic actions or tests, Saadia thought that the root cause of a high $\Delta p$ could be "damaged alumina adsorbent" causing the $\Delta p$ across the bed to be too high. One factor that could cause such damage would be excessive temperature in the regeneration. This might be caused by an incorrect TRC T3 sensor that reads 177 °C but it really is 245 °C, for example. Such damaged adsorbent would also have less adsorbent capacity.

**Test A**. Check/calibrate T3.

**Test ?**: Another test would be to open up the adsorbers and sample the adsorbent, checking the adsorption loadings and checking for decomposition. However, this is expensive and takes the unit off-line. She preferred to do other simple on-line tests first.

**Test II and III**: Gather a set of data that can be used for four other tests, called **B**, **C**, **D** and **E**. Since this is a batch-cyclical process, most of tests require data that are gathered every 15 minutes over the whole cycle and with bed A being both adsorber and regeneration mode. Since the steam trap and the separator traps discharge below grade, data about the steam/condensate flowrate are not easily accessible. However, the trap on the separator can be isolated and condensate collected from S2.

In **Test II**, 6-hour cycle with the proportional valve closed, measure the exit moisture concentration in the air at S4, collect condensate at S2, measure temperature T4, plus all other usual readings.

In **Test III**, a 6-hour test with the proportional valve open and a corresponding small flow of air through the regenerator heater and the condenser-separator, measure the exit moisture concentration in the air at S4, measure temperature T4, plus all other usual readings.

Such information could then be used to do four other tests:

**Test B**. Mass balance on moisture. Excessive water coming out might suggest a steam leak.

**Test C**. Compare adsorption specs on alumina with moisture adsorbed in one cycle. Low adsorption/ mass of adsorbent means the cycle time is not long enough to load the adsorbent or the alumina is lacking the adsorption capacity.

**Test D.** Compare time plots of temperature T4, liquid collected S2 and Dew Points from S1, S4. She sketched the time plots she expected. These are shown in Figures 4-1 and 4-2.

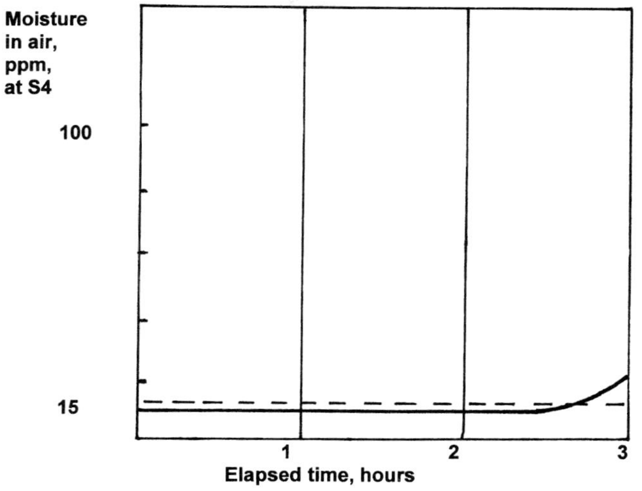


Figure 4-1 Prediction of the adsorption cycle.

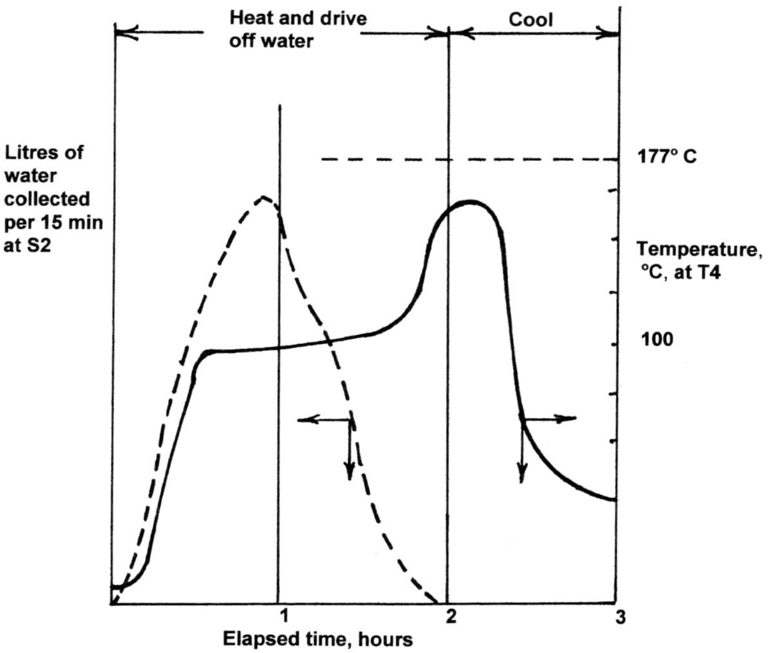


Figure 4-2 Prediction of the regeneration.

Figure 4-1 shows a typical loading with the exit concentration below the target 15 ppm. Near the end of the cycle, the concentration may increase as the start of the breakthrough curve appears. That's what I would expect to see for all the tests.

Figure 4-2 shows what I would expect for the regeneration and condensation of the water from the separator. During the hot regeneration period, the temperature on T4 should increase and level off on a plateau around 105 °C and then rise to the inlet temperature, 177 °C when the bed is almost regenerated. If regeneration was started before the bed was fully loaded, then the alumina near the end of the bed will be dry, then wet and then dry as the desorption band sweeps through the bed. If the bed loading is small (or the gas velocity high) then the length of the temperature plateau is short. The condensate collected should spike before the cool-down period. For the cool down, the effluent temperature T4 should continue to rise but will drop after the high-temperature wave from the heated, regenerated adsorbent passes through the bed.

**Test E**. Compare pressures on P1–P4 for four different conditions: a) the heating cycle for regeneration versus b) the cooling cycle (to see $\Delta p$ over heater) and d) when the proportional valve is shut versus e) partially open to see the $\Delta p$ in the adsorbing bed versus including the $\Delta p$ for regenerating bed and cooler-separator. These tests should be done with both A and B beds being adsorbers.

Before I order the tests, let me check. She reflected on how she had handled the program so far, her hypotheses and the proposed tests. Satisfied, she planned the tests.

*What were the strengths and weaknesses of Saadia's approach so far?*
*What would you have done?*
*Should Saadia have just replaced valve V2?*
*Should Saadia have called in the vendor?*
*Should she have opened both beds and tested the alumina?*
*What would you do next?*

Saadia was pleased that she had resisted the urge to jump in and change Valve 2, test the alumina and to call in the vendor. The 2 days of tests and analysis of data should clarify many issues and dispel incorrect hypotheses before such major action is taken. Here are the results.

**Test A:** TRC 1 responds to change and calibrates OK. Conclude: the regeneration air temperature is 177 °C. *Hypothesis 10* that the regenerated air temperature is hotter than 177 °C is disproved.

**Test B:** Mass balance on moisture. **Test II**, for the 6-hour cycle with the proportional valve "closed"; exit gas flowrate, FR 1 = 2830 $Ndm^3/s$; loading bed B for the first 3 hours and collecting the condensate from the regeneration of B during the second 3 hours:

79.7 kg in =? 66.1 kg out.

For bed A:

79.7 kg in =? 74.3 kg out.

The balance isn't within the 10% that I like to see, but this suggests that steam is not a source of water coming into the system. The condensate collected is less than the apparent bed loading. Reject *hypothesis 1* that there is a steam leak and perhaps consider poor separation in the separator with resulting carryover of entrained water to the adsorber.

**Test C:** Check adsorption loading of water/ mass of dry adsorbent.

From my files for activated alumina: 0.14–0.22 kg water/kg dry adsorbent.

Calculated from the vendor specs: 0.089 kg water/kg dry adsorbent for air flow 7300 $Ndm^3/s$

Calculated from **Test II** for the 6-hour cycle with proportional valve "closed"; exit gas flowrate, FR 1 = 2830 $Ndm^3/s$; loading bed B with subsequent regeneration: 0.016 kg/kg.

*Notes*: for the **Test II** conditions the flowrate is below vendor design specs so the moisture coming to the unit is less. If this incoming moisture is increased in proportion to the flowrates, however, the loading on the adsorbent becomes $0.016 \times 7300/2830$ $Ndm^3/s$ or 0.0412 kg/kg. This suggests that the adsorbent is not adsorbing the amount of water expected from the vendor design specs (0.09 versus actual 0.04 or about 1/2) nor that expected based on file data about activated alumina (0.14 versus actual 0.04). Conclude that *Hypothesis 4* might be true.

**Test D:** Plots of data are given in Figure 4-3 for **Test II**, the 6-hour cycle with the proportional valve closed. The data are the exit concentration S4 from the adsorbing bed for beds A and B respectively, in Figures 4-3 a and b. What a surprise! For both beds A and B on the adsorption cycle, there is a spike of water in the middle of the cycle. A comparison with the water evolution from the concurrently occurring regeneration, in Figure 4-4 a and b, shows consistency in the spike of water. This suggests leakage across valve V2. Saadia estimated the amount of gas leaking through the valve based on the times of the peak condensate collection to be about 1.5–3 $dm^3$ / s. I conclude that *Hypothesis 3* is correct.

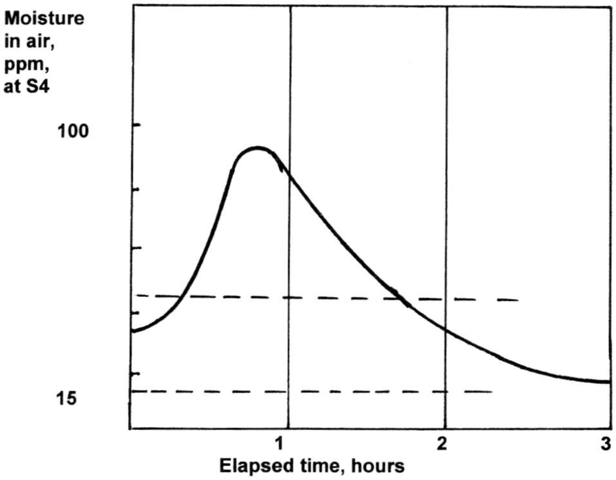


**Figure 4-3a** Test II, bed A adsorb with proportional valve closed.

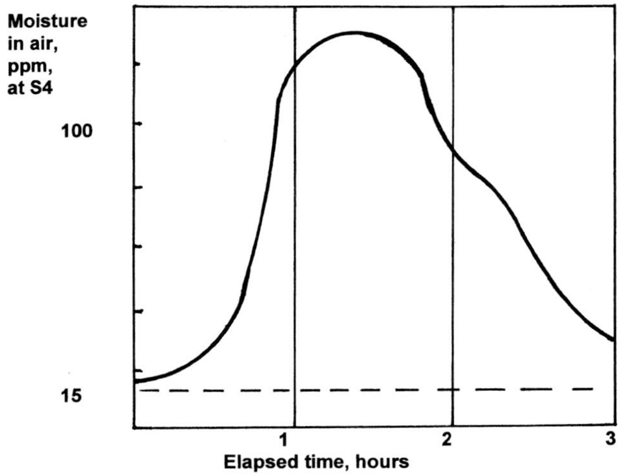


Figure 4-3b Test II, bed B adsorb with proportional valve closed.

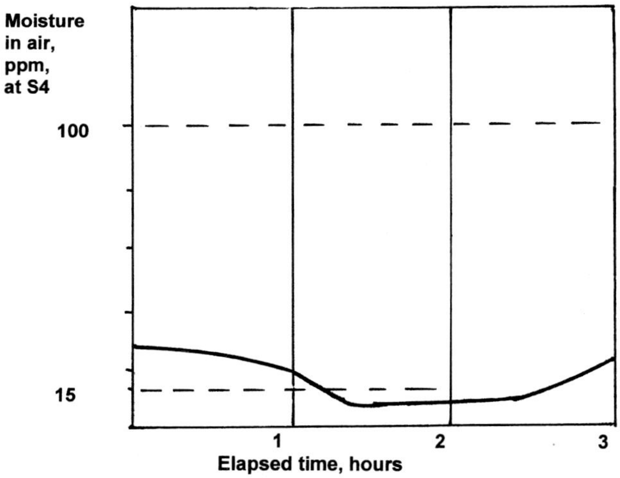


Figure 4-3c Test III, bed A adsorb with proportional valve open.

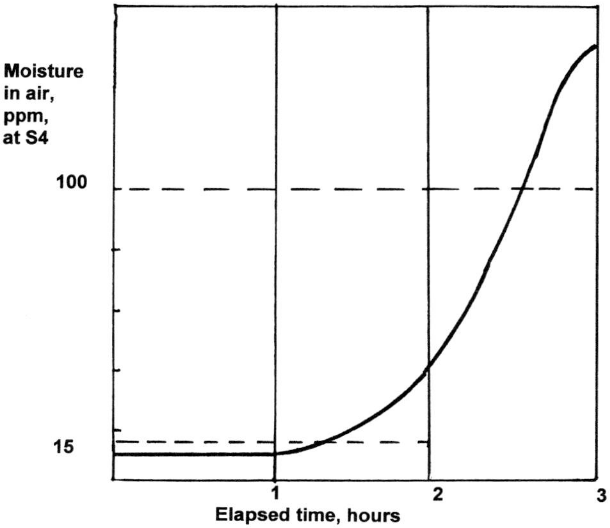


**Figure 4-3d** Test III, bed B adsorb with proportional valve open.

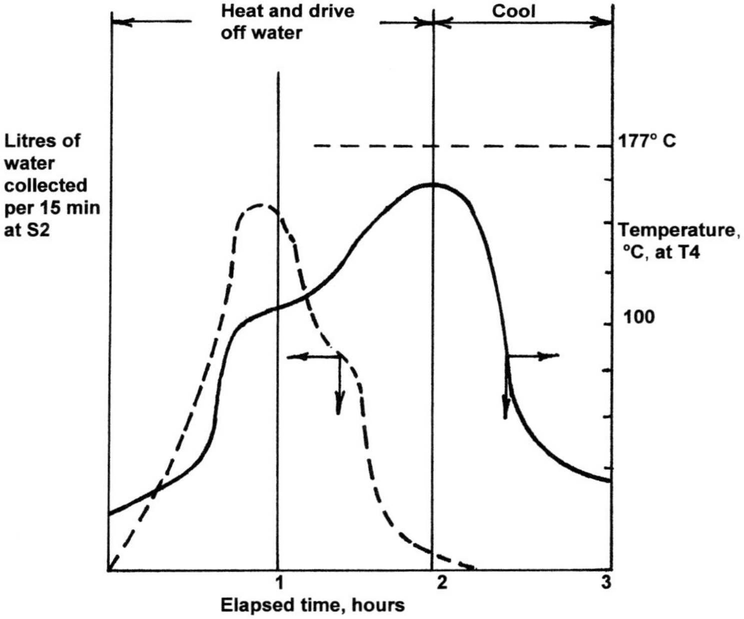


**Figure 4-4a** Test II, bed B regenerate (while bed A is adsorbing) with proportional valve closed.

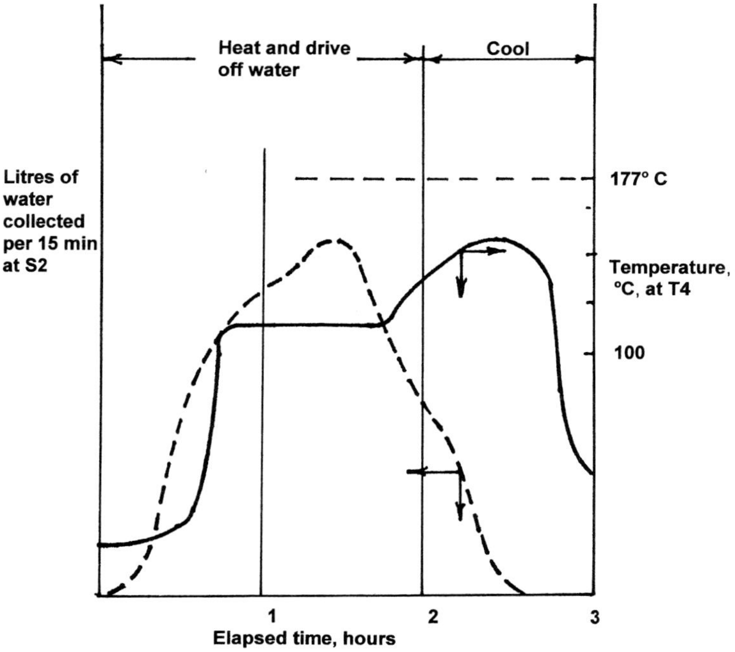


**Figure 4-4b** Test II, bed A regenerate (while bed B is adsorbing) with proportional valve closed.

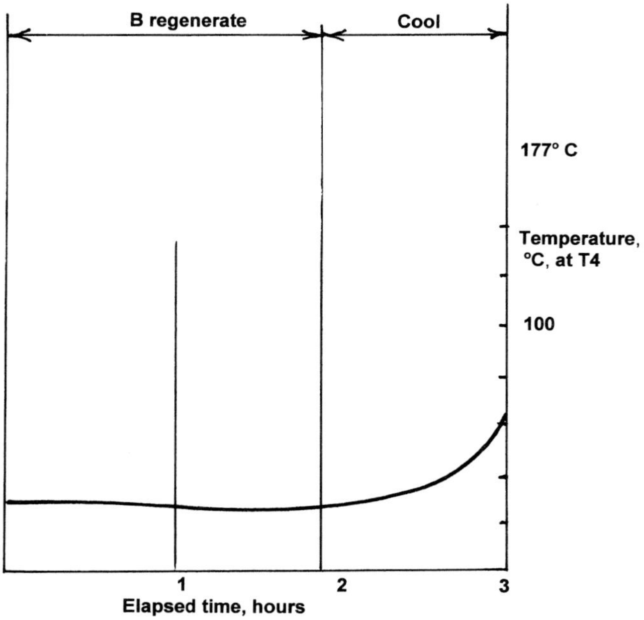


**Figure 4-4c** Test III, bed B regenerate (while bed A is adsorbing) with proportional valve open.

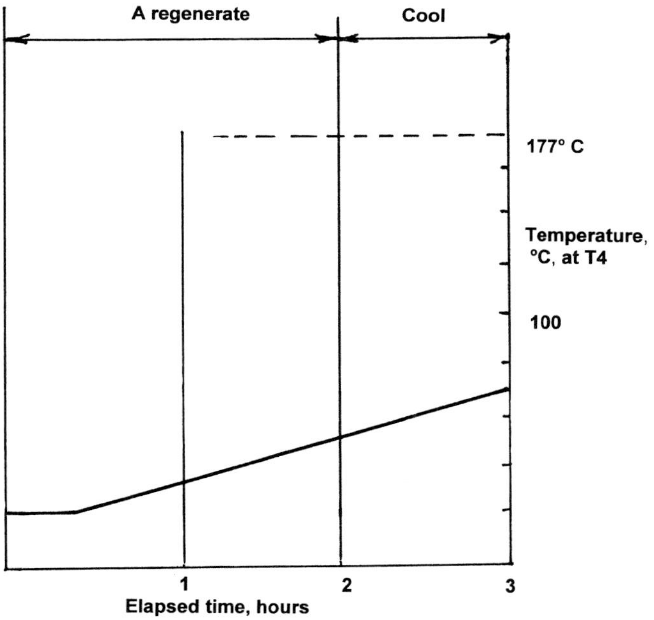


**Figure 4-4d** Test III, bed A regenerate (while bed B is adsorbing) with proportional valve open.

**Test III**, a 6-hour test with the proportional valve open and a corresponding small flow of air through the regenerator heater and the condenser-separator. The adsorption results are given in Figures 4-3c and d. Bed A seems to show the appropriate adsorption behavior. Bed B exhibits breakthrough of the water front. No condensate from S2 was gathered from this run so a mass balance could not be done. Since the flow through the cooler is relatively small, I am beginning to think that the separator may not be operating well, based on high concentrations of water found at S3 of > 24 000 ppm. The temperature profiles during regeneration are given in Figures 4-4 c and d. These profiles are not what I expected even though Bed B adsorbed only a small amount of water.

**Test E**. Pressure drop profile: P1–P4. *Vendor.* = 14 kPa. *Simple spot test I* with proportional valve closed = 27 kPa. *Test II:* proportional valve closed: regeneration of bed B and loading of bed A: $\Delta p = 62$ kPa and remained the same during the cooling cycle. I conclude that the pressure drop across the heater is negligible because there was negligible change in $\Delta p$ when the heater was bypassed via the 3-way valve V1. For the regeneration of bed A and the loading of bed B: $\Delta p = 125$ kPa with negligible change when the heater was bypassed. *Test III:* proportional valve is open with most of the air going directly to the adsorbent bed: for bed A adsorbent $\Delta p = 34$ kPa; with bed B adsorbent $\Delta p = 62$ kPa. These results suggest that there is a significant $\Delta p$ across the cooler-separator and that the adsorbent in bed B, with the larger $\Delta p$ and early breakthrough behavior, should be tested.

Saadia used the data to perform an energy balance over the regeneration operation and found, to her surprise that the energy would balance only if the flow through the "closed proportional valve" was 1270–1360 $Ndm^3/s$. The balance could also be closed if 60% of the water entering the separator was entrained.

Saadia arranged for the 4-way valve, V2, to be changed, and she contacted the vendor. For the new valve she recommended the replacement of the iron core with a stainless steel core. She also was concerned that the temperature of the gas leaving regeneration exceeded the plateau temperature of about 110 °C.

*Discussion*

Overall rating of the difficulty of this problem 9/10, relatively difficult because many faults are present. Saadia took several days to identify the fault in the valve and to define a series of concerns about this unit. Overall, Saadia drew on her experience with adsorption, especially in designing the tests and predicting what she expected to see before the test results were obtained. She astutely chose to follow the Worksheet because this setup was different from her experience. This was a "*problem*" and not an *exercise* for Saadia and she handled it that way.

**Problem Solving**

Monitored, especially by predicting the shape of the adsorption and regeneration curves before doing the tests. She was organized and systematic. She checked and double checked. Her dominant P style showed through in that she delayed changing the valve and calling in the vendor until she had more data. Overall a rating of A.

**Data Handling**

Actions were based on fundamentals. She related the test to the hypothesis. Good reasoning displayed. Overall rating B+.

**Synthesis**

She listed a variety of hypotheses. Considered many viewpoints. She had large masses of data to consider; she managed this well. Overall rating A.

**Decision Making**

She seemed to continually try to prioritize. No obvious bias was apparent. Criteria were present for some of the decisions, but for others, the criteria were not noted explicitly. Overall rating B+.

*Strengths:* Recognized the need to problem solve, used fundamentals, systematic, managed large volumes of data well, rich set of hypotheses, did not get discouraged.

*Areas to work on:* provide more explicit details for decisions.

## 4.5
## Case #7: The Case of the Reluctant Crystallizer

Frank has 25 years of experience. He is a process engineer responsible for several processes on site including the crystallization unit. The phone call comes in from Phil, the operator in the control room. "Something wierd is happening on the vaccum crystallizer, Frank. We've been operating the VC for about two hours now. The first hour with only two ejectors turned on, the operation was OK. We turned on the booster ejector and it "held" for the first 1/2 h or so and then it "kicked out". While the booster ejector was "holding" the liquid level in the crystallizer drops at a fantastic rate. What's going on here? We just can't produce quality crystals under these conditions."

"I agree, it sounds wierd. I'll be right out," responds Frank.

Frank has developed a good working relationship with all the operators so it doesn't surprise him to receive the call and for Phil's succinct summary. He took out the diagram for the process and mentally reviewed his thoughts. The vacuum crystallizer, VC, is the core of this operation. Pregnant liquor feed enters the vacuum crystallizer at 55 °C where it is concentrated and cooled to precipitate the product. This is a batch process. To start, we pull a vacuum on the VC (using the two-stage steam ejectors with city water to the interstage condenser), open the valves from the feed tank and syphon feed into the VC until the unit is 2/3 to 3/4 full, as seen on the sight glass. The feed valves are then shut and the eight-hour batch process begins. While operating with the two-stage ejector we estimate the absolute pressure to be about 6.5 kPa abs. After about an hour, the temperature of the liquor has decreased to 40 °C and the liquid level has dropped by about 10 cm. At that time, the booster ejector and the second barometric condenser are turned on to decrease the pressure further to about 2.5 kPa abs. At the end of the 8 hours the liquid level has dropped about 40 to 50 cm. The temperatures of the water in both barometric condensers are monitored to ensure that the water temperature does not exceed 26 °C. The operators have some control over this temperature by judiciously mixing cold city water with warmer water from the bay. Experience has shown that if the booster ejector is turned on too soon, it will not "hold" but rather "kicks out". "Kicking out" produces a very distinctive sound. This jargon has been used ever since Frank started on the unit. "Kicking out" has been interpreted as being when the steam controlled by valve S3 goes directly into the VC instead of through the ejector nozzle and thus does not pull the required vacuum.

Frank headed out to the control room. Here are some of his thoughts. Steam ejectors…. they are so sensitive to upstream steam pressure. Frank mulled over his experience: "*unstable operation or loss of vacuum*": steam pressure < 95% or > 20% of design/ steam superheated > 25 °C/ wet steam/ inlet cooling-water temperature hot/ cooling-water flowrate low/ condenser flooded/ heat-exchange surface fouled/ 20–30% higher flow of non-condensibles (light end gases, air leaks or leaks from fired furnaces)/ seal lost on barometric condenser/ entrained air in condenser water/ required discharge pressure requirement high/ fluctuating water pressure. Hmm.. It was a hot summer day.. so the bay water is likely hotter than usual; the maintenance turnaround was only a month ago and the barometric condenser was inspected and looked OK. Everything had operated fine since the turnaround. We usually operate the steam to the booster ejector with the valve half open. Hmm, lots of possibilities. By this time Frank had zeroed in on five hypotheses:

- steam to the booster ejector was at too low a pressure,
- steam superheated or wet,
- cooling-water inlet temperature too high,
- loss of vacuum caused maybe by loss of seal in the barometric leg or leaks in VC,
- pressure in the barometric condenser higher than usual.

In a moment he would be in the control room. He formulated some questions he wanted to ask Phil before he went out on the plant.

*What were the strengths and weaknesses of Frank's approach so far?*
*How would you have handled the call from Phil?*
*Would you go directly to the plant before seeing Phil so that you are better prepared to ask questions?*
*Would you have phoned the boiler house first to see if they had upsets?*
*What questions would you ask Phil?*
*What are your hypotheses?*
*What is the evidence so far?*

Frank realized that he worked better when he wrote things down but he wanted to do that with Phil's help. Furthermore, one of the reasons Frank got along so well with plant operators is that he always went to the control room first, before ever venturing out on the plant, and he always respected their experience and ideas.

After greeting a worried Phil, Frank and Phil sat down to put their thoughts on paper, the way they usually did in situations like this. Phil knew that Frank liked to use the IS and IS Not approach to summarizing the evidence. "Check me if I'm wrong," said Frank, "but this is what we have so far." He wrote:

IS: what and when: operating OK during startup, feed intake and first hour of operation before the booster ejector started. The booster ejector operates on "*hold*" without "*kicking out*" for an hour.

IS NOT: what and when: is not operating as usual when booster operating as "*hold*" with evidence that the liquid level (on the sight glass) drops "*at a fantastic rate*".

Phil looked at the list and said "Looks OK." and added "When the liquid level started to drop I went up and listened to the booster ejector and it sounded fine. The only thing I noticed was the pressure gauge on the bay water line to the booster condenser was fluctuating wildly. I also knew that you attribute most vacuum-system malfunctioning to steam pressure to the ejector. I checked the steam pressure gauges on P1, P2, P3 and P8. They are all 550 kPa g and *rock steady*. Gauge P4 was 725 kPa g, as usual, and steady."

Frank said, "Good thinking. I need some clarification. What did you observe when the liquid level drops *at a fantastic rate*?"

"The level in the sight glass dropped about 1 cm every two seconds. You could visibly watch it go down."

"Thanks. When you checked the pressure gauge P8, did you tap the gauge?"

"Yes, I tapped it and it seems to be working OK. The pressure on all steam gauges was 80 psi."

Frank mentally converted this to 650 kPa abs, that was normal steam pressure. "The pressure on the bay water line usually reads 205 kPa g; today it fluctuated wildly. How extensive was the variation, Phil ?"

"We don't normally check that pressure. It isn't shown in the control room. Out on the plant that gauge needle was jumping back and forth between 140 and 550 kPa g. Wild!"

Thinking about today's hot August temperature Frank asked "What about the temperatures on the barometric legs?"

"The temperatures on both legs were usual, certainly less than 27 °C; T2 was about 24 °C and T1, about 22 °C."

"Just for the record, what were the readings on the instruments you checked on the plant and how do these compare with the usual valves?"

"The steam pressures I talked about earlier; pressure on the city water P6 was the usual: 310 kPa g (45 psi) before the valve and P5 was 0 to 35 kPa g (5 psi) after the valve. I didn't note the temperature of the liquor in the VC. The level had disappeared below the sight glass. That's when I called you."

Frank added that evidence to his charts and thought about his hypotheses now in the light of the new evidence:

- Steam to the booster ejector was at too low a pressure. But all the pressure gauges read the same usual values, and they are steady! Perhaps the pressure gauge P8 is faulty.
- Steam superheated or wet. Can't tell, but if it was then the other ejectors should malfunction as well. They don't; so this is unlikely.
- Cooling water inlet temperature too high. Still possible but the exit temperatures in the barometric legs suggest this is not a cause.
- Loss of vacuum caused maybe by loss of seal in the barometric leg or leaks in VC. Perhaps; relatively easy to check, so why not?
- Pressure in the barometric condenser higher than usual. No gauge is available on the condenser or downstream of valve W5. Not easy to tell but perhaps this is connected to the wild variation in pressure in the bay water coming in. Perhaps valve W5 has a worn valve stem that is vibrating and causing the oscillations in pressure. Slugging flow of water? Is the bay water pump behaving itself? Should it be checked?

So to sum up, Frank is down to three hypotheses: 1) steam pressure to the booster is low or fluctuating but the pressure gauge is broken. Check the pressure gauge. 2) changes in the vacuum with leaks; questionable but let's check. Frank can't see how the fluctuating pressure gauge on the bay water is connected to rapid evaporation in the VC. However, he decides to change his focus to be "Why is gauge P7 fluctuating?" He hypothesizes that P7 is fluctuating because of 3) fluctuations in the pump exit pressure from the bay water pump or oscillations of the valve stem in valve W5. What's going on here?

*What were the strengths and weaknesses of Frank's approach so far?*
*What would you have asked Phil?*
*What are your hypotheses now?*
*What has Frank missed?*
*Is Frank exhibiting* pseudodiagnosticity *or* fixation *when he keeps looking for a steam fault to the booster ejector?*
*Should Frank have considered the evidence of "kicking out" of the booster ejector?*
*What would you do next?*

Frank makes a list of his next steps.

1. While the plant is still operating, visually inspect the VC for a leak. Attach a gauge to the VC via the valve and nozzle at the top and read the vacuum/pressure noting it every 10 minutes for two hours. Hypothesis, if there is a leak, the pressure will gradually increase. If the pressure remains constant, then there is no leak.
2. Shut down the process, and remove and visually check valve W5; while we are at it check valve W4. Frank realizes that this is not going to confirm any of his hypotheses, except perhaps his wild idea of a vibrating valve stem in hypothesis 3. However, he still can't see any connection between the oscillating pressure and the rapid evaporation. However, since the *change* is occuring near the valve, let's check out the valve.
3. To test hypothesis #1, either replace the pressure gauge at P8 or recalibrate the existing pressure gauge P8.

Frank hoped that the results of these tests would resolve the mystery. The results were:

- Hypothesis #1) steam pressure to the booster is low or fluctuating. Evidence is that gauge P8 is working and calibration is correct. Hypothesis denied.
- Hypothesis #2) changes in the vacuum with leaks. Evidence is that a visual inspection and the vacuum test could be interpreted as being *no leaks.* Hypothesis denied.
- Hypothesis #3) fluctuations in the pump exit pressure from the bay water pump or oscillations of the valve stem in valve W5. Frank received telephone confirmation from Utilities that the exit pressure on the bay water pump is constant, steady and the usual value. Valve W5 (and valve W4) were removed, visually checked as OK. A frustrated Frank decided to replace W5 with a new valve anyway. The process is started up. A number of other engineers have gathered in the control room to see if Frank has solved the mystery. The system behaves the same as it did before the changes: the booster ejector "held" for the first 1/2 h or so and then it "kicked out". While the booster ejector was "holding" the liquid level in the crystallizer drops at a fantastic rate.

Frank revisits his starting hypotheses and notes that perhaps the water temperature is too high in the barometric condenser. After all it's a really hot day. Frank follows his dominant J behavior and decides to take action and "correct the fault" rather than gather more data. Frank repiped the system with a temporary hose that sent city water into the bay water feed line so that only cold city water went to the condenser. The use of the hotter bay water was eliminated. Unfortunately this repiping makes no improvement!

Frank returns to his office. "It's back to basics! I've never seen anything like this before. I can do this!" *What can cause the liquid level in the crystallizer to drop at such a rate?* Brainstorm: liquid is leaking out, the change is in the sight glass and not in the vessel; something is sequestering the liquid, the vacuum is much higher than expected and the liquid is flashing off, the temperature is higher than expected, the absolute pressure is much lower than expected; steam flow into the booster is higher than expected; the steam flow is oscillating but its oscillations are so fast that the changes aren't picked up on the steam gauge P8 but are picked up on the water load and reflected on P7; the fluctuating water flow into the condenser is causing a vacuum that is pulling off more vapor than expected; liquid is going back into the feed tank; the discharge line is open, liquid leaks around the mixer shafts; the temperature gauge on the VC is broken so the temperature is much higher than it reads; liquor from the VC is entrained and not evaporated; liquor goes out through the booster ejector.

Now let's refocus and brainstorm on the *causes of wildly fluctuating gauge P7 when the bay water pump exit pressure is stated as being steady and the usual value.* Frank wrote down the following: change in water pressure, upstream oscillations, oscillations in the steam, oscillations in the liquid level, oscillations in condenser, city water hitting the bay water in condenser, oscillating liquid level in barometric leg, vibrating baffles in condenser, vibrating valve stem on water, water bypassing periodically directly into the barometric leg, corrosion particle in the pressure tap, unstable bourdon tube to pressure gauge, bay water pressure is not steady.

"I recall the case of the filter press operation was affected by the upstream oscillations of a centrifuge", murmured Frank to himself. "An idea that links the rapid evaporation and the fluctuating pressure gauge could be oscillations in the steam flowrate going into the booster ejector." Frank hasn't quite given up on the idea that the steam into the ejector is the key. Frank headed back out to the control room. Frank and Phil went to scrutinize the steam line into the booster ejector. The valve was a globe valve that was operated partly open. When Phil carefully adjusted the valve he felt some vibration. "Maybe that's it," exclaimed Frank. They shut down the troubled plant, removed the steam valve and found that the valve stem had been severely eroded. They replaced the steam valve S3 with a tapered plug valve. The replacement of the valve solved the problem. Case dismissed.

*Discussion*

The overall degree of difficulty of this problem is 10/10, very difficult. Frank took a week to solve this problem. The costs in loss of product and loss of pregnant liquor carryover into the hot well was extremely high. Let's reflect on Frank's approach following the guidelines in the feedback form. Frank started this problem using an exercise approach although even here he showed excellent problem-solving characteristics: IS and IS NOT, systematic, patient.

**Problem Solving**
Systematic and organized even though he was very frustrated. Patient and takes time to understand the system (even though he had worked many years on this plant). Some monitoring. Some checking. He's active and wrote things down. Overall rating B+.

**Data Handling**
Based on fundamentals–especially after he recognized it as a *problem* and not an *exercise*. Very systematic in planning and gathering evidence. Used simple tests. Overall rating A–.

**Synthesis**
Used a variety of hypotheses. Flexible. Kept an open mind even when nothing seemed to be working out. Overall rating A–.

**Decision Making**
Prioritized. His bias about the low steam pressure actually paid off. Decisions tended to be made intuitively with few explicit criteria. Overall rating B.

*Strengths:* excellent interaction with the operators, very systematic, didn't panic, flexible, time well spent exploring and understanding the problem, patient.

*Areas to work on:* more monitoring and more explicit thought process especially with criteria for decisions.

## 4.6 Reflections about these Examples

In each of the cases, the engineer "solved the problem." Each used a different style. The less- experienced engineers, Michelle and Dave, used the Trouble-Shooter's Worksheet with varying degrees of comfort. The more-experienced engineers, Pierre, Saadia and Frank, tended to resort to the Trouble-Shooter's Worksheet *as needed*. Pierre used it to remind himself to try the *Why? Why? Why?* approach. He might have tried this without resorting to the Worksheet. Indeed, the problem he was addressing seemed to be one of the few where that technique was useful. Saadia recognized the uniqueness of her problem early on and, although she had a lot of experience with adsorption, she wisely disciplined herself to work slowly and systematically through the Worksheet. Of the five engineers she did the best job of using the "hypothesis-symptoms-actions" chart. Frank treated the problem like an *exercise* until he discovered it really had more unusual components than he had usually encountered. Before it was too late he started to use the Worksheet.

What components of the Trouble-Shooter's Worksheet seemed to be used more effectively?

All tried to slow down the process and work carefully in the **Engage** process, although Pierre's efforts were not the greatest. This was astute of the trouble shooters because most mistakes are made right at the start. Dave almost got caught here. Pierre was stymied here for a frustrating few moments.

The **Define the stated problem** stage was handled better when all five engineers used the IS and IS NOT tactic effectively.

For the **Explore** stage, many used a variety of approaches and, indeed, were selective in the elements they included. Dave used the *What if?* elements effectively. The less experienced engineers drew on the symptom-cause information in Chapter 3 to help them overcome their limited practical experience. Fortunately, for the problems they were working on some data were available. For some equipment, the published information is lacking.

The "hypotheses-symptoms-action" chart was only used well by Saadia. The others, especially Frank, tended to use the ideas intuitively, but such an intuitive approach has pitfalls. Saadia also excelled at using her fundamentals to predict, *ahead of time*, what she expected to see from the results of the tasks.

Frank seemed to be the only one who was comfortable working with and drawing on the experience of the operators. This didn't come without patience in developing trust and a good working relationship with Phil. Michelle, and Dave, should polish their skills in this area.

# 5
# Polishing Your Skills: Problem-Solving Process

Five components are useful in developing your skills: becoming comfortable talking aloud about your thought processes, identifying a strategy and monitoring the stages you use, defining the problem, creativity and self-assessment. Activities to develop your awareness, skill and confidence in these skills are given[1]. To gain the most, select the areas you want to work on and complete the activities. Developing skill is not a spectator sport. Participate.

## 5.1
## Developing *Awareness* of the Problem-Solving Process

Awareness is the ability to describe what goes on in our mind as we solve problems and make decisions. Such awareness is a prerequisite for the triad improvement activities in Chapter 8.

Such awareness also helps us improve our problem solving – and trouble shooting – skill because:

- we can explicitly identify "where we are in a problem",
- we can compare "how we do it" with how "others do it",
- we can identify our strengths and areas to work on to improve our problem solving,
- we can get ourselves "unblocked" if we cannot seem to solve a problem,
- we need to be able to describe our thoughts for team problem solving,
- we can describe our thought processes to another so that we can improve,
- we build on this awareness to develop skills with strategies, in Section 5.2.

One of the more successful techniques to quickly acquire confidence and skill in "awareness" is the Whimbey pairs or Talk Aloud Pairs Problem Solving, TAPPS, approach. Two people are needed. One plays the role of a "*talker*" or problem solver; the other plays the role of the "*listener*".

1) The materials in this chapter are from the MPS program, copyright Donald R. Woods 1982 ff and used with permission.

*Successful Trouble Shooting for Process Engineers.* Don Woods

ISBN: 3-527-31163-7

In this section we list the target skills that are developed by doing the activities in this section, describe the roles, give the activities and the feedback forms.

### 5.1.1 Some Target Skills

Research in problem solving has shown that successful problem solvers exhibit the following characteristics (related to awareness and the thought processes and the activities in this section): They

- are aware of their thought processes and use that awareness to identify where they are in the process and get themselves unstuck,
- are skilled in describing aloud their thoughts as they solve problems,
- pause and *reflect* about the process and about what they are doing,
- accept that their particular style works for them; others may have a different preferred style,
- are active as opposed to passively trying to remember stuff. They are active by writing things down, making charts and diagrams. Being active helps overcome the space limitations of Short-Term Memory (the portion of the brain where problem-solving tasks are done. This portion can only "hold" five to sevens bits of information at a time.)
- focus on accuracy and not on speed,
- accept that problem solving is a social process; we interact with others,
- know that self-assessment is about performance and not about them as a person,
- know that self-assessment is based on evidence and not on gut feelings or wishful thinking, and so to assess progress in developing skills they will use feedback forms and written evidence and reflections.

In addition, this activity provides an opportunity to

1. Acquire some skill at listening,
2. Acquire some skill in self-assessment,
3. Acquire some skill in giving and receiving feedback,
4. Through self-awareness, begin to improve self-confidence.

### 5.1.2 The TAPPS Roles: Talker and Listener

**Talker/ problem solver**: In pairs, with a partner as a listener, the *talker* reads the problem statement aloud, and talks aloud as he/she works on/solves the problem. The goal is not to get the "right" answer to the problem. The goal is to talk aloud continuously about the process. The goal is to have, in a ten-minute period of talking, fewer than two silent periods of more than 10 s duration. The goal is to focus on accuracy. The goal is to be active and write things down.

Your listener will not give you any hints about how to solve the problem. Your listener will help you to talk about what you are thinking so that the listener can follow and understand what you are saying. More specifically, here is what you do in this role:

1. Sit side by side; have paper and pencils available.
2. The talker starts by reading the problem statement aloud.
3. Then start to solve the problem *on your own*. *You* are solving the problem. Your partner is only listening to you. He or she is not solving the problem with you or for you.
4. Talking and thinking at the same time is not easy. At first you might find it hard to think of the right words to use. Do not worry. Say whatever comes to your mind. No one is testing you or marking you. You are playing the role.
5. Go back and repeat any part of the problem you wish. Use such words as "I am stuck! I do not know what to do! Maybe I should read the problem statement again."
6. Try to solve the problem no matter how easy it is. You are learning to talk about your thinking methods. We use simpler problems to help make this activity as easy as possible.

**Listener**: You have an important and difficult role to play. You are to help the *talker* talk. You are *not* to solve the problem nor to give hints to the *talker* about how you would solve the problem. This is the problem *talker*'s turn. You will get a turn later to be *talker*.

Encourage them to talk aloud. At the same time, you are to monitor their thinking. Can you understand what they are saying? Can you follow the path that their mind is following? Could you describe what they are thinking to others? You are to help them to talk about the mental processes they are using–no matter how silly or incorrect they might be. You must not laugh at them. You must not criticize them and tell them that they are wrong. If you think they made a mistake, then say "Can you check that?" or "Are you sure?" Do not tell them what they should be doing. Do not tell them what you think is the "correct" answer.

1. Help the *talker* to see that you are not a "critic". Instead, you are a *helper-for-talking*. You might say " Please keep talking." or "I was not able to understand or follow what you just said; would you please explain." "Can you tell me what you are thinking now?" "Do not worry about how it sounds – just say an idea about what you are thinking." "Can you check?" "Are you sure?" "OK" "Ahmmm". (For more suggestions about listening, see Section 7.1.2.)
2. You might tell the *talker* that your role is to:
   - Remind them to keep talking,
   - Help them improve the accuracy of their talking about their thinking by saying *"Can you check?" "Are you sure?"*
   - Be able to understand and follow each step of the talker's thinking.

3. Do not turn with your back to the *talker* and try to solve the problem on your own. Do not solve the problem on your own and then tell the *talker* what they should do.
4. Do not let the *talker* continue if:
   – you do not understand what they have done. Say "I didn't quite understand that, could you please elaborate on your thoughts, Thanks."
   – you think that a mistake has been made. Do not say, "You have made a mistake." You might say "Are you sure?" or "Do you want to check that?"

Your goal, when you receive feedback on how well you played the listener role, is that you should be within two scale ratings of "about what I wanted" on both the "degree of interaction" and the "tone of the interaction."

### 5.1.3
### **Activity 5.1**: (35 minutes)

This activity will take 35 minutes. 3 minutes getting set up; 10 minutes talker A, 5 minutes reflection, complete feedback forms; switch roles, 2 minutes getting set up, 10 minutes talking, 5 minutes reflection, complete feedback forms.

- Getting set up. (3 min) Find a partner, flat table, two chairs side by side, pencil and paper. One person starts as talker, the other is the listener. Read over your role (Section 5.1.2).
  *Talker* selects an "exercise" (or perhaps a "problem") from among Tasks 5.1A to E.
- First 10-minute talk (10 min): *talker* talks for 10 minutes. If the *talker* completes the selected task early, then the *talker* selects another task. Do not change roles! The *talker* talks for 10 minutes.
- Reflections, discuss and feedback forms (5 min). As individuals, write down reflections about the activity, use Worksheet 5-1. (1.5 min). In pairs, discuss what you experienced. Don't continue working on a Task. (1.5 min). *Listener* completes feedback form in Worksheet 5-2 and signs it; gives it to the *talker* as evidence. *Talker* completes feedback form in Worksheet 5-3 and signs it; gives it to the *listener* as evidence. Worksheets 5-1 to 5-3 are given in Section 5.1.4.
- Switch roles and get ready (2 min) Read the instructions about your new role (Section 5.1.2). *Talker* select a Task from among Tasks 5.1.A to E.
- Second 10-minute talk (10 min). See instructions above.
- Reflections, discuss and feedback forms (5 min) as above.

**Task 5.1.A:** Content-general

A.1 (based on Lochhead and Clement, Univ. of Massachusetts) At Mindy's restaurant, for every four people who ordered cheesecake, there are five who ordered strudel. If C represents the number of cheesecakes ordered and S represents the number of strudels ordered and M represents Mindy's restaurant, write an equation using these variables to represent this situation:

a) M: $4C = 5S$
b) $4C \times M = 5S \times M$
c) $4C = 5S$
d) Total $= 4C + 5S$
e) $(5/9)\ S = C$
f) M: $5C = 4S$
g) $5C \times M = 4S \times M$
h) $5C = 4S$
i) $S = (4/9)\ C$
j) other

A.2 (reprinted courtesy of Art Whimbey) There are two clocks, A and B. Clock A keeps perfect time but clock B runs fast. When clock A says 4 minutes have passed, clock B says that 6 minutes have passed. Both clocks are set correctly at 5 a.m. What is the correct time when clock B shows 9 p.m.?

**Task 5.1.B** Tasks related to science and engineering

B.1 (based on Lochhead and Clement, University of Massachusetts)

A pillar supports a floor. Is the pillar doing any work?:

a) Yes, of course it is doing work; otherwise the floor would come crashing down.
b) No.
c) Yes, a definition of work is a force acting over a distance; the force is the force pushing up on the floor above and the distance is the height of the pillar
d) Yes, we know this because of energy considerations. The potential energy is the mass of the slab or floor times the height of the pillar that is supporting it. If the pillar wasn't there, the potential energy would turn into kinetic energy and would eventually be released in the form of heat and sound as it crashed down.
e) Other

**Task 5.1.C** General tasks more related to trouble shooting.

C.1 A researcher predicted that if part of a leaf is in the shade and the other part of the leaf is in the sun, then equal amounts of starch will be found in both parts of the leaf. Which of the following hypotheses is the researcher most likely assuming is true?

1. Chlorophyll is present in both the shaded and sunny parts of the leaf.
2. In the shaded part of the leaf photosynthesis is increased.
3. Carbon dioxide and water can enter the leaf cells in both the sunny and the shaded parts of the leaf.
4. By shading part of the leaf, photosynthesis is increased in the sunny part of the leaf.
5. Starch can move to different parts of the leaf.

**Task 5.1.D** Engineering related

D.1 Oversized condenser

The overhead condenser on the distillation column is oversized by 38%. That is, it has the correct number of tubes to promote turbulent flow inside the tubes. However, the length of the tubes has been increased. The baffle spacing on the shell side and the baffle window has been sized for the design overhead rate. This is in addition to the usual allowances for fouling. The condenser is horizontal. Which of the following observations are consistent with this situation when the condenser is first put into service:

a) The exit cooling-water temperature will be colder than expected.
b) The pressure drop on the shell side will be $1.38^2$ more than we expected.
c) The reboiler will act as though it were undersized and the column operation will be unstable with vapor blanketing in the reboiler because film boiling now occurs.
d) Nothing unexpected; the controller action will account for the overdesign.
e) Increased power required on the cooling-water circulation pumps.
f) The feed location should be changed to be closer to the reboiler.
h) Other

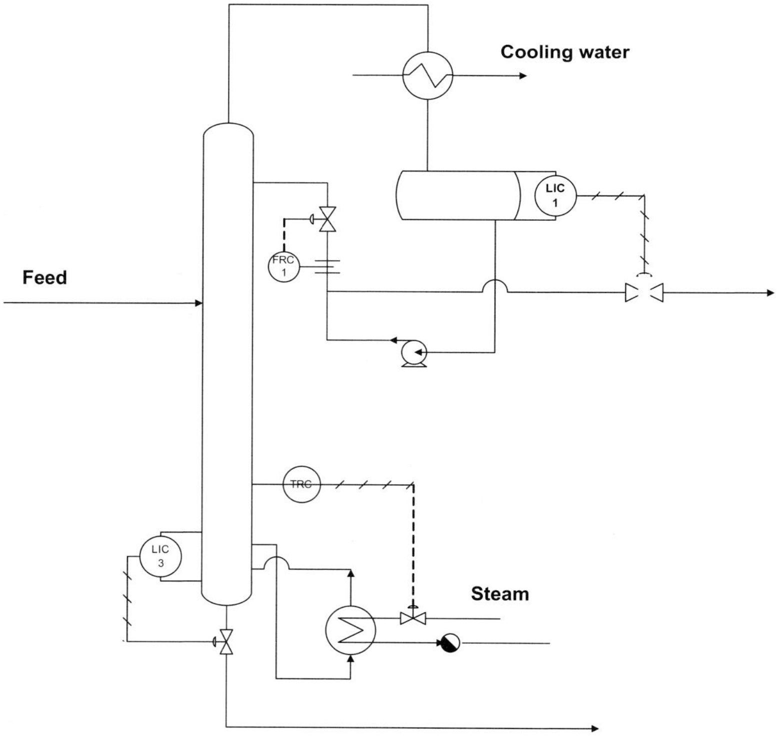


**Figure 5-1** A column.

**Task 5.1.E** Technical tasks related to methane-steam reforming

E.1 In the methane-steam reformer shown in Figure 5-2, the total feedrate to the reformer is 25% higher than the flow instrument reads. The other instruments usually read reformer outlet temperature T1 = 454 °C (850 °F) with the exit methane concentration of 9.5 mol %. The tubewall temperature is usually 960 °C (1765 °F). The design $\Delta p$ = 345 kPa (50 psig). The reformer works on outlet temperature control. Which of the following observations accurately describe the situation:

a) Nothing different; everything works fine because the controllers adjust to yield the same exit temperature.
b) The tube wall temperature will be about 30 °C higher because the controllers adjusted to yield the same exit temperature; the exit methane concentration is 7.5 mol %.
c) The tube wall temperature will be about 23.5 °C lower than usual because of the cooling effect of the larger mass of gas flowing through.
d) The inlet gas temperature will increase by 38.2 °C because of the controller action.
e) With controller action, the methane at the exit will increase to 10.5 mol % and the tube wall temperature will remain 960 °C.
f) With controller action, the $\Delta p$ will be about 500 kPa.
g) With controller action, the exit product gas temperature will be 20 °C lower than design.
h) Other.

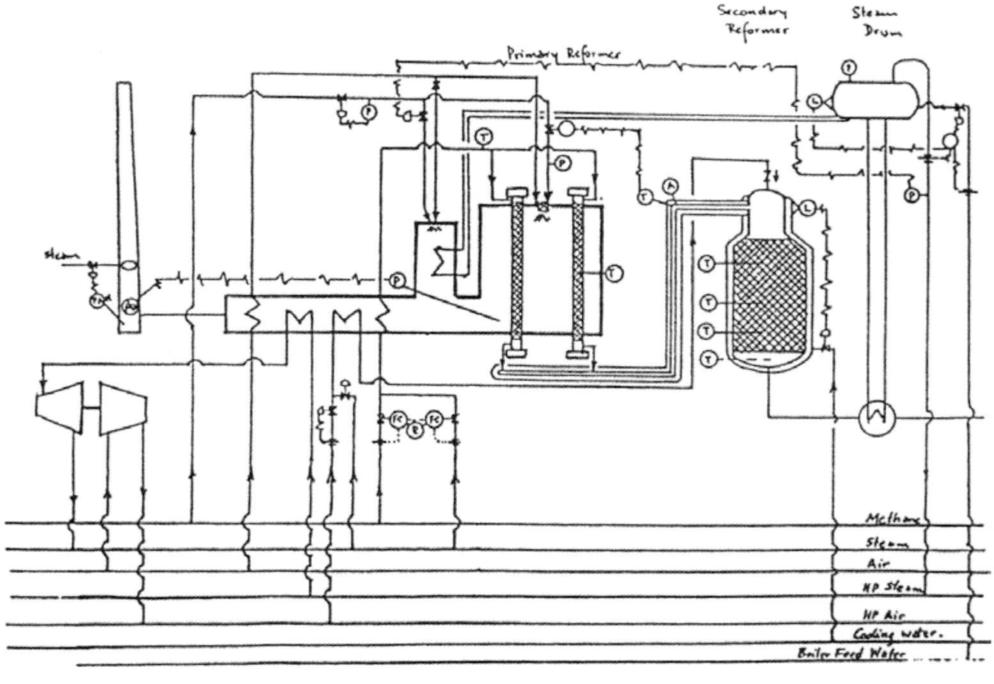


**Figure 5-2** A steam reformer.

### 5.1.4
### Feedback, Self-Assessment

Self-assessment is based on written evidence. Throughout this activity there are several times when you are asked to write reflections and complete forms. This is a necessary part of the skill development process.

The evidence can include:

- the problem statement you used when being the talker and the marks, underlines and notations you made directly on this.
- the paper you worked on.

Other, more structured evidence includes:

- your reflections you made after you did the task, Worksheet 5-1;
- the feedback from the listener about the task, Worksheet 5-2.
- the feedback from the talker to the listener about the role playing, Worksheet 5-3.

**Worksheet 5-1:** Reflections: A place for you to record your ideas about the TAPPS Method:

Being the **talker**: What did you enjoy most? What was most difficult about the task? What did you discover by being the talker? What did you discover from interacting with a listener? What were your strengths? Focus on accuracy? Very few silent periods? Being active? Good communication?

________________________________________
________________________________________
________________________________________
________________________________________
________________________________________
________________________________________

Being the **listener**: What did you enjoy most? What was most difficult about the task? What did you discover by being the *listener*? What did you discover about problem solving by comparing the talker's approach to yours? What were your strengths as listener? Quality of your prompts and degree of interaction? Tone of interaction? Good communication? Non-intrusiveness?

________________________________________
________________________________________
________________________________________
________________________________________
________________________________________
________________________________________

**Worksheet 5-2:** Feedback from the listener to the talker about the process used.

Awareness ______________________ ______________________

problem listener

| | | | | | | | |
|---|---|---|---|---|---|---|---|
| Number of silent periods | 0 | 1 | 2 | 3 | 4 | 5 | >5 |
| Number of checks, double checks | >5 | 5 | 4 | 3 | 2 | 1 | 0 |
| Amount of writing/ charting | >5 | 5 | 4 | 3 | 2 | 1 | 0 |

Comments:

______________________________________________

______________________________________________

Validated by: ______________________________________________

**Worksheet 5-3:** Feedback to the listener:

problem ______________________ listener ______________________

I found the listener:

- The quality of the comments:

–10 –8 –6 –4 –2 ☺ –2 –4 –6 –8 –10

too passive little too passive about right a little interruptive too interruptive

- The attitude displayed:

–10 –8 –6 –4 –2 ☺ –2 –4 –6 –8 –10

too passive little too passive about right a little interruptive too interruptive

- The listener's emphasis was listening to me □; helping me verbalize □; helping me solve the problem □; solving the problem for me □.

validated by talker ______________________________________________

## 5.2 Strategies

A strategy is an organized approach used in solving a problem. Such an organized approach identifies steps or stages for different parts of the process. Based on a survey of the cognitive literature and a critique of over 150 published strategies, the MPS 6-stage strategy, given in Figure 2-3, is a good working strategy to use. A strategy is important because:

- we all usually use one,
- a strategy helps us to be organized and systematic,

- having a strategy helps calm us down if we become anxious when we are given a very difficult problem to solve,
- having a strategy helps us to "monitor" our mental processes.

One of the more effective ways to build our skill and confidence with the use of strategies is an extension of the Whimbey TAPPS approach used in Section 5.1. This is an extension of that experience in that the *talker* is also asked to move a marker to indicate the stage in the strategy in which he/she is working and asked to explicitly say monitoring statements frequently. The *listener's* role is expanded to include recording the minutes the *talker* spends working in each of the stages in the strategy and recording monitoring statements made by the *talker*.

In this section we list the target skills, describe the roles, give the activities and the feedback forms.

### 5.2.1 Some Target Skills

Research in problem solving has uncovered the following behaviors of skilled problem solvers as they relate to the use of a strategy:

- Spend time reading the problem statement (up to three times longer than unsuccessful problem solvers).
- Define the problem well; do not solve the wrong problem. Are willing to spend up to half the available time defining the problem. Most mistakes made by unsuccessful problem solvers are made in the define stages.
- The problem that they solve is their mental image of the problem; such a mental image is called the internal representation of the problem.
- Differentiate between exercise solving and problem solving (that were illustrated in Figures 2-1 and 2-2.)
- Use an organized strategy that focuses on defining the real problem, whereas unsuccessful problem solvers tend to search for an equation that uses up all of the given variables or given information (regardless of whether it applies to the situation or not).
- Define the real problem with a focus on key fundamentals whereas unsuccessful problem solvers tend to memorize and try to recall equations and solutions that match the description of the situation.
- Break "Defining the problem" into three separate activities to avoid errors. These are 1) listen, read, get initial information and manage both distress and panic (called **Engage**), 2) classify the initial information into the "stated goal or task to do", "constraints and criteria", and "description of the situation", without trying to "define the real problem" (called **Define the stated problem**), and 3) create a rich, internal image of the problem as seen from many different perspectives and evolve a definition of the real problem (called **Explore**).
- Use a strategy so as to be systematic and organized, whereas unsuccessful problem solvers tend to take a trial and error approach.

- Are aware that a strategy consists of a series of about six stages with each stage using different thinking and feelings. This strategy is *not* used serially (following rigidly one step after another). Rather it is used flexibly; applied many times while solving a single problem with frequent recycling from one stage to another.
- Problem-solving skill draws on subject knowledge (needed to solve the problem) and with the sample solutions (from past-solved similar problems) as was illustrated in Figures 2-1 and 2-2.
- *Monitor* their thought processes about once per minute while solving problems.

In summary, the goal of this activity is to develop your skill and confidence in

1. Extending and reinforcing the skills addressed in Section 5.1 on Awareness.
2. Recognizing patterns in the problem-solving process.
3. Realizing that a "strategy" is not applied *linearly* and sequentially; that it is used flexibly.
4. Recognizing the difference between *problems* and *exercises.*
5. Understanding the relationship between subject knowledge, past solutions to problems and problem solving.
6. Acknowledging the importance of *defining* problems and to recognize this as a three-step process.
7. Acknowledging the importance of *reading* the problem statement.
8. Realizing that *problem solving* is not "doing some calculations." Conversely, to correct the misconception that if you are not "doing some calculations" you are not solving problems.
9. Acquiring skill in explicitly monitoring the process.

### 5.2.2
### The Extended TAPPS Roles: Talker+ and Listener+

**Talker/problem solver+**: This is an extension of the pairs activity with roles described in Section 5.1.2. Do all of the activities described in Section 5.1.2 *and* now the *talker* also

1) moves a marker on a strategy board, please use Figure 2-3, p. 21, to indicate which of the 6 stages you think is being addressed. By considering the location of the marker, the *listener* should agree 80% of the time that the activities you describe are consistent with the stage represented by the marker. You should need prompting no more than three times in a ten minute period.
2) tries to say frequently such monitoring statements as "*Where am I?*" "*Have I finished this?*" "*If I calculate ..., what will that tell me?*" "*If I ask this question ..., what will that tell me?*"" *Where do I go next?*" "*Can I check this?*" *If a hypothesis is shown to be wrong or if you calculate a "strange answer." then ask "OK, What did I learn from that?"* You should exhibit four verbal management statements during a ten minute period of problem solving.

This is difficult to do. Be patient with yourself. You may not completely understand the meanings of the stages yet. You may use different stages than the ones on the Strategy Board. Please, do your best. The listener will not move the marker for you. The listener will not tell you what stage you are in. The listener might ask you "Are you still in the "***Explore***" stage?" Remember to keep talking, to be active, to use pencil and paper, and to check and check again. Before you start, go over the meanings of the six different stages in the McMaster-6-Step strategy with the *listener*. Agree on the meanings of the words.

1. Sit side by side; have paper and pencils available, have the Strategy Board and the marker.
2. The *talker* moves the marker to the ***Engage*** part of the Strategy Board and starts by reading the problem statement aloud.
3. Then move the marker to whatever stage you are going to work on next and start to solve the problem *on you own*. Keep talking aloud. *You* are solving the problem. Your partner is only listening to you. He or she is not solving the problem with you or for you.
4. It is not easy to talk, think and move the marker at the same time. You might forget to move the marker. That is OK. Do the best you can.
5. Go back and repeat any stage of the strategy you wish.

**Listener+**: This is an extension of the listener role described in Section 5.1.2.

Here you are to record the amount of time the *talker* spends in each of the stages on the Strategy Board. Do not correct them; do not argue with them about which stage the *talker* is working on. Do not move the marker for them. You may have to ask *"Are you still in the ... stage?"*

In all that you do, your interventions will be judged by the *talker* to be *helpful*, and not judged to be *disruptive*. The worksheet to be used to record the evidence for the *talker* is given in Worksheet 5-4; some example data are given in Figure 5-3. Add a ▼ to the chart whenever a monitoring statement is said.

### 5.2.3
### Activity 5.2: (time 35 minutes)

Allow the same timing as used in Activity 5.1, in Section 5.1.2. The physical arrangement for the ten- minute talk period is illustrated in Figure 5-4.

The Tasks may be selected from Tasks 5.1.A to E or from Task 5.2.A including options from Appendix F, Sections 5.1 and 5.2.

In pairs, one be a *talker*, the other be a *listener*. The *talker* plays the role for 10 minutes and "solves" problems during all the allotted time. Do not change roles.

Use Worksheet 5-1 for reflections and Worksheet 5-5 for feedback about listening. The Worksheet 5-4 will be validated by the *listener* and given to the *talker* as evidence.

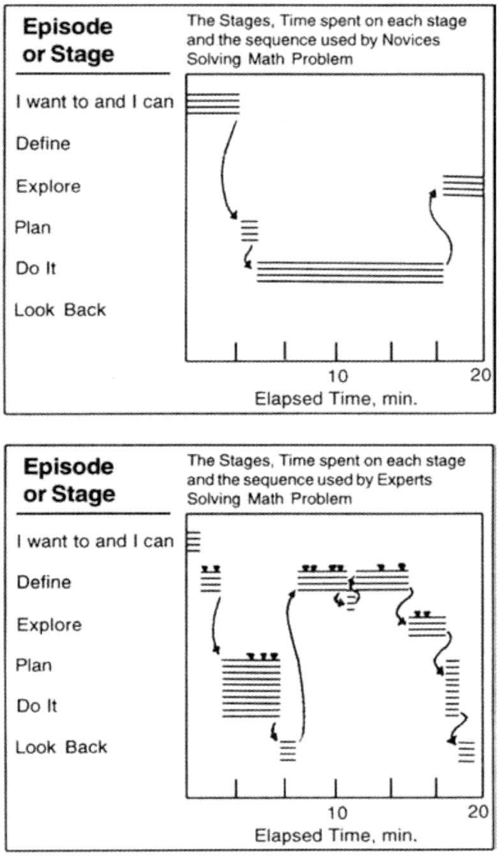


**Figure 5-3** (top) Some example data on the Worksheet.

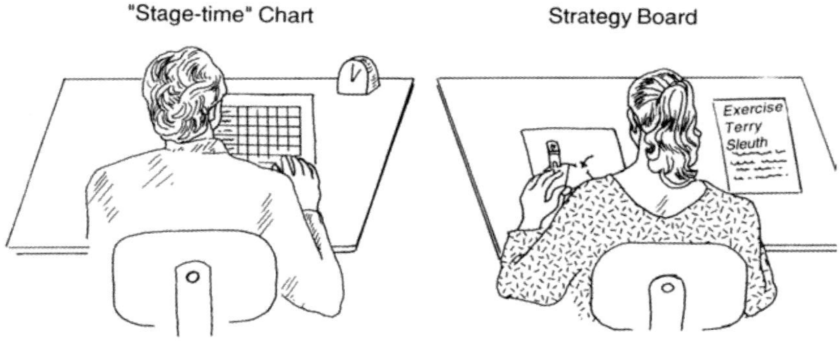


**Figure 5-4** (bottom) The physical arrangement: the listener is shown on the left; the talker, on the right.

**Worksheet 5-4:** Record of the talker's strategy with ▼ for monitoring statements.

Talker ____________________ Case __________ Listener ______________________

| Stage | | | | | | |
|---|---|---|---|---|---|---|
| **Engage**: *"I want to and I can!"* | | | | | | |
| **Define-the-stated problem**: Sort the given problem statement | | | | | | |
| **Explore** the problem to discover what the problem really is | | | | | | |
| **Plan** | | | | | | |
| **Do it** | | | | | | |
| **Look back**: elaborate, check | | | | | | |

0 2 4 6 8 10 12

Time, minutes

**Task 5.2.A:** Terry Sleuth in the Poly Room

Terry Sleuth ventured into the polymerizer room on the way to the R&D lab for an appointment with Bill Wright. Seeing Terry, Charlie, the poly-room engineer, called "Hi Terry, please come over here and help us sort out this mess. Look at these reactors. We're continually losing quality product in reactors two and three but not on reactor one."

"Tell me more," encouraged Terry.

"The problem is that just near the end of the run, the motors driving the mixers overload and cut out. They stop! Then the whole batch is ruined because the heat transfer is insufficient," explained Charlie.

"Does it happen only on reactors two and three and not on reactor one?" asked Terry.

"Yes. Reactors two and three have calandria coolers and marine propeller mixers. Reactor one has an internal coil cooler and a turbine mixer. The mixing flow patterns are chosen specifically for the heat-exchanger configuration: the propeller moves the liquid up through the tubes with the coolant on the shell side of the internal calandria; the turbine shoots the liquid out onto the tubes with the coolant inside the tubes. Perhaps the calandria tubes are plugged with polymer," continues Charlie.

Terry thought for a moment and then asked "Are both cooling systems about the same surface area and serviced by the same cooling water?"

"Yes."

"Has this problem ever occurred before?"

"This is the first time we have processed this product on any of these reactors; we have processed other products for years on all three reactors and with great success," beamed Charlie.

Terry smiled. Terry had all the information needed to identify the cause. What did Terry say?

**Task 5.2-B:** Terry Sleuth and the Case of the Delinquent Decanter

Ring ... The incessant ringing of the telephone disrupted the steady hum of the engineering office. Betty answered, looked perplexed, looked across at Terry Sleuth's desk and knit her brow as she carried on an animated conversation. Terry could almost guess from Betty's reactions that the call was about the hexane-water decanter on the soybean-miscella still. The Ministry of Environment had been after us for the last week to get the concentration of hexane in the wastewater from the decanter down to acceptable levels. Unfortunately, the decanter wasn't working well. Betty hung up, donned her hard hat and headed toward Terry's desk. "It's the decanter again, isn't it!" said Terry. "You've said it", replied Betty disheartenly. "Before we head out to see the beast, let's review what we know", suggested Terry encouragingly. "OK", Betty said and went on "the decanter is a vertical cylinder with a conical bottom; the feed enters about midpoint with exit lines at the top for the hexane and at the bottom for water. The feed is a mixture of about 60% hexane and 40% hot water. The foam or droplet layer fills the center band of the decanter; the drops coalesce to form pure layers of hexane (that rises to the top) and a pure layer of water. There is a cover on the decanter that has a vent. The last time we talked we realized that some hexane dissolvers in the water and some droplets of hexane may go out with the water because the drops haven't coalesced. How's that?" Terry replied, warmly, "Very good. However, we were unsure as to whether the feed is hexane drops in water or water drops in hexane. Since there is more hexane, it is probable that the feed is hexane containing droplets of water. So it is the water drops that are coalescing." "Yes, but I don't see that that makes any difference whether it is hexane drops coalescing or water drops coalescing. What I have done, since we last talked, is a lot of research about decanters. Mizrahi and Barnea report in a refereed article that for every 10 °C increase in temperature, the coalescence time will be faster by a factor of two. I think I'll rig up a heater on the feed line so that the incoming temperature of the feed is 70 °C instead of 60 °C. That should fix this baby!" gloated Betty. Terry paused, checked the Handbook of Chemistry and Physics and said, "Don't be too sure." What did Terry look up and why did Terry cast doubt on Betty's idea?

### 5.2.4
### Feedback, Self-assessment

Self-assessment is based on written evidence, not on intuitive feelings. Throughout this activity evidence has been gathered. The evidence includes:

- the problem statement, with all your underlining and notations,
- your paper you worked on,
- the time-stage-monitoring evidence of Figure 5-4,
- the reflections, similar to Worksheet 5-1 but with prompts related to this activity,
- feedback about your role as *listener*, Worksheet 5-5. You might elect to provide additional feedback to the *talker* from Worksheet 5-2 about the process used.

**Worksheet 5-5:** Evidence for listening: Feedback to listener ☐ the listener will encourage verbalization, an emphasis on accuracy, active thinking and encourage the problem solver to move the marker correctly on the strategy board. Your interventions will be judged by the problem solver to be *helpful*, and not judged to be *disruptive*.

Activity 1: Talker ____________ Case ____________ Listener ____________

| | | | | |
|---|---|---|---|---|
| encourage verbalization: | not needed | interruptive | OK | really helped |
| encourage emphasis on accuracy: | not needed | interruptive | OK | really helped |
| encourage active thinking | not needed | interruptive | OK | really helped |
| interventions: | not needed | interruptive | OK | really helped |

Comments:

________________________________________________

________________________________________________

## 5.3 Exploring the "Context": what is the *Real* Problem?

During the ***Explore*** stage, we may wish to place the problem in a larger context. A very useful technique is Basadur's *Why? Why? Why?* technique (Basadur, 1995). In this technique, we start with the initial "problem", ask *Why?* and redefine the problem in a broader context. The process is repeated. After we have completed several levels of context, we then can address "which level of generality is best, given the current constraints and contexts."

This approach was used very successfully by Pierre in the **Case** #4, the case of the Platformer Fires, Section 4.3. Consider first an example and then an activity.

### 5.3.1 Example

Here we illustrate the application of the approach for **Case** #7, the Case of the reluctant crystallizer, problem 1.5 in Chapter 1 and considered in Frank's approach in Section 4-5.

Beginning symptom: *"the liquid level in the crystallizer is dropping at a fantastic rate."*

Rephrase as *Why do I want to* stop the liquid level in the crystallizer from dropping at a fantastic rate. OK. Perhaps your answer is "so that I can produce quality crystals from the crystallizer." OK. Now ask *Why?* Now your answer might be "*So that I* have quality crystals to sell to my customers." This process continues. It is convenient to summarize this are follows:

*Why?* so that have Happiness and Bliss

↑

*Why?* so that have me a rich and productive life;

↑

*Why?* so that have challenging and exciting employment;

↑

*Why?* so that keep the company profitable;

↑

*Why?* so that keep sales healthy,

↑

*Why?* so that have quality crystals to sell to my customers;

↑

*Why?* so that produce quality crystals from the crystallizer

↑

Start: → stop the liquid level in the crystallizer from dropping at a fantastic rate.
*Why?*

In this case, perhaps the most important problem to address is "to have quality crystals to sell to my customers." That being the case, Frank should have addressed his efforts toward obtaining the quality and quantify of crystals, even if it means buying them from a competitor. Thus, this technique helps us to see the problem in a bigger context and prompts us to ask "what is the *real* problem?".

### 5.3.2
### Activity 5-3

For **Case #9** The bleaching plant, do a *Why? Why? Why?* analysis.

**Case #9:** The bleaching plant.
Our process makes margarine. One step in the process is to remove the odors and color bodies from the edible oils used to make margarine. This removal is done in the bleacher illustrated in Figure 5-5. The light volatiles are removed by subjecting the oil to high vacuum. The color is removed by adsorbing the color species on powdered adsorbent that is subsequently filtered from the oil. The bleacher is shown in the diagram. The procedure, as set out in the startup manual, is to charge the vessel with edible oil, isolate the bleaching vessel (by turning the appropriate valves), draw a vacuum by means of the steam ejector system until the absolute pressure is 20 kPa abs on the vessel, open the valve connecting the bleach vessel to the hopper filled with the adsorbing powder, Fuller's earth, and suck (or pneumatically convey) the powder into the bleach vessel. The physical arrangement, including the approximate elevation, is shown in the diagram. This is a batch process.

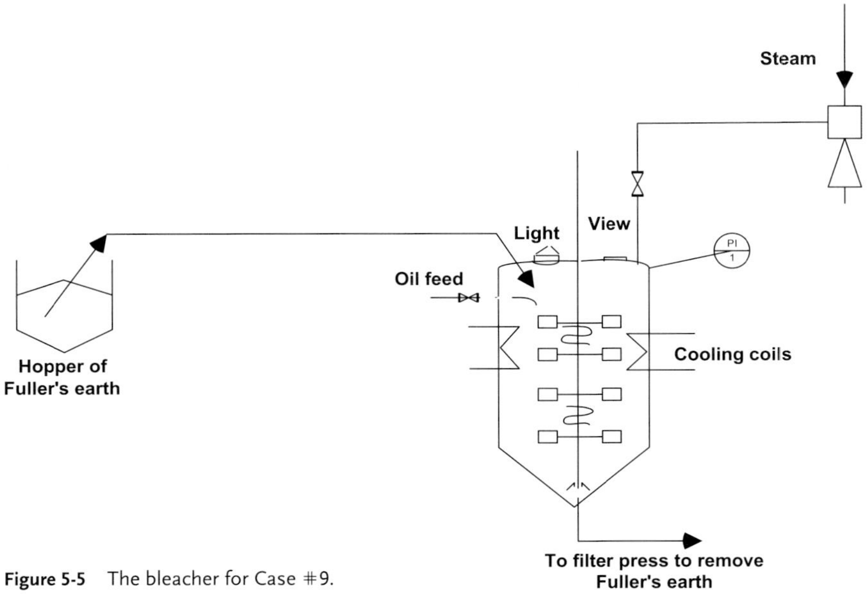


**Figure 5-5** The bleacher for Case #9.

The gauge on the vessel read 13 kPa abs, but when we opened the valve to convey the powder into the bleacher, nothing happened. That is, we expected to see, through the view port, the powder dumping into the liquid in the bleacher. This is the first time this plant has ever started up. The product has already been sold and because of previous delays in startup we are now losing $15 000 per day.

____________________

*Why?* ↑

____________________

*Why?* ↑

____________________

*Why?* ↑

____________________

*Why?* ↑

____________________

*Why?* ↑

Start → ____________________

**Reflect** on this approach. Was it easy to do? What insights did you gather? What might you address as the *real* problem? When might you use this approach?

## 5.4 Creativity

Whether solving exercises or problems we need creativity to generate different points of view (for the creation of a rich internal representation), to create hypotheses about change and to create hypotheses based on the basics.

### 5.4.1 Some Target Skills

Skilled creative thinkers:

1. Defer judgment;
2. Are succinct;
3. Can list 50 ideas in 5 minutes;
4. Create a risk-free environment;
5. Encourage free and forced association of ideas;
6. Can *piggy back* on previous ideas;
7. Use triggers, such as those listed in Table 5-1, to maintain the flow of ideas;
8. Aren't discouraged. In the last two minutes of a five-minute brainstorming session, over 85% of the ideas are not practical. But, they spend time to identify the treasures among the 15%;
9. Are positive;
10. Manage stress well. Manage any negative *self-talk*;
11. Use impractical and ridiculous ideas as "stepping stones" to innovative, practical options.

*After* they have generated a large list of ideas, skilled creative thinkers then list measurable criteria and select at least five technically feasible hypotheses.

**Table 5-1** Checklist of triggers for brainstorming.

| Name of trigger to change point of view | | Elaboration of how it is to be used | Comments |
|---|---|---|---|
| To be used for objects<br>"How to improve ion-exchange resin" | Function | What is the function of this object? How else can we achieve this function? What function *cannot* it do? | Objects for engineers are "products" and hardware. |
| | Physical uses | What are its physical properties and characteristics? What are they *not*? How else might we obtain or use these physical properties? | Probably the most useful perspective; use first. The negative view is often extremely illuminating. |

**Table 5-1** Continued.

| Name of trigger to change point of view | | Elaboration of how it is to be used | Comments |
|---|---|---|---|
| | Chemical uses | What are its chemical properties? How can these be exploited? What are they *not*? | Easy for us to use; use second. |
| | Personal uses | What personal uses can we make of this object? | These three give a unique perspective that is often overlooked. |
| | Interpersonal uses | What interpersonal uses can be made? | |
| | Aesthetic uses | How can we create pictures? music? artistic creations with the object? sculpt? weave? | |
| | Mathematical or symbolic properties | What are the mathematical or symbolic uses? What can they not be used for? What are they often confused with? | Only applicable for certain objects or ideas. |
| Situations<br><br>"I need to design an ion-exchange unit"<br>"The reformer is not functioning; fix it." | Checklist | Of the many checklists published SCAMPER is probably the most effective:<br>**S**: substitute who? What? Other processes? Other places?<br>**C**: combine purposes? Ideas? Appeals? Uses?<br>**A**: adapt what else is like this, new ways?<br>**M**: modify, maximize, minimize?<br>**P**: put to other uses, other locations<br>**E**: eliminate,<br>**R**: reverse, rearrange | Effective for objects and some situations. Try each viewpoint. Easy to do. |

**Table 5-1** Continued.

| Name of trigger to change point of view | | Elaboration of how it is to be used | Comments |
|---|---|---|---|
| | Wildest fantasy | Think of the craziest idea or use. | Need to establish self-confidence and group confidence before really outlandish ideas are presented. Build confidence in the merit of this approach by later bridging to unique ideas. Brings laughter and tension relief. |
| | How nature does it | Identify the situation and then think of how nature fulfills this function. Bridge to engineering reality. | Great potential; depends on the situation. Successful in designing new bridges; new bog machines. |
| | What if? In the extremes | Extrapolate from the unfamiliar to a simplified version | Extension of problem-solving skill in defining simple problems. Relatively easy to apply. |
| | Boundary exploration | Identify the constraints and remove them | Much easier for some to do than for others. |
| | Functional analogy | How else is the function achieved? | |
| | Appearance analogy | How else might we get the appearance of this situation? | |
| | Morphology | Break the problem situation into a series of parts. For each part list 10+ options. Then, systematically combine one option from each part and ask why not? | More mechanical; easy to computerize; most of the work is in setting up the parts and the options. Fun and surprising to see some of the results. |
| | Symbolic replacement | Replace the original problem by an interesting idea generated in the session; refocus. | Occurs naturally in many brainstorming sessions. Add this trigger explicitly if needed. |
| | Juxtaposition | Bring in three completely random words and bridge to current situation. Example "refrigerator, light switch, clock." | Very effective. Don't fake it by using words "you think will help". Use random words. |

**Table 5-1** Continued.

| Name of trigger to change point of view | | Elaboration of how it is to be used | Comments |
|---|---|---|---|
| | Personal analogy | Imagine yourself-as part of the situation. Describe your feelings and what you are experiencing. Be imaginative. | Tricky. Works for some people; not for others. Effective with fluid-dynamic problems. |
| | Reversal | Do the reverse. | More challenging than one thinks. Focus systematically on the reverse of different elements of the situation at a time. |
| | Book title | Create a title for a "best-selling" novel. The title should sum up the current situation. "Will exchanging bring happiness to Mary Lou?" | Interesting. Worth a try for 1 minute. |
| | Letter, word, sentence. | Focus on three different levels of detail corresponding to the big view (a sentence), an intermediate view (a word in the sentence), and a letter (in a word). In ion exchange this might be "the ions, the resin, the packed bed, the separation". | Very effective for most of our problems. Consider a 5-minute brainstorm at each level. |
| | Visual image | Look at three or four famous paintings. Describe aloud what you see and make connections. | At first glance this sounds too bizarre. However, for some this is one of the more effective ways of seeing the situation from a new perspective. Difficult to do out on the plant. |

### 5.4.2
### **Example: Case #10:** To dry or not to dry! (based on Krishnaswamy and Parker, 1984)

Our product is dry crystals of calcium nitrate. The crystals are precipitated in a continuous crystallizer. The exit flow from the crystallizer – consisting of a slurry of crystals and mother liquor – goes to a fixed, parabolic screen (DSM) to separate the large, product crystals from the undersized crystals that are recycled. The undersized are pumped continuously through a hydrocyclone and returned to the continuous crystallizer. As screening progresses the screen gradually *blinds* because the finer material that gets stuck in the screen cloth causes too many fines to be carried over with the large product crystals. Hence the screen is operated batchwise. The feed to the screen is stopped; the screen is washed, and then the screen is put back in service.

The large crystals from the screen go to a scraper-discharge centrifuge. The operation of the centrifuge is continuous but the moist crystals from the centrifuge are discharged batchwise because the centrifuge goes through a filter-wash-peel/discharge cycle. The exit from the centrifuge goes directly to a continuous, steam-tube rotary dryer via a chute and a conveyor screw. In the past, the crystal washing in the centrifuge was not very efficient. They modified the cycle to provide more wash water in the washing cycle of the centrifuge. The system is shown in Figure 5-6.

"The crystal product from the rotary dryer has a moisture content of 4.5% whereas the design value is 1.5%. Cake seems to be building up in the dryer feed chute, in the feed screw and on the steam tubes at the feed end of the rotary dryer." Fix the problem. Get the dryer working again so that the exit crystal moisture content is always below 1.5%.

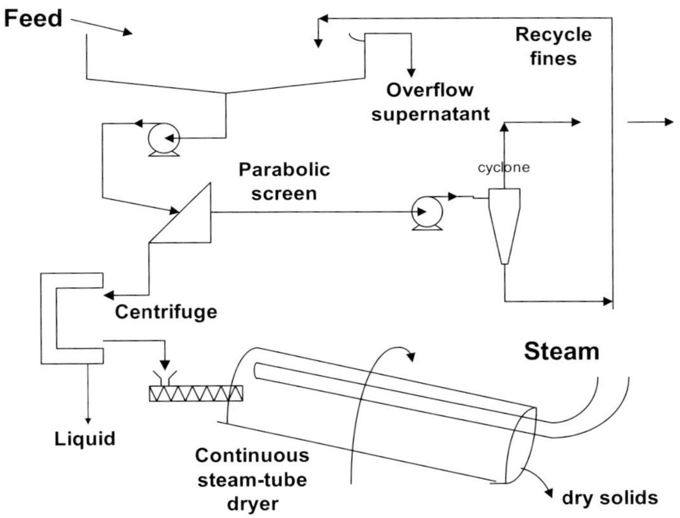


**Figure 5-6** The system used to dry the crystals; Case #10.

*Example results of brainstorming:*
Too much water in wash cycle for dryer to handle, cake buildup on steam tubes decreases heat transfer, cake dries up too early causing buildup while the rest of the cake is very wet, not enough steam supplied, wrong centrifuge, wash water carryover from the centrifuge, steam leak into dryer, it's raining, cycle wrong in centrifuge, cycle from screen not coordinated with cycle from centrifuge, crystals too wet from the screen, washing from the screen contaminates the feed to the centrifuge, feed

from the screen too wet, fines carryover to the centrifuge causing blinding in centrifuge, peel cycle too short, peel cycle too long, filter cycle too short, filter cycle too long, cycles now out of synchronization between the screening and filtering, feeder from screen to centrifuge broken, feed screw to dryer rotating too fast, rotating too slow, the rotational speed of the dryer has increased, the residence time in the dryer has decreased, the steam pressure has changed, the quality of steam from the boiler house has changed, the condensate trapping on the dryer is backing condensate up and reducing the heating area, air is blanketing the heat-transfer area in dryer because air was not bled off before restarting dryer, pitch of the dryer has changed, operating procedure has changed, pump from the screen bottoms malfunctioning and water spilling back into feed to centrifuge, cycle on the parabolic screen has been increased, more fines into peeler centrifuge causing poor filter cycle.

*Try trigger "craziest"*: operators not reading instruction correctly, sandwich dropped in line and plugging flow of crystals, wash down water used to clean the floors is splashing onto feed screw, the dryer is turning backwards, the peeler cycle is reversed to peel-wash-filter, the feed to the centrifuge is going directly to the dryer, feed from the crystallizer is going directly to the dryer, for this phase of the moon the vampires are out *(is that crazy enough?)*.

*Try trigger "reversal"*: focus on *the product from the dryer is too dry*: possible reasons are: long residence time, higher steam pressure/temperature, excess drying area, feed to the dryer from the centrifuge is dryer than expected, centrifuge rpm faster than usual, washing of the cake in the centrifuge is done with new solvent that evaporates faster, feed to the centrifuge from the screen is dryer than usual. OK, for the current problem take the opposite of all these.

*Try trigger "juxtaposition" using random words elastic bands, stapler and coffee cup.* The process I will use is *to take characteristics of this idea and find links to possible feasible ideas: I'll show that as arrows → and keep going until I get an idea that I think is feasible.*

Elastic bands: stretches → returns to original shape → screen becomes compressed and later pops back

Elastic bands → made of rubber → gaskets leaking?

Elastic bands → long and narrow → crystal size changes

Stapler: batchwise operation → dryer should be run batchwise too

Stapler: runs out of staples, need to refill periodically → centrifuge operates but has run out of feed

Coffee cup: holds liquid → liquid condenses near feed end of dryer and drops onto feed; recycle of wet damp gas in dryer.

Coffee cup: handle → holds → liquid retained in crystals by interstitial forces

Coffee cup: handles hot liquids → wash water is too hot; hot condensate clogs dryer tube

Coffee cup: status symbol with logos embossed on it → words → instructions wrong.

Coffee cup with metal: can't put in the microwave → contamination from rust, plugging lines

Coffee: brown liquid → dirty wash water

Coffee: tasty → micro-organisms growing in centrifuge, in wash water, in crystallizer

Coffee: justification for breaks → operators don't pay attention when they should; breaks in the cycles between the screen and the centrifuge.

Coffee: social event → water clings to crystal, not water anymore

*Select the craziest as a "stepping stone" or "piggy back" to feasible idea.":* Crazy idea: "for this phase of the moon the vampires are out" *The process I will use is to take characteristics of this idea and find links to possible feasible ideas: I'll show that as arrows → and keep going until I get an idea that I think is feasible.*

Phase → sinusoidal motion → cycles → periodicity → the cycle of the screen interferes with the cycle of the centrifuge → periodic pulses of wet stuff to dryer.

Moon → night → cool at night → is there a temperature effect on the feed? the feed to the dryer is too cold; is the steam too cold? is the wash water in the centrifuge too cold?

Moon → pale light in the dark → hard to see → difficult to see the moisture content in the feed because it changes so fast → things are happening on this plant that are hard to see → slow cycles and feedrate down to each section of the plant in turn and try to "see better"

Moon → changes shape from crescent to circle → the crystals change shape so filtering is different

Moon → changes shape from crescent to circle → crescent → curve → the curve of the parabolic screen is too steep → decrease in capacity

Moon → changes shape from crescent to circle → changes → same ideas as listed above for Phase

Moon → land on the moon → long distance away → something a long distance away from the dryer is causing problems → utilities? steam? wash water?

Moon → land on the moon → long distance away → something a long distance away from the dryer is causing problems → feed to crystallizer? pregnant liquor? crystallizer? screen? hydrocyclone? pump to hydrocyclone? screw feeder to dryer? centrifuge?

Moon → land on the moon → long distance away → something a long distance away from the dryer is causing problems → head office giving new policies and stressing out employees → spouses stressing out the operators? → events stressing out the operators → operators making mistakes

Moon → land on the moon → long distance away → something a long distance away from the dryer is causing problems → vendors supplied faulty equipment? lousy wash system?

Moon → land on the moon → different environment on the moon, lack of oxygen → oxygen in wash water affecting crystals? contaminant in wash water causing poor washing? is it the washing cycle? of the drying cycle in the centrifuge?

Vampire → person → operators error?

Vampire → person → operators → operating instruction error?

Vampire → teeth → sharp objects breaking up crystal size → slower filtering and slower drainage rate in centrifuge.

Vampire → drinks blood → blood → operator injured? water is not water, it's contaminated with low vapor pressure.

**Comment:** This session was characterized by:

- mix of symptoms and roots causes;
- lots of duplication;
- most feasible ideas in the first 32 ideas;
- a lot of impractical ideas, especially later in the session but there were some new feasible ideas amongst these.

Total ideas: 98 with 32 from the initial burst, trigger "craziest" = 8; trigger "reversal" = 8; juxtaposition= 20; stepping stone= 30. The stepping stone seemed to work well for me on this problem.

From this are selected the following possible hypotheses:

1. not enough steam,
2. wash water carryover from the centrifuge,
3. cycle from screen not coordinated with cycle from centrifuge,
4. feed crystals too wet from the screen,
5. rotational speed to the dryer has increased,
6. more fines into peeler centrifuge causing filter cycle to be too long; fines carryover to the centrifuge causing blinding in centrifuge,
7. centrifuge rpm faster than usual,
8. crystal size change; crystals change shape so filtering is different,
9. centrifuge operates but has run out of feed,
10. dryer feed too cold,
11. vendor supplied faulty equipment.
    - slow down the cycles and feedrate down to each section of the plant, in turn, and try to see better what is happening.
    - "slower filtering and slower drainage rate in centrifuge" symptom! use as potential brainstorming idea.
    - "periodic pulses of wet stuff to dryer" symptom! not a root cause but might be profitable as symptom.

5.4.3
### Activity 5-4

For **Case #9**, the bleacher (described in Section 5.3) brainstorm 50 possible causes and write these down. Suggested time frame: initial write 6 min.

- Say to yourself-"I can do this." reread the problem statement: 2 min.
- Try trigger "wildest fantasy" 1 minute
- Try trigger "what if? in the extremes. 1 minute
- Try trigger "juxtaposition" using the words *pencil, hat,* and *perfume.* 1 minute each word.
- Try trigger "reversal" 1 minute.

Stretch.

Look over your list and select the "craziest one". Use this idea as a "stepping stone" to a technically feasible one. By this we mean, take the properties of the crazy idea and use these in a feasible idea.

Reflect on what happened. Complete the feedback form given in Figure 5-7.

### 5.4.4 Feedback, Self-Assessment

Self-assessment is based on written evidence, not on intuitive feelings. Throughout the brainstorming evidence has been gathered. The evidence includes the number of ideas produced and the succinctness of the ideas.

In addition we can write reflections, using a form similar to Worksheet 5-1. Figure 5-7 gives a convenient summary form specific for brainstorming.

Number of Ideas generated ________ Activity or problem: ________

**In the session:** rate the following out of 10:

| | 0 | 2 | 4 | 6 | 8 | 10 |
|---|---|---|---|---|---|---|
| How much elaboration? | Extensive | | | | | None, all succinct ideas |
| How much criticism? | Extensive | | | | | None |
| How many silences? | Extensive | | | | | Few |

**Triggers used?** No ☐ Yes ☐ give name(s)

craziest boundary personal function juxtaposition checklist morph
book title replacement reverse senses fishbone wave other

Strengths ________

Areas to work on ________

from D.R. Woods, "How to Gain the Most from PBL", 1994

**Figure 5-7** Feedback form for brainstorming.

## 5.5 Self-Assessment

Your skills and confidence as a trouble shooter will improve the most if you create a formal approach to self-assess your current skill, set goals and gather evidence about your progress to achieving the goals. Skilled self-assessment develops objective awareness of how you perform a task and develops self-confidence. In this section we define the target skills for self-assessment, in Section 5.5.1, and then propose activities for growth in Section 5.5.2.

### 5.5.1 Some Target Skills

Skilled assessors realize that assessment is:

1. about performance; it is not about personal worth.
2. based on evidence; it is not based on wishful thinking or gut feelings.
3. essential for growth.
4. not possible without published unambiguous goals and measurable criteria.
5. based on a wide variety of different types and forms of evidence.

### 5.5.2 Activity for Growth in Self-Assessment

Attitude and skill – these are the key elements to work on.

Assessment involves a change in *attitude*. Many think of exams, performance reviews and evaluations as being stressful and something to be avoided. We need to realize that self-assessment helps develop growth and is one of the most positive activities we could do to improve trouble-shooting skills. Assessment is about performance! Assessment is based on evidence and is a judgment done in the context of published goals and with measurable criteria.

The three *skills* needed relate to creating goals, measurable criteria and designing forms of evidence

a) Ability to identify and create observable and unambiguous goals. Examples have been given in each section. In general unambiguous means that we can *observe* the performance. Such words as “know”, “create” are unacceptable because we cannot observe someone demonstrating that they “know this”. Words like “list”, “write out” are an improvement.
b) Ability to identify and create measurable criteria related to the goals. This task is difficult, boring and tedious – but necessary. For example, in Section 5.4 the goal was to list ideas. To make that measurable we need a number and a time: 50 ideas in 5 minutes.
c) Ability to write out, gather and evaluate evidence as it relates to the goals. As demonstrated previously in this book, evidence in the form of reflections, the worksheets and feedback forms are all useful. Usually, once the goals and criteria are created, it is relatively easy to create the forms of evidence.

**Activity 5-5** Self-assessment

Based on the self-assessment in Section 1.3 create goals for growth, measurable criteria and pertinent forms of evidence. Begin gathering a collection of evidence and systematically go over the evidence and self-assess your progress.

### 5.5.3 Feedback About Assessment

Figure 5-8 provides a feedback form related to the assessment process.

**Goals**: Content is well identified, goals are challenging and achievable, goals are written in observable terms, goals are unambiguous, the "given" conditions are specified.

| None of these behaviours | | Few of these behaviours but major omissions | | Most features demonstrated | | All of these behaviours |
|---|---|---|---|---|---|---|
| ☐ | ☐ | ☐ | ☐ | ☐ | ☐ | ☐ |
| 1 | 2 | 3 | 4 | 5 | 6 | 7 |

**Criteria**: Criteria are consistent with the goals and are measurable and practical. The criteria are challenging and achievable.

| None of these behaviours | | Few of these behaviours but major omissions | | Most features demonstrated | | All of these behaviours |
|---|---|---|---|---|---|---|
| ☐ | ☐ | ☐ | ☐ | ☐ | ☐ | ☐ |
| 1 | 2 | 3 | 4 | 5 | 6 | 7 |

**Evidence:** The type and quality of evidence gathered is consistent with the goals and criteria. The evidence has been gathered conscientiously over a long enough period of time. The evidence is well organized. The quality and extent of evidence is sufficient to allow me to judge the extent to which the goals have been achieved.

| None of these behaviours | | Few of these behaviours but major omissions | | Most features demonstrated | | All of these behaviours |
|---|---|---|---|---|---|---|
| ☐ | ☐ | ☐ | ☐ | ☐ | ☐ | ☐ |
| 1 | 2 | 3 | 4 | 5 | 6 | 7 |

**Process:** The assessment process has been applied and as an independent assessor I concur with the decision as to the degree to which the goals have been achieved.

| None of these behaviours | | Few of these behaviours but major omissions | | Most features demonstrated | | All of these behaviours |
|---|---|---|---|---|---|---|
| ☐ | ☐ | ☐ | ☐ | ☐ | ☐ | ☐ |
| 1 | 2 | 3 | 4 | 5 | 6 | 7 |

| Strengths | Areas to work on |
|---|---|
| ______________________ | ______________________ |
| ______________________ | ______________________ |
| ______________________ | |
| ______________________ | |
| ______________________ | |

from D.R. Woods, "How to Gain the Most from Problem-based Learning" (1994)

**Figure 5-8** Feedback about the assessment process.

## 5.6
## Summary and Self-Rating

In this chapter we identified five skills related to problem solving. These were awareness or the ability to describe your problem-solving processes; strategies or the ability to see patterns in the process; exploring the context using the *Why? Why? Why?* process; being creative and being skilled in self-assessment. For each target skills were listed, some examples were provided, activities to develop the skills were described and forms of evidence were given.

Reflect on what you have experienced in this chapter:

______________________________________________________________

______________________________________________________________

______________________________________________________________

______________________________________________________________

| Rate | I think I am skilled | I have some evidence and have some skill | I am confident. I know the goals the criteria and have a variety of evidence | Not sure this is for me |
|---|---|---|---|---|
| Awareness: can describe, focus on accuracy, am active | ☐ | ☐ | ☐ | ☐ |
| Strategies: see patterns, monitor | ☐ | ☐ | ☐ | ☐ |
| Explore the context using Why? Why? Why? | ☐ | ☐ | ☐ | ☐ |
| Am creative: 50 in 5 min, use triggers, | ☐ | ☐ | ☐ | ☐ |
| can step from crazy ideas to feasible ones | ☐ | ☐ | ☐ | ☐ |
| Self-assess: | | | | |
| attitude about performance; | ☐ | ☐ | ☐ | ☐ |
| needed for growth, evidence-based skill: create observable goals; | ☐ | ☐ | ☐ | ☐ |
| create measurable criteria; | ☐ | ☐ | ☐ | ☐ |
| identify pertinent evidence; | ☐ | ☐ | ☐ | ☐ |
| valid judgment in this context; | ☐ | ☐ | ☐ | ☐ |

# 6
# Polishing Your Skills: Gathering Data and the Critical-Thinking Process

Critical-thinking skills are needed by trouble shooters. These skills include selecting and designing tests, checking for consistency, classifying sets of ideas or data, recognizing patterns and reasoning and drawing valid conclusions.

Here are some examples of where some of these critical-thinking skills are used in the task of trouble shooting. We:

- See/receive information that suggests something it wrong: compare conditions with unsafe conditions and decide from among a) emergency shutdown, b) change to "safe-park" or c) continue current conditions. Select the symptoms. Separate *fact* from *opinion*. The thinking skills needed are consistency and classification.
- Gather more information about the current situation and compare and contrast this with the "expected" performance to obtain a rich mental image of the situation. The thinking skill needed is how to compare and identify differences.
- See if there are any patterns in the data. The thinking skill needed is pattern recognition.
- Brainstorm possible faults or causes that could cause the symptoms. Alter those ideas that are "symptoms" to be "root causes". Classify the list and apply criteria to prioritize the list of hypotheses/causes that are consistent with practical experience. The thinking skills needed are classification and consistency.
- Select hypotheses/faults that are consistent with the symptoms. The thinking skills are using cause–effect information and classification.
- Create tests; the tests must be pertinent to the hypothesis, and not suffer from confirmation bias. The thinking skill needed is how to select valid diagnostic actions.
- See if there are any patterns or trends in the evidence collected. The thinking skill needed is pattern recognition.
- Check that the results of the tests confirm and disprove the hypothesis. The thinking skill needed is reasoning.

In this chapter we consider first, in Section 6.1, how to select valid diagnostic actions. In Sections 6.2 to 6.5, we look at the critical-thinking skills of checking for

*Successful Trouble Shooting for Process Engineers.* Don Woods

ISBN: 3-527-31163-7

consistency, classification, seeing patterns and reasoning. Each is considered in turn and illustrated through examples.

## 6.1 Thinking Skills: How to Select Valid Diagnostic Actions

First, we offer criteria for selecting diagnostic actions; then we provide a structured list of possible actions. Then we consider how to perform and interpret the actions. In Section 6.1.3 we explore how to become sensitive to personal preferences, style and biases that might interfere with the selection and use of diagnostic actions. Consider each in turn

### 6.1.1 How to Select a Diagnostic Action

In selecting an action to take, the following criteria are useful:

- Safety first!
- Keep it simple. Pertinent, easy-to-gather information should be gathered first.
- The results should provide the accuracy needed.
- Safety and time are critical.
- Select actions that will prove or disprove hypotheses.
- There is an expense associated with any action or lack of action.
- Stopping production to "inspect" or "change equipment" is usually very costly.

### 6.1.2 Select from among a Range of Diagnostic Actions

The actions taken differ widely from immediate emergency response, to gathering information to better understand the problem, to testing hypotheses, to immediate correction of the suspected cause. In general, the actions selected depend on the type of TS problem, the TS strategy you elect to follow and your personal style. A structured list of options includes: a) is it an emergency? b) put the situation in the context of recent events, c) understand what should be happening, d) check what is happening now, e) test hypotheses and, perhaps, f) take immediate corrective action. Each is considered in turn.

**a. Safety first! Is it an emergency? What's the general background information about this situation?**

- Invoke emergency shutdown? Although trouble shooters should know the hazards related to the process, information about hazard, safety, MSDS sheets can be obtained from Chapter 3, Section 3.12, the MSDS sheets on file or from two web sites for MSDS data http://www.msdsxchange.com or http://www.msdssearch.net.

- Is "safe-park" a required option?
- Tabulate IS and IS NOT data.
- Note details about the weather and offsite services: air temperature and humidity, rain-snow, ice-fog? bay water temperature, river temperature, city water properties/temperature, tower water, steam and air plant; wastewater treatment plant.

*Comment:* Start here and put the process into perspective.

**b. Should I be using a TS strategy best suited for "change"? What has happened recently that might affect the process?**

What and when did changes occur:

- changes made to operating instructions, to feed, to specifications, to flow-rates,
- work done on the plant,
- what was done during the maintenance turnaround,
- what was done in routine maintenance.

*Comment:* Easy to do. You may know this already but check out the details of exactly what was done. This provides the basis for selecting your TS strategy and subsequent diagnostic actions.

**c. What should be happening: expected, "usual" performance?**

This information is usually obtained in the office from files and records, from simple calculations and from simple, astute *What if?* exploration.

*Files and records:* consult key files and records and predictions of performance:

- design files and simulation: design basis, fouling factors, assumptions made, specifications and predicted conditions throughout the plant.
- vendor files for the equipment: performance expected, sometimes trouble-shooting information* (* often available in both the engineering office and the control room).
- operating procedures*.
- commissioning data.
- recent P&ID. Often not available, or if available, not up-to-date.
- recent tests and internal reports.

Data from handbooks and texts in your office.

Trouble-shooting files: data relating symptom to root cause with the probabilities or likelihood, similar to that given in Chapter 3.

*Calculations or estimations that can be done before special tests are done.*

- calculated/estimated pressure profiles: these can be done from simulations or from order-of-magnitude estimates. Note the pressure difference across barriers to determine the direction of flow if a leak occurs.
- mass balance: predicted by simulation or estimated.
- energy balance: heat in = heat out across exchangers, furnaces, condensers, reboilers. Steam usage with 5 kg organic evaporated by1 kg steam, flame temperature,% excess air.
- thermodynamics: equilibrium conversion, vapor-liquid equilibrium, energy changes for flow across turbines, ejectors and valves.
- rate: use the temperature difference in the reboiler to estimate whether boiling is in the nucleate, film or transition regime.

Equipment performance calculations: check that the sizing and general operability conditions are met. Will the installed equipment do the job expected?

For process control, check the location and type of sensor, type of control used.

*Exploration from simple "What if?"* questions.

Sometimes we can quickly eliminate extraneous tests by asking such questions as "*What if* you could actually look at the flow patterns inside the vessel?" "*What if* you could look at the catalyst surface?" "*What if* there was no insulation? "*What if* there was infinite insulation?"

*Comment:* This should be a valuable, third level of data acquisition. However, the usefulness of these actions depends on the quality of the documentation available (often it is poor), and of your files, on your skill with order-of-magnitude estimates; and on the information available in the case-problem statement.

**d. What is currently happening?**

We start with information in the control room and from the operators.

In the control room, here are some options:

- read the displayed information and check past records of same variables. Look for trends. Look for interconnects (two or more events that seem to be happening together: temperature increase and conversion decrease).
- check the operating procedures*.
- scan the vendor files including performance data and trouble shooting*.
- talk to the operators about changes, about this shift versus previous shifts, about their hypotheses, about the facts.

*Comment:* This is a vital source. Meet the operators; find out the details; read the instruments displayed. This is a *must* visit before venturing out onto the plant.

**e. What is currently happening and gathering information to test hypotheses?**

We learn more about what is currently happening i) by putting the process in context; ii) by using our senses to gather information from out on the plant, iii) by gathering data for calculations, iv) by checking out the sensors and controllers, v) from quick checks with specialists, vi) from more ambitious tests and vii) shut-down activities.

i) **Put the process in context.** Check with the operators of the utilities and of upstream and downstream plants.
ii) **On the plant visit.** Out on the plant we use our senses, do simple tests, check for consistency, and look for trends.

*Use your senses.* Read the instruments, listen, smell, look. Note the position of the valve stem, look for steam leaks, look at the discharge from steam traps, read the motor amps, check the condition of the insulation, look for rust around a flange, leaks around a gland, look at flows (if they are visible). Hot smells? Listen for noise: indicating flow, cavitation or liquid level in vessels. Use your intuitive feelings.

*Simple on-site tests:* Shift to a more quantitative perspective: estimate/measure:

- the temperature: use the glove test or a laser or surface pyrometer.
- the humidity or dew point.
- the response of sensors: check the response to a change in set point.
- the signals to controllers/valves: inlet pressure/signal; pressure/signal to the valve.
- diameter of insulation/ of pipes.
- the liquid flows in accessible drains.
- the exit pressure for zero flow for centrifugal pumps.
- if the bypass valve is open or shut.
- condition of the valves: turn and seal.

*Comment:* Your senses and simple on-site tests are easy to use and often provide excellent, key information.

*Check for consistency:* do multiple sensors agree? are the lab data consistent with the results from on-stream analyzers? temperature–pressure–composition agreement? phase-rule agreement? mass-balance agreement? make-sense agreement (fluids flow from high to low pressure? thermal energy flows from high temperature to low?) physical-thermal data agree with measurements (pressure enthalpy data for refrigerant)?

*Check for trends:* have the temperatures, pressures or yields been changing gradually over the weeks? Is there a trend every 5 minutes? every 15 minutes? every hour? What is the frequency and amplitude of cycling?

*Check that the P&ID agrees with the actual process configuration.*

iii) **Gather data for calculations.** Fundamentals underpin the process. Check a mass and energy balance. Gather data to estimate performance of equipment.
iv) **Sensors and controllers.** Put the controller on manual. Is there evidence that the sensors are working and are accurate? Use temporary instruments to check measurements. Request specialists to calibrate the sensors or tune the controllers.

v) **Quick checks with specialists.** *Consultants:* call vendors, original designer, consultants, licensee of process and/or suppliers of raw materials, adsorbents, catalysts.

*Comment:* the operative word is "quick". Although it's better to talk by phone than e-mail, the key people may not be available to answer your call.

vi) **More ambitious tests.** *Samples of gas, liquids or solids for analyses:* obtain the samples: Are the samples representative? Valid? Is there a time dependency? Are all the samples correctly labelled? Can the sampling be done safely? Type of analysis: often chemical or particle-size analysis.

*Specialists:* Should a velocity profile be measured? gamma scan? tracer study?

Perhaps a well-designed set of experiments should be run and the results analyzed statistically.

*Comment:* although the process can continue to operate, these tests take time and may be expensive.

vii) **Shut-down-type activities.** *Open and inspect:* shut the process down, safely isolate the piece of equipment, and clear of the process fluids and vapors. Know what you are looking for, but be prepared for surprises.

*Comment:* this requires that the process be stopped. This is expensive. Try to leave this action as the last resort. Much can be learned from the previous activities.

#### f. Possible corrective action

Some trouble shooters, especially those who prefer *action* to *patiently gathering evidence* (dominant J behavior as described in Section 6.1.3c), take "corrective" action prematurely. Others might astutely have identified the cause early; others, are lucky. In general, be cautious in taking corrective action before simple tests of hypotheses can be made. Usually "*Take corrective action*" requires the expensive shutting down of the process and making alterations.

## 6.1.3
## More on Gathering and Interpreting Data

Here we consider the guidelines for designing experiments to test hypotheses, and list the resources needed for gathering data from simple tests. We also explore the implications of your personal style in planning and selecting tests.

### 6.1.3.1 Guidelines for Selecting and Designing Experiments to Test Hypotheses

The tests should be simple, inexpensive and should provide positive and negative validations of a hypothesis. Usually we try to:

- avoid the temptation to "open the equipment and see". Usually some very simple tests can keep the process going and provide the insight needed to identify the fault.
- check one hypothesis/variable at a time.

- isolate variables.
- test the "most likely" fault based on probabilities of past failures. Table 6-1, for example, lists some data related to instrument failure, relative to "equipment failure".

**Table 6-1** Failures per annum for instruments and equipment (from Woods, Process Design and Engineering Practice, originally published by Prentice Hall, 1995, © Donald R. Woods).

| Instruments | Failures | Equipment |
|---|---|---|
| Analyzer: GLC | 20.9 failures/annum | |
| Analyzer: $CO_2$ | 10.5 | |
| Analyzer $O_2$ | 7 | |
| Analyzer: *general* | 6.2 | |
| Analyzer: pH | 4.3 | |
| Flow: $\Delta p$ transducer | 1.9 | |
| Liquid level $\Delta p$ transducer | 1.8 | |
| Liquid level float type transducer | 1.6 | |
| Liquid level *general* | 1.6 | |
| Flow: *general* | 1.1 | |
| Purge system | 1 | |
| Pressure: *general* | 1 | |
| Temperature: transducer | 0.9 | |
| | 0.9 | turbine |
| | 0.6 | centrifugal pump |
| Temperature sensor | 0.3 | |
| | 0.15 | distillation column, reactor |
| | 0.1 | exchanger |

The tests should be based on a good understanding of the equipment and the system.

*Example 6-1:*
For cavitation of a centrifugal pump, we could a) listen for the crackling noise typical of cavitation, and b) reduce the flow through the pump by partially closing the line on the discharge line. This should cause crackling noise to subside. Fundamentally, at lower flowrates, less NSPH is required and the friction loss on the suction side is reduced by the $(\text{velocity})^2$.

**Activity 6-1:** Selecting tests
We hypothesize that there is an obstruction in a 10-m length of pipe, that has numerous bends and fittings, through which liquid is flowing. Andre suggests the following tests:

a. stop the operation, open the pipe at the various fittings and look.
b. increase the flowrate and note the difference in pressure drop.
c. decrease the flowrate and note the difference in pressure drop.

d. estimate the pressure drop and compare with measured values.
e. stop the operation, open one end and send a plumber's worm through the line.
f. and maybe there might be some others.

What would you select and why?

**Activity 6-2:** Selecting tests

In a refrigeration cycle we hypothesize that impurities got into the refrigerant. Comment on the appropriateness, limitations and assumptions of each of the following tests:

1) check the pressure difference between the refrigerant side and the process side.
2) sample the refrigerant and analyze for impurities using GLC.
3) read the temperatures and pressures and compare these with the data on pressure–enthalpy charts for the refrigerant.
4) read the pressure gauge on the compressor suction.

**Activity 6-3:** Case #6

For **Case #6**, Saadia completed the following chart on the Trouble-Shooter's Worksheet for the case of the utility dryer (details are given in Section 4.4). Complete the chart below by matching the actions to the hypothesis. Comment on the appropriateness of each action.

Symptom a. exit air "wet": 3 × higher than specifications
b. pressure drop double expected value

**Activity 6-3:** Hypothesis chart for worksheet.

| Working Hypotheses | Initial Evidence | | | | | Diagnostic Actions | | | |
|---|---|---|---|---|---|---|---|---|---|
| | a | b | c | d | e | A | B | C | D |
| 1. Steam leak | S | N | | | | | | | |
| 2. Excessive moisture carryover from the separator | S | N | | | | | | | |
| 3. Valve S2 leaking | S | N | | | | | | | |
| 4. Adsorbent lacking adsorption capacity. | S | N | | | | | | | |
| 5. Absorber on-line too long: breakthrough | S | | | | | | | | |
| 6. Not enough regeneration time | S | | | | | | | | |
| 7. Condenser not cooling sufficient | S | | | | | | | | |
| 8. Instruments wrong, pressure | N | S | | | | | | | |
| 9. Absorbent broken down | ? | S | | | | | | | |
| 10. Temperature TRC, T3 reads low | S | S | | | | | | | |

Diagnostic actions:

A. Test calibration of temperature T3
B. Test calibration of pressure gauges P1 and P4
C. Calculate a mass balance on moisture
D. Calculate the expected removal of adsorbent from adsorption/kg adsorbent data
E. Sample adsorbent to determine if damaged
F. Read pressure drops across different parts of the system and when different parts are "on-line"
G. Predict water removed via regeneration, temperatures and moisture in exit gas profiles and compare with data taken over several cycles
H. Vary the regeneration time allowed

#### 6.1.3.2 Resources for Gathering Data

Here we consider the types of equipment to take, and the time required to do some tests.

**a)** Simple equipment and stuff to take with you on-site

Here are some of the things I used to take with me on a trouble-shooting mission:

- notebook and calculator: keep accurate records.
- leather gloves: especially important to allow me to sense the temperature on either side of a steam trap.
- stethoscope: very useful to magnify the sounds inside a vessel, steam trap or valve. This helps to identify, for example, the typical flow through a thermodynamic steam trap; to listen for vibrations.
- string: in trying to estimate the diameter of a pipe or insulated pipe I found it much easier to measure the length of string around its circumference and then calculate the diameter.
- tape measure.
- clamp-on ammeter.
- flashlight.
- stopwatch.
- knife.
- pens and marker, tape, labels, sample bags.
- tachometer.
- magnifying glass.
- brick: to help me estimate a mass or energy balance I often needed to know the flow of water in the drains. On some of the plants these were easily accessible. A brick could be placed in the rectangular drain to create a dam or "weir" and then from a measure of the height of liquid above the "weir" the flow could be estimated.

- pail and stop watch: measure other types of flows.
- long needle: allowed me to estimate the thickness of insulation.
- spray bottle of soap solution and tape: to identify leaks around valve stems, spray soap solution. For flanges, tape around the flange and then spray on a single hole you make in the tape.
- a surface pyrometer or laser pyrometer may be a useful addition, especially if steam and steam traps are causing problems.
- camera.

In your journey as a trouble shooter, you too will collect your favorite collection of simple things to make it easy for you to uncover the secrets of the process.

**b)** Time it takes for gathering data

Here are some examples of the time it takes to gather certain types of data and to make alterations to the process.

- Laboratory analysis: routine analysis: 2 h; special test for impurities 8 h.
- Instruments: install orifice plate 10 h; rotameter, 5 h, level gauge 8 h.
- Agitator: hook up, 40 mh; blower, hook up 25 mh; pump, hook up 50 mh.

#### 6.1.3.3 **Personal biases and style in collecting evidence and reaching conclusions**

Trouble shooters create hypotheses and then select cues, evidence and tests to confirm or disprove the hypotheses. They check that the evidence/ data/ cues actually support their hypothesis. Sounds straightforward. However, most mistakes occur here.

Each person has a personal style. Furthermore, mistakes and biases affect how data are collected and conclusions reached. Consider each in turn.

**a)** Personal style in trading off data gathering versus taking action

Each of us has a preferred style of making decisions. Some prefer to be active, to make choices even though they might be wrong. Others, want to gather data and really understand the situation before action is taken. The **P-J** dimension of the Jungian typology, or Myers Briggs Type Indicator (MBTI) may provide you with insight as to your preferences. For example, a dominant **P** style is characterized by wanting to collect detailed information and data before making a decision; a person with a dominant **J** style wants action and may use an approach of changing a suspected cause before doing simple tests to check whether that really is the cause.

*Example 6-2:*

For the **Case #10**, to dry and not to dry, and described in Section 5.4, the hypotheses selected by Jason and Heather are 1) wash water carryover from the centrifuge, 2) cycle from screen not coordinated with cycle from centrifuge, 3) feed crystals too wet from the screen, 4) condensate trap on the dryer is malfunctioning causing the condensate to back up in the tubes and reduce the heat transfer area, 5) more fines into the centrifuge causing the filter cycle to be too long; fines carryover to the centrifuge causing blinding in centrifuge, and 6) crystal size change; crystals change shape so that filtering is different.

Jason, after consulting with the vendor of the dryer, wants to replace the steam trap on the dryer (since the vendor suggests that this is a likely cause and it's relatively easy to change). "I'm convinced it is hypothesis #4. Let's act." Heather, on the other hand, wants to sample the crystals going into and out of the centrifuge every 30 s for three cycles of operation. "The samples can be analyzed for water content and for particle-size distribution. This is relatively easy to do and this will help us test hypotheses #1, 3, 5 and 6. On the other hand, Jason, if you really feel strongly that it is the steam trap, let's go out and check out the trap first."

*Comment:*
Here we have apparent disagreement. Jason is showing predominant **J** behavior; Heather, predominant **P**. Provided Jason and Heather see their disagreement as "Hurrah! we balance out our different styles," then a better result will occur. If Jason and Heather had both been dominant **J**, then the steam trap would probably have been changed, only to discover that the steam trap was not the fault.

**b)** Common biases in collecting evidence

The four major types of error are pseudodiagnosticity, confirmation, availability and representative biases.

- *pseudodiagnosticity* or *overinterpretation*: actively seek worthless data and change opinion based on irrelevant data; treat noncontributory cues as relevant. This happens about 30% of the time and is the most common bias.
- *confirmation bias*: actively support a favored hypothesis even though all the evidence points elsewhere. Seek confirming information and ignore disconfirming cues.
- *availability bias*: prefer to use data that are more readily available.
- *representativeness bias*: see similarity that doesn't exist between two events. Even when given the correct underlying knowledge trouble shooters consistently disregard that knowledge in favor of stereotypes about how "representative" these particular data/characteristics are.

**c)** Common biases in reaching conclusions

These biases tend to relate to personal preference and the amount of training and experience.

Personal foibles include: premature closure and anchoring may occur despite experience and training.

- *premature closure*: the conclusion is not justified by existing data. Tend to be a Jungian typology "dominant J" (discussed in part i).
- *anchoring*: adhere to a preconceived belief even though the evidence refutes it.

Three other biases that occur usually with inexperienced trouble shooters who lack the training and experience with the process equipment are inadequate synthesis, underinterpretation and misinterpretation.

- *inadequate synthesis*: unjustified conclusions are drawn. The trouble shooter fails to use knowledge effectively in interpreting and making inferences from the data.
- *omission of clue or underinterpretation*: an important clue is ignored. This occurs about 2% of the time.
- *wrong synthesis or misinterpretation*: the available data contradict the conclusion. This occurs about 6% of the time.

**Activity 6-4:** Confirmation bias

Consider the task of testing the hypothesis that "Every card that has a vowel will have an even number on the other side". The four cards shown below in Figure 6-1 are available to test this hypothesis. (From Johnson-Laird and Watson, 1970). Which card or cards do you need to turn over in order to test the validity of the hypothesis?

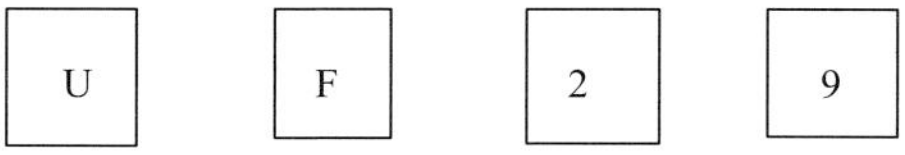


**Figure 6-1** Four cards.

**Activity 6-5:** Feedback about your style

To give yourself-some feedback about your style, consider the following case.

Assume that the information given in the scenario is factually correct. A code letter is given at the end of separate bits of evidence in the account of the murder, for example "Tom Dayton had many enemies (a)." where the code letter is (a).

**Worksheet 6-1:** The Tom Dayton murder (adapted from Sherlock Holmes).

Tom Dayton had many enemies. (a) He was a scalawag and a prankster who never passed up an opportunity to embarrass someone through a practical joke (b). It was Tom who invented the joy buzzer and the whoopee cushion, and some even credit him with having originated fake vomit.

It is well known that Tom's favorite target was his old headmaster, Stanley Bosworth, (c) at Bromley School. Stanley Bosworth was the victim of some of Tom's most elaborate pranks. Tom's eventual marriage to Bosworth's daughter, Melissa, was considered by many to be Tom's ultimate joke (d) on the respected headmaster.

Among the more prominent victims of Tom Dayton's past pranks were Judge Walter Brighton (e), Lord and Lady Morton (f) of Westchester, banker Mortimer Fawcett (g), Doctor Fabian Peerpoint (h), and tobacco merchant Dawes Flescher (i).

All of the above were present at the dinner party held on the Bosworth yacht in honor of Stanley Bosworth's 60th birthday (j).

Following an uneventful dinner, most of the guests retired to their staterooms to freshen up. The clock in the dining room struck 10 pm (k) when a shot rang out (l). Most later claimed they heard a second shot (m). All aboard the yacht, including the yacht's captain, Jonas Fenton, (n) and cook, the curvaceous Mildred Weekson (o), arrived at Tom Dayton's stateroom to find him dead – shot in the forehead (p).

A smoking revolver lay near the doorway (q); Tom's body lay on the floor across the room (r), just below an open porthole (s). Scrawled in the dust near Dayton's body were the initials SB (t). In the corner of the room was a suitcase filled with $500,000 (u). No bullets were found in the walls, ceiling or floor of Tom's room (v). Who killed Dayton?

During the investigation the following factual evidence was produced:

w. Although just about everyone claim they heard two shots, Tom Dayton had one bullet in his head.
x. Lady Morton was being blackmailed.
y. The clock in the dining room was 15 minutes slow.
z. The $500 000 in the suitcase was counterfeit and was accompanied by a withdrawal slip for $1 000 000 from Mortimer Fawcett's bank.
aa. Bosworth angrily stated at dinner that he felt Tom was mistreating his daughter.
bb. The actual murder weapon was found under water below the porthole.
cc. Dr Fabian Peerpoint advocates mercy killing.
dd. The smoking revolver in the room belonged to Stanley Bosworth.
ee. Tobacco merchant Dawes Flescher is a talented mountain climber.
ff. Jonas Fenton, the yacht's captain, went to school with Tom Dayton.
gg. Dr Fabian Peerpoint revealed that Tom Dayton was terminally ill with only a few months to live.
hh. The bullet in Tom's head did not come from the smoking revolver on the floor in Tom's room.
ii. Melissa says that she was visiting her father in his stateroom from 9:45 pm until she heard the shot.
jj. Mildred Weekson was having a secret affair with Tom Dayton for the past two years; and with Lord Morton.
kk. The shot that killed Tom Dayton was fired from outside the porthole. Directly below the porthole is water.

Based on the evidence so far would you:

1. accuse ____________ of murder based on evidence (list the letters of the evidence supporting your conclusion) ____________________________.
2. accuse Dr. Peerpoint of mercy killing based on evidence (list the letters) ____________________________________________________________

3. conclude that Tom died from ________ based on evidence.
   (list the letters) ________________________

or… .

4. require that the following information is needed before any conclusion can be drawn:
   ll) what pranks had Tom played on ____________
   mm) check for poison in Tom's body
   nn) other ________________________

Feedback about your style is given in Appendix G.

**d.** Cautions about interpreting data

All data have errors. We should have an idea of

a. What is an acceptable error in all the target or specification conditions.
b. What is an acceptable error for all the measurements.
c. When is there a significant enough deviation that we recognize that something is wrong.

The evidence may be relayed to us as fact, opinion or opinionated fact. Actually it is difficult for people to relate just the facts; they want to infer and give their own interpretation to the information. As trouble shooters we need to separate carefully *fact* from *opinion*. We need to concentrate on getting the *facts* and the focus on relating the *facts* to the cause.

Maybe time variation occurs. Perhaps the data have to be collected over a time cycle.

*Normal instrument error or a fault?* All instruments have errors. Judgment is needed to tell whether the following set of measurements are about what we expect from the instrument? Or that there is trouble on the process?

*Example 6-3:*

Here are records from the operator's log book of the digital printout for the temperature at the top of the reactor. 932.56; 938.64, 930.28, 935.67, 932.19, 937.52.

If the expected temperature is 936 are the values within acceptable error or is something wrong?

Evidence that trouble exists usually comes with error associated with it. We have to distinguish whether the variation in the data represent "expected error" or a fault.

*Example 6-4:*

The laboratory reports that the analysis of the recycle gas shows 3.4% methane. It should be 0.3%. All the other instruments on the process read normal.

Follow-up action: Another sample was taken; the laboratory reported 3.5%. All other instruments still read normal.

Next: No action was taken by the process operators. They continued to operate the plant as usual. Each day, the laboratory reported values of methane in the range 3.5 to 3.65%.

Later: Two weeks later the head chemist returned from holidays and notes that while she was away, all the analyses had been taken on the "calibration mode". When the instrument is set on its correct settings, the concentration of methane in the recycle line was 0.25%.

Too often we *hope that the data* are applicable. A colleague, in designing a petrochemical plant was unable to locate the physical properties of the organics. He decided to assume they were the same as water and hope that they would work out. Just a short time spent in a critical assessment of this assumption would have saved six months of wasted work.

Too often we accept data from the published literature; yet about 8% of data published are mistakes. "The temperature into the hydrodealkylation reactor is >1150 °C" states one reference. This should read >1150 °F. Another example is that a major handbook published an incorrect value of the heat of vaporization through several editions. Check the data coming from computer programs and simulations. Check the physical property package estimates.

### 6.1.4
### Summary

Trouble shooters gather information to solve the problem. Information can be gathered for six different purposes: a) safety check plus provide background information about the situation, b) put the situation in the context of recent events, c) understand what should be happening, d) check what is happening now, e) gather data to test hypotheses and, perhaps, f) take immediate corrective action. Criteria are given for how to select which type of information to gather. In general, start simply, isolate variables, test the "most likely" faults first and have a purpose for gathering each particular piece of information. We need to realize that each of us has a preferred style (and perhaps bias) in how we gather and interpret data. Two inventories were used to help identify preferences and biases. These ideas were illustrated by a range of examples and activities including revisiting **Cases #6** (the utility dryer) and **#10** (to dry and not to dry).

## 6.2
## Thinking Skill: Consistency: Definitions, Cause–Effect and Fundamentals

Scriven (1976) emphasizes that the main criterion used in critical thinking is *consistency*. Indeed, we trouble shoot in a world defined by consistent terms, fundamentals and concepts. We work with *facts*. A clear distinction needs to be made between *facts* and *opinions*. The equipment with which we work has clearly defined symptoms that are generated by each fault. The behavior of fluids and materials follows fundamental principles. We communicate in English, with its defined rules; we work with mathematics that follows rules. In this section we remind ourselves of those rules and framework in which we check for consistency.

### 6.2.1 Consistent Use of Definitions

Probably the most crucial concept is **Facts**. To effectively separate facts from opinion we use definitions based on Obform coding developed by Johnson of the Journalism Department at the University of Wisconsin.

**a. Facts.**

Johnson defines three sources of facts as *factual data, conclusions* and *background information. Factual data* may be accepted as fact if we can attribute an observer with being able to hear, feel, smell, taste or see the observations recorded or stated. For example, *"the gas evolved from the anode is oxygen and that from the cathode is hydrogen."* The facts we can observe are that colorless and odorless gases are evolved from the anode and the cathode. We can infer something about the nature of the gases from further tests. However, the facts are not that "oxygen is evolved from the anode", etc.

*Conclusions* from factual data may be taken as facts if the reasoning is correct: 1) the validity of each step in the sequence of reasoning is proven, 2) there are sufficient steps and 3) the nine steps for valid reasoning (given in Section 6.5) are followed.

*Background information* is factual if references are given for the direct quotes and if statements made by people are direct quotations. For the latter, the message that the speaker said might be factually incorrect, but it is a fact that the person said those words.

**b. Opinion.**

This is information other than Facts.

**c. Opinionated facts.**

This is factual information that contains opinion. For example, *The temperature is as high as 45 °C*. The fact is the temperature is 45 °C. The opinion is that this temperature is "high". Put the two together and we have opinionated fact.

*Example 6-5:* Examples of Facts and Opinions

| Example statement | Factual statement by people | Factual data about operation |
|---|---|---|
| John said "The gauge reads 50 kPa." | Yes | Yes |
| John said "The pressure is 50 kPa." | Yes | No |
| John said that the pressure gauge reads 50 kPa. | No | Yes |
| John said the pressure gauge reading is too high. | No | Opinion via words "too" |
| John said that the pressure is too high. | No | Opinion |
| John said "The pressure gauge reads 50 kPa, the gauge was calibrated yesterday, the sampling line is clear. If I increase the pressure slightly, the gauge reading increases slightly. I infer that the pressure is 50 kPa." | Yes | Yes |
| John said "The pressure gauge reads 50 kPa, the gauge was calibrated yesterday, the sampling line is clear. If I increase the pressure slightly, the gauge reading increases slightly. I infer that the pressure is 50 kPa. I conclude that pressure is too high." | Yes | Yes + Opinion |

**Activity 6-6:** Facts and opinions
Analyze the following passage and classify it according to facts, opinion and opinionated facts.

Here is an accurate account of what happened.

The telephone rang! "Trouble out on the ethylbenzene unit," said Bill. Harry said that he would be right out as he slammed down the phone. As Harry approached the unit Bill came out to meet him and said, "I'm sure that the heat transfer is insufficient in the reboiler to the product column; I'll show you what I mean."

Harry glanced at the rotameter and saw that the flow to the column was the usual amount of 3000 gpm; the pressure gauge read 150 psig and the bottoms temperature was 140 °C. Rounding the column, he saw that the liquid level in the bottoms level gauge was rising at a rate of about 3 cm/min. The liquid level disappeared out of the top of the level gauge. After about two minutes the level reappeared in the sight glass and disappeared out of the bottom of the sight glass within several minutes. "See," said the operator, "we have lost all the bottoms out of the column just like that!" "It has gone off to the storage tank," offered John. "No, it has gone through the reboiler and straight up the column. You can see by the instabilities in the pressure gauges that occur just after the level disappears out the bottom of the sight glass," said Bill.

**Activity 6-7:** Based on the account given in **Activity 6-6,** which of the following statements are True (T), false (F) or can't tell (?).

1. Bill said that there was trouble out on the ethylbenzene unit. T F ?
2. Harry said "I'll be out immediately." T F ?
3. The heat transfer is insufficient. T F ?

| | | | |
|---|---|---|---|
| 4. The trouble is in the reboiler. | T | F | ? |
| 5. The flow to the column is 3000 gpm. | T | F | ? |
| 6. The pressure was 150 psig. | T | F | ? |
| 7. The bottoms temperature was 140 °C. | T | F | ? |
| 8. The level in the bottom of the column was building up and then suddenly dropping. | T | F | ? |
| 9. John said, " The bottoms have gone off to the storage". | T | F | ? |
| 10. When the level in the bottoms of the column drops the pressure gauges show instabilities. | T | F | ? |

*Example 6-6:* **Case #3:**
Consider the account of Michelle working on **Case #3**. The case of the cycling column (presented in Chapter 4, Section 4.1). At one stage, Michelle is trying to decide if the control system is at fault. Her diagnostic action was to "put the control system on manual."

*When the set point was increased manually, the valve stem on the steam moves up, the liquid level appears in the sight glass and continues to drop but shortly thereafter the level appears in the glass and is rising. I conclude that the control system is not at fault.*

For Michelle's thinking reproduced above, has Michelle focused on facts? opinion? opinionated facts?

a. *When the set point was increased manually, the valve stem on the steam moves up.*

These are both observable, and therefore called "facts".

b. *When the set point was increased manually, the liquid level appears in the sight glass and continues to drop but shortly thereafter the level appears in the glass and is rising.* These three are observable and therefore are "facts".

*Comment:*
Michelle did a good job here. She might have incorrectly said, "*When the set point was increased manually, the steam flow increased.*" Since Michelle could not have seen the steam flow, saying the *steam flow increased* would have been an opinion.

In summary, both facts and opinions are used to trouble shoot. However, we need to know which are facts and which are opinions. A definition of facts is given. Examples illustrate how to use that definition consistently. We revisited **Case #3**, the cycling column.

### 6.2.2
### Consistent with How Equipment Works: Cause → Effects: Root Cause-Symptoms

Equipment is fabricated and works according the fundamental principles of science and engineering. Therefore, if a fault occurs in the equipment, certain symptoms will appear. We define a *symptom* as something that can be observed, heard or felt related to the performance of equipment or a system of equipment. The degree to which a *symptom* can be explicitly observed by the trouble shooter depends on the sensors and configuration for the equipment layout. For example, for the usual configuration of a pump the *symptoms* we might be able to observe include the flowrate (if there is a flowmeter), the exit pressure/head (if there is a gauge), the power on the drive (if the amperage is measured), the "crackling noise" (if we can get close enough to "hear") and the temperature of the motor/drive and of the suction line (if we can "feel" or use a surface pyrometer to measure the surface temperature). It is important for us to document cause–effects or cause–symptoms for different types of equipment in terms of the usual types of instrumentation that is available.

**a. Cause → symptom consistency**

An important task is to ensure that we are familiar with cause → symptom data for different pieces of equipment.

*Example 6-7:* Causes (and symptoms)

Three of the common faults with centrifugal pumps (and the symptoms) are: operates at very low capacity (vibration and noise, and pump overheats and/or seizes), rotor not balanced (short bearing life, vibration and noise, short mechanical seal life, pump overheats and/or seizes and stuffing box leaks excessively) and impeller partially clogged with solids (either no liquid delivered or flow lower than expected, power demands higher than expected).

**Activity 6-8:** Listing symptoms for causes or faults.

For the depropanizer shown in **Case #8**, for the following faults/causes,

a. list the symptoms that would be observed, heard or smelt.
b. estimate the order of magnitudes of the deviations (or the extent of the symptoms).

1. The vortex breaker in the overhead drum V-30 is welded such that the cross-sectional area for flow has been reduced to 15% of the pipe internal cross-sectional area.
2. The vortex breaker in the overhead drum V-30 has corroded away. The exit pipe is fully open.
3. In the overhead condenser, E-25, the inerts have been inadequately vented from the shell side (from the tube side) before startup.
4. In the reboiler, E-27, the inerts have been inadequately vented from the shell side (from the tube side) before startup.
5. For the depropanizer, the trays are bent so that the downcomer clearance is $^1/_2$ what it was supposed to be.
6. For the depropanizer, the trays are bent so that the downcomer clearance is double what it was supposed to be.
7. For the depropanizer, corrosion as increased the diameter of the holes in the sieve trays by 10%.
8. For the depropanizer, tray 5 has collapsed because of inadequate support.
9. For the depropanizer, trays 5, 13, 20 and 25 are not level. They are 30° to the horizontal.

More examples of typical faults/causes are given in Appendix H. Appendix I lists the symptoms for some of these.

We can refer to the cause → symptom statements as *If ... then* statements and invoke the rules of logic. The characteristics of *If ... then* statements are (we include the terms antecedent and consequent from the reasoning literature):

- If the "cause" (antecedent) is positive, then the "symptom" (consequent) is positive.
- If the "symptom" (consequent) is negative, the "cause" (antecedent) is negative.
- If the "cause" (antecedent) is negative, we cannot conclude anything about the "symptoms" (consequent). For example, *If* the pump impeller is not turning backwards, *then* we cannot conclude anything about the flowrate and head.
- If the "symptom" (consequent) is positive, then we cannot conclude anything about the "cause" (antecedent). For example, *If* the flowrate and head are abnormally low, *then* we cannot say the impeller is turning backwards. Other reasons can cause this.

This last characteristic causes frustration for the trouble shooter because typically we are not given the cause, rather we are given the symptom and expected to deduce a cause. In other words, symptom ← cause. Furthermore, trouble shooting is also confounded because there may be more than one cause that could be contributing the symptoms. "Symptoms" are not necessarily caused by one and only "root cause". Nevertheless, documenting cause → symptom information is a good starting point for trouble shooting from which we can develop information about symptom ← cause.

**b. Symptom ← cause consistency**

The starting point for trouble shooting is a list of symptoms. The challenge is to create hypotheses as to the probable cause that are consistent with the symptoms. This is the *reverse* of the data discussed in section **a**. Some example symptom ← cause information is given in Chapter 3 for different equipment. The challenge is, that unlike cause-effect data that are "true", the symptom–cause data are "not necessarily true".

**Activity 6-9:** Symptoms and causes

Consider **Case #11**, the symptoms and the hypotheses/causes. Are they consistent?

**Case #11:** The Lazy Twin (courtesy of W. K. Taylor, B. Eng. McMaster, 1966)

The situation: Pump A is usually running. Pump B, identical to pump A, is the spare pump. The operators of the process notice that the flow meter, FRC-100, shows that not enough flow is going to the process. So they switch over from pump A to pump B by shutting off the power to the drive motor for pump A and turning on the power to the drive motor for pump B. Now the flow to the process comes back to where it should be. What is wrong? The pump configuration is shown in Figure 6-2. Data about the flow and setting of the valves are given in Figure 6-3.

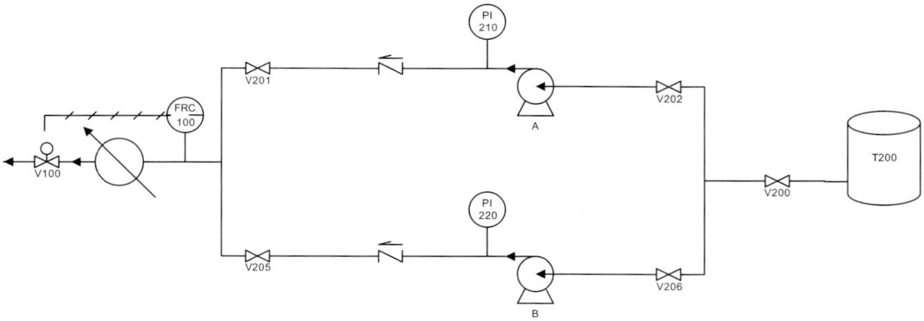


**Figure 6-2** Operator's sketch of the equipment for Case #11.

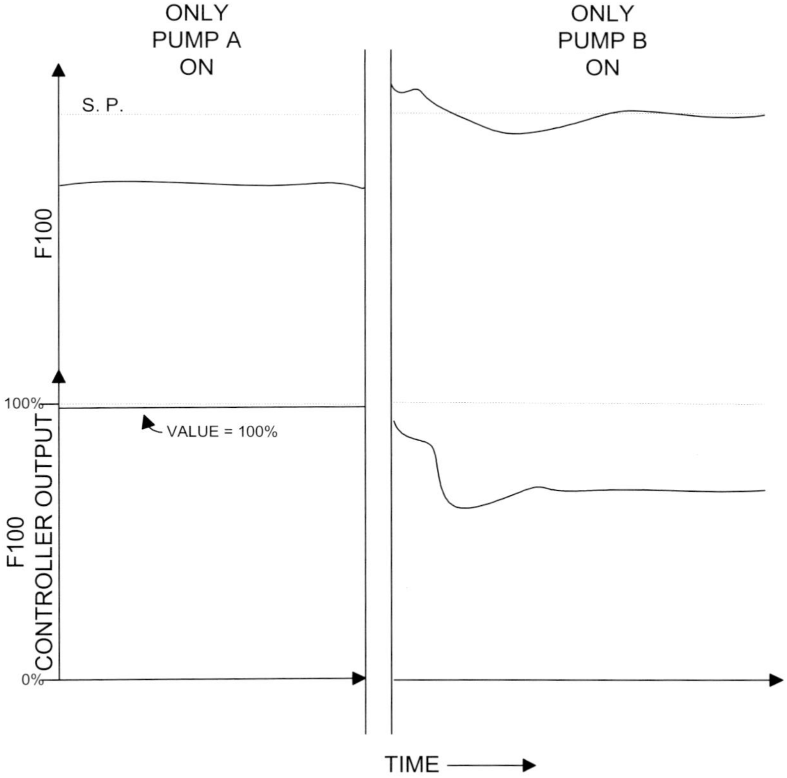


**Figure 6-3** Flow and valve-setting data for Case #11.

For this **Case #11** we might summarize the symptoms as:

a. when pump A is running, the flow meter shows that not enough flow is going to the process
b. when pump B is running, the flow meter shows the correct flow is going to the process

In the following table, column "a" represents symptom "a", "b", symptom "b". Consider each hypothesis and in columns "a" and "b" place a code **S** supports; **D** disproves and **N** neutral or can't tell this hypothesis or fault.

**Table 6-2** Worksheet for hypotheses.

| Working Hypotheses or cause | Initial Evidence: symptoms | | | | | Diagnostic Actions | | | |
|---|---|---|---|---|---|---|---|---|---|
| | a | b | c | d | e | A | B | C | D |
| 1. Flowmeter is reading wrong. | | | | | | | | | |
| 2. Pump A has a sandwich stuck in the impeller. | | | | | | | | | |
| 3. Pump B is much bigger capacity than we need | | | | | | | | | |
| 4. Full electrical current does not go to pump A. | | | | | | | | | |
| 5. A fuse is blown in the circuit to pump A. | | | | | | | | | |
| 6. Liquid is vaporizing in the line to pump A and so pump A cannot produce the expected flow. | | | | | | | | | |
| 7. The operator cannot read the flowmeter correctly. | | | | | | | | | |
| 8. Some corrosion products are clogging the flow-meter. | | | | | | | | | |
| 9. There is not enough Net Positive Suction Head for the Pump A. | | | | | | | | | |
| 10. There is not enough liquid in the tank to the pumps. | | | | | | | | | |
| 11. The signal is not getting from the controller to the valve. | | | | | | | | | |
| 12. The flow-control valve is stuck partway shut. | | | | | | | | | |
| | | | | | | | | | |

**c. Symptom or "root cause" dilemma**

In brainstorming possible causes (that are consistent with the symptoms) sometimes the idea is another symptom, instead of a "root cause".

*Example 6-8:*

Chapter 3 suggests that for distillation if *all temperatures are falling simultaneously* then the cause might be *low boilup*. Is "low boilup" a root cause? No. The root cause is something that *causes* low boilup, which (for a thermosyphon reboiler) could be such causes as condensate flooding/ inadequate steam supply/ steam valve closed/ superheated steam/ boiling-point elevation of the bottoms/ inert blanketing/ film instead of nucleate boiling/ increase in pressure on the process side/ undersized reboiler/ control system fault/ fouling on the process side/ low liquid level/ high liquid level/ heavies in the feed/ pipe lengths< design/ pipe diameter > design/ process fluid level < 30–40% of tube length.

To test if we have a symptom or a root cause, test the postulated "cause" by asking "What would cause this?" If there is no new answer, then we probably have the root cause.

**Activity 6-10:** Root cause
For **Case #11**, The Lazy Twin, and the possible causes listed, which of these causes are root causes and which are symptoms?

In summary, this is perhaps the key section in this chapter. Consistency between cause and symptom; symptom and cause; hypotheses and symptom are key to trouble shooting. It is simplest to start with typical causes/faults and listing symptoms that are produced. Then we move to the trickiest part, namely the reverse: identifying possible causes from a given set of symptoms. A new **Case #11**, the lazy twin, was introduced. The challenge of working with "root causes" was emphasized.

### 6.2.3 Consistent with Fundamental Rules of Mathematics and English

Our reasoning and actions need to be consistent with the rules of mathematics and English.

For *Mathematics* consider the following:

*Example 6-9:*
Please check the reasoning in the following:

$$\sqrt{\frac{-1}{1}} = \sqrt{\frac{-1}{1}}$$

$$\sqrt{\frac{1}{-1}} = \sqrt{\frac{-1}{1}}$$

$$\frac{\sqrt{1}}{\sqrt{-1}} = \frac{\sqrt{-1}}{\sqrt{1}}$$

$$1 = \sqrt{-1}\sqrt{-1}$$

$$1 = -1$$

What went wrong? Here we encounter an incorrect answer so we *know* that we were inconsistent with the rules of mathematics.

But what happens when we don't have a surprise? We need to condition ourselves to check for mathematical consistency in all the calculations we do as we trouble shoot. We do have guidance in Sections 6.2.4 and 6.2.5 because our answers should be mathematical answers that are consistent with fundamental laws and with experience. To some extent Saadia followed these principles in **Case #6.** She calculated the expected results ahead of time. Then when she saw the results she would know if there were surprises.

For *English*, we have the rules of grammar and the meanings of words. Words only have meaning in people. Therefore, we need to use words that are "understood" by all. Consider the following examples:

*Example 6-10:*
An open window, a gust of air, glass on the floor, water on the floor, Mary's dead. Why?

*Example 6-11:*
A man is walking along. He tears his sleeve on a sharp corner. Within minutes he is dead. Why?

In these two examples, if we realize that Mary is a goldfish and that "walking along" means walking on the bottom of the ocean, then these are not puzzles and we all *understand*.

When we are trouble shooting, some of the words that are ambiguous and that interfere with communication include:

*"The instrument looks OK!"*

*"I followed the usual practice."*

**Activity 6-11:** Ambiguous words
Make a list of ambiguous words that you encounter when you trouble shoot. Then think of ways you can overcome this ambiguity.

In summary, critical thinking requires that we are consistent in our use of Mathematics and English.

### 6.2.4
### Consistent with Fundamental Principles Of Science: Conservation of Mass, Energy, High to Low Pressure, Properties of Materials

Mass and energy balances, pressure profiles and an understanding of the unique properties of materials are probably the most useful source of information in trouble shooting. Yet, trouble shooters often neglect these.

### 6.2.5
### Consistent with Experience

Having a wide range of "rules of thumb" and memorized experience factors help us to validate and check answers and ideas for consistency. Sources include Brannan (1998), Walas (1988), Woods (1995) and Woods (2001).

### 6.2.6
### Summary

Consistency is the key to critical thinking. Here our focus was on consistency with

- definitions: to sort facts from opinions;
- cause → effect information for equipment: to guide in the creation and testing of hypotheses;
- the rules of mathematics and English: to help us focus on accuracy;
- the fundamentals: to remind us of the basics to check;
- experience: to validate our thinking, calculations and assumptions.

Most of the effort was spent on the first two topics: *facts* versus *opinions* and sorting out cause → effect relationships.

## 6.3
## Thinking Skills: Classification

Classification is dividing the whole into parts such that there is a meaningful relationship among the parts. The classification is done for a purpose, and for each level of classification there must be one and only one criterion. Each level of classification should be complete. There should be no single subclass. The classification should be consistent in the amount of detail given and the classification should have neither faulty coordination nor faulty subordination. These are the general characteristics of a good classification.

Skill in classification is needed to:

- classify the starting information during the stage of **Define the stated problem**, discussed in Chapter 2.
- classify the possible causes generated during the brainstorming session as part of the **Explore** stage, discussed in Chapter 2 with the brainstorming session being illustrated in Chapter 5.

### 6.3.1
### Classify the Starting Information

For the purpose of understanding the problem, the starting information should be classified using the criterion "what are the key parts to the trouble-shooting problem statement". Usually the key parts are:

- the situation or system,
- the symptoms that suggest a fault,
- the triggering event,
- the criteria for success (either stated or inferred),
- the constraints.

The focus here is to identify the "symptoms". The symptoms are defined as "evidence stated by the operators, those displayed by sirens or alarms, reports from the laboratory, or the analyzer showing non-specification behavior, or complaints from customers describing *unexpected or unusual* behavior." Sometimes the symptom is accompanied by a triggering event. The triggering event is often expressed as "*When* ...". Sometimes there is no apparent triggering event. Sometimes the triggering event is not related to the symptom; it is coincidental. However, when we start trouble shooting the problem we may not realize the coincidence. Some examples of triggering events include:

When we increased the flowrate to new higher capacities ...
When we started up for the first time ...
When we started up for the first time after maintenance ...
When the operator increased the pressure in the column above the usual value ...

The mental TS process is to scan all the given, starting information and apply the definition of "symptom" to identify this portion of information.

*Example 6-12:* Identifying symptoms.
For **Case #9**, the symptom is stated ambiguously as "nothing happened." Before we can proceed, we should identify why this is unexpected behavior. From the context, we can rephrase this as, the vacuum inside the deodorizer should have been sufficient to suck the Fuller's earth into the deodorizer when the connecting valve was opened. However, the Fuller's earth was not sucked into the deodorizer because we could not see powder dumping in when we looked through the view port.

*Example 6-13:* Identifying pertinent triggering events:
Several *When*'s occur in this situation in **Case #9**: "*When* the valve was opened;" "*when* the vacuum had stabilized"; "*when* we started up this plant for the first time". Are any of these triggering events? The most likely triggering event is *when* we started up this plant for the first time.

### 6.3.2
### Classifying Ideas from Brainstorming

As illustrated in Chapter 5, usually we obtain 50 to 100 ideas in a brainstorming session. Of these often 80% are non-sensible. That's OK because it is worth a few minutes to generate such ideas since hidden among the crazy ideas are some real winners. After the brainstorming session, however, we need to classify for the purpose of listing at least six feasible hypotheses. The first level of classification would be "technically feasible" versus "non-sensible" or it could include an additional class of "interesting". The next level of classification would be to sort the technically feasible ideas according to underlying type of cause.

## 6.4
## Thinking Skills: Recognizing Patterns

Patterns in the information could include increasing or decreasing trends, cycling data, series or unexpected systematic external changes that affect performance.

Here are three examples.

*Example 6-14:*
One process operator might be experiencing great personal stress at home. Whenever this operator is on shift the process does not run smoothly. If we were unaware that the operator was under stress it would be hard to identify the cause.

*Example 6-15:*
Whenever there is a severe thunderstorm some of the instruments on the plant malfunction.

*Example 6-16:*
Everything seemed to be working OK although gradually the performance was moving off specification. The cause was a minor leak in a heat exchanger.

In these three examples, the trick was to be able to see a pattern between stress and poor performance; thunderstorms and malfunctioning instruments; and gradual changes in performance and cross contaminant because of a leak.

The general guidelines and criteria for identifying patterns are not very helpful. Keep an open mind. Plot (and be sensitive to) time variation in data. Note the possible interaction between different systems and especially the interaction between cycling processes.

Consider now more about the patterns in the symptoms and in how we collect data.

### 6.4.1
### Patterns in the Symptoms

The unexpected and outside factors that can affect performance include the weather:

- showers and rain; extreme cold; hot sunny weather affects cooling towers and air-cooled condensers.
- electrical storms affect sensitive electronic/electrical instrumentation.
- atmospheric pressure affects pumps that pump liquid from tanks that are open to the atmosphere.
- extreme cold can freeze vents shut; freeze bucket steam traps.
- hot weather plus steam tracing can cause vaporization and vapor binding.

Cycling processes, that might interfere with other cycling processes include:

- batch distillation and batch reactors,
- adsorption and ion exchange,
- centrifuges and filters.

When cyclical processes are used, the data should be reported for different parts of the cycle and not reported just as average values. Averages mask the trends and patterns.

When cyclical processes occur, the frequency of the cycle should be determined. For example, cycling steam flow or column pressure for a distillation (30 s to several minutes) coincides with a specific set of probable causes. On the other hand, thermodynamic steam traps should have about six cycles/minute.

*Example 6-17:* The cycling level in the bottom of the column.
Engineer Ted Tyler used the cycle to help him predict that the collapsed tray, causing the cycling, was tray 13 to 20 from the bottom.

*Example 6-18:* Another cycling level in the bottom of the column.
In another cycling column, the coordination between the cycle of the level and the cycle of the steam trap, as determined by listening with a stethoscope to the trap, helped pinpoint the steam trap as the fault.

*Example 6-19:* The bombastic bagging machine.
PVC powder was pneumatically conveyed to a hopper located above the bagging machine. The dust was filtered from the conveying air. The problem was that periodically there would be loud thumping in the hopper that sounded like an avalanche of powder suddenly arrived in the hopper. Yet all the evidence suggested that the powder was being fed to the hopper smoothly and continually. This periodic appearance of slugs of powder caused problems with the continual bagging operation. What was discovered was that the bags in the dust filter were being periodically cleaned. The design fault was that the powder-laden air flowed *up the inside* of the cylindrical, vertical bags so that the blow-ring cleaners could travel on the outside of the bag. As the blow ring cleaned a bag it blew the powder into the center of the cylindrical vertical bag and the powder was held in the up-flowing conveying air. The powder became fluidized in some of the bags. When the mass of fluidized powder became excessive in the central core of one of the bags, the whole mass from the fluidized bed would avalanche down into the hopper. Which bags became fluidized beds and when the bed became unstable seemed to be completely random. The corrective action was to replace the configuration so that the feed air flowed down the central core of the bag. In this example the cyclic operation was the cleaning of the filters.

The gradual buildup of solids, of corrosion products or of trace contaminants can cause trends in performance. The cause might be that the solids are too wet, the amount of purge is insufficient or a leak. Gradual changes that cause trouble are often the most difficult to detect. Good records are needed to note the trends.

Unexpected pockets of water or condensate in low-lying sections of pipe can cause patterns in performance.

**Activity 6-12: Case #7:** The reluctant vacuum crystallizer
The data given in this case are cyclical! Analyze the approach Frank took, as described in Section 4.5 of Chapter 4, and look for patterns in the data that Frank should have spotted.

**Activity 6-13: Case #10:** To dry and not to dry
This case is described in Section 5.4. The operation consists of a continuous crystallizer, a batch screen, a batch centrifuge and a continuous rotary dryer. The symptoms are "The crystal product from the rotary dryer has a moisture content of 4.5% whereas the design value is 1.5%. Cake seems to be building up in the dryer feed chute, in the feed screw and on the steam tubes at the feed end of the rotary dryer." Critique these symptoms.

### 6.4.2 Patterns in the Evidence

When we suspect a pattern, then we should collect data that will make the pattern easy to spot.

*Example 6-20:* In **Case #6**: The utility dryer
Saadia sampled every 15 minutes for the 120-minute regeneration cycle. She also astutely sampled such that she could have data for bed A adsorbing and then being regenerated (and for B adsorbing and being regenerated).

**Activity 6-14: Case #10:** To dry and not to dry
Before the change in washing, the centrifuge and screen cycle were 10 minutes online and 3 minutes water wash. What sampling would you recommend to try to identify a pattern?

## 6.5 Thinking Skill: Reasoning

In evaluating evidence and making decisions about the hypotheses, the reasoning that we use should be sound. In this section we outline an organized nine-step approach to evaluate the reasoning.

In general, an overall nine-step approach for critical thinking is:

1. Classify all the given information into the key parts.
2. Write the conclusion.
3. Identify the context. What are the stated and the inferred contexts? Are there other pertinent contexts?
4. Check the definitions; identify and clarify ambiguous terminology.

4a. Change the argument to show the relationship between the conclusion and the evidence.

5. Consider the evidence. What is the quality of the evidence? Is it sufficient? Relevant? acceptable? What are the facts? What are the opinions? Is the correct type of data gathered for the purpose? Usually the purpose is to test a hypothesis. Check the evidence. The evidence is the basis of an argument.
6. Formulate the assumptions.
7. Assess the quality of the reasoning.
8. Assess the strengths of the counterarguments.
9. Evaluate the consequences and implications.

To illustrate this process consider some of Michelle's reasoning as presented in Chapter 4, Section 4.1 as she tries to resolve **Case #3**, the case of the cycling column. At one stage, Michelle is trying to decide if the control system is at fault. Her diagnostic action was to "put the control system on manual."

*When the set point was increased manually, the valve stem on the steam moves up, the liquid level appears in the sight glass and continues to drop but shortly thereafter the level appears in the glass and is rising. I conclude that the control system is not at fault.*

### 6.5.1
### Step 1: Classify the Information

The given information includes the context, the conclusions, the evidence related to the conclusion, the stated counterarguments and the stated assumptions. Later we will add inferred assumptions and counterarguments.

The context: the situation. Supply WHO, WHAT, WHERE, WHEN IS and IS NOT details plus the equipment and system.

The conclusion: that is usually expressed as either what is the cause or what is not the cause.

The evidence: collected as described in Section 6.1.

The stated counterarguments: evidence that disproves a statement or hypothesis that you are trying to prove.

The assumptions: a statement for which no proof or evidence is offered.

The qualifier: a constraint or restriction or limiting condition on the conclusion.

*Example 6-21:* Classify the information given in the short segment of **Case #3**, the case of the cycling column:

*Context*: a level in the sight gauge at the bottom of a distillation column is cycling. The frequency is about once per 2 minutes. More details will be given in Step 3. The control system was "put on manual."

*Conclusion:* I conclude that the control system is not at fault.

*Evidence:* When the set point was increased manually, the valve stem on the steam moves up, the liquid level appears in the sight glass and continues to drop but shortly thereafter the level appears in the glass and is rising.

*Stated counterarguments:* none in this passage.

*Stated assumptions:* none in this passage.

*Inferred assumption:* the control system actually was on manual; the level in the sight glass is related to the level in the bottom of the column.

*Stated qualifier:* none.

### 6.5.2
### Step 2: Write the Conclusion

It is important that the focus of the reasoning is stated clearly. To locate the stated conclusion look for words like "therefore," "because", "I conclude", "so" and "thus". Usually the conclusion is near the end of dialogue. There may be several conclusions each building on another. Identify the main conclusion. Write it down so that we can focus on the conclusion. The conclusion can be in many different forms.

- We could be identifying the context (as in Section 6.5.3) and our conclusion is "Therefore, we have correctly identified the context." But have we?
- We could be monitoring our trouble-shooting process (as was recommended in Chapter 2) and we conclude "Thus, we have completed the hypothesis generation stage, now let's move to the data collection stage." But is this correct?
- We may conclude that "In summary, we have selected a *reasonable* set of hypotheses from the brainstorming activity in Chapter 5." But have we? Are the causes *root causes* or are some of them symptoms?
- We might have listed the symptoms and the hypotheses and concluded that "Therefore, hypothesis #1 is consistent with the evidence." But is this a valid conclusion?
- We might have gathered some evidence and concluded that a hypothesis has been confirmed, denied and can't tell.

*Example 6-22:* Michelle's main conclusion is "*I conclude that the control system is not at fault.*" The question is, *Is her conclusion valid?*
Label each conclusion.

### 6.5.3
### Step 3: Identify the Context

What are the stated and the inferred contexts? Are there other pertinent contexts? The context depends on the conclusion. I find it helpful if I sketch a cross section of the equipment to remind me of the internal possibilities. I draw a line around it to define the system showing the ins and outs. Label specific contexts.

*Example 6-23:* in **Case #3** the context is that the level in the sight glass cycles. The distillation column is as shown in the Case and, for this portion of the evidence, the control system was placed on manual. Whether this is included as a qualifier or context depends on how detailed you want to present the argument. A sketch of the context is illustrated in Figure 6-4.

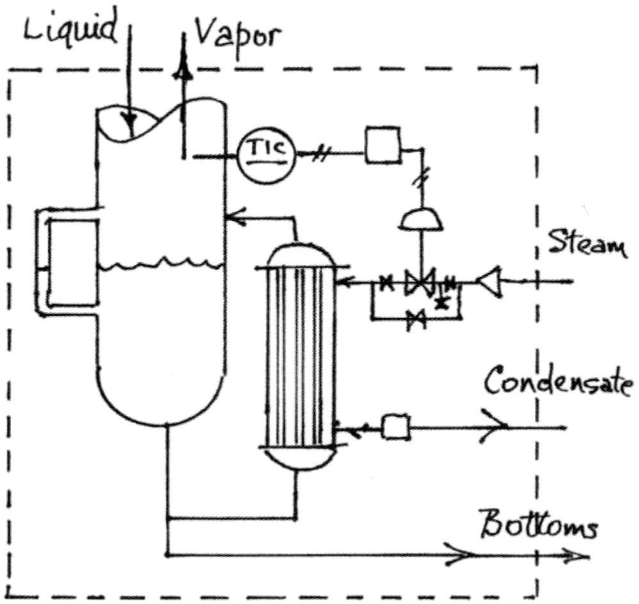


**Figure 6-4** Sketch of the system.

This sketch includes more detail around the steam control valve; it includes the bottoms draw-off and uses the dotted line to indicate the system.

### 6.5.4
### Step 4: Clarify the Meaning of the Terminology

Identify and clarify ambiguous terminology. Write out your interpretation of any ambiguous words and any unstated but intended implications. Check that you understand all the words and terminology. Often, on process plants the personnel create jargon words for operations and pieces of equipment. For example, on one plant the condensate receiver was called "the pig". If you are unfamiliar with the terminology you need to ask for clarification. This takes courage. Use the principles of consistency outlined in Section 6.2, especially Sections 6.2.1 and 6.2.3.

*Example 6-24:* in **Case #3** the words that Michelle should know include "the set point", "control system" and "on manual". *On manual* = signal from the sensor no longer alters the steam-valve setting; manually change the set point and the valve should respond.

### 6.5.5 Step 5: Consider the Evidence

In trying to decide on the validity of Michelle's conclusion the factors to consider include: What is the quality of the evidence? Is it sufficient? relevant? acceptable? What are the facts? What are the opinions? Are the correct types of data gathered for the purpose? Usually the purpose is to test a hypothesis. Check the evidence. A diagram usually helps.

#### 6.5.5a Identify the Evidence

Label each assertion. Sometimes several assertions appear in the same sentence. Separate each one. Don't give a separate number to the same assertion even though the same assertion might be stated several times. Give numerals to conclusions, counterarguments and assumptions.

*Example 6-25:* For **Case #3**

[*With the control system on manual,* 2].
[*When the set point was increased manually,* 3]
[*the valve stem on the steam moves up,* 4],
[*the liquid level appears in the sight glass and continues to drop but shortly thereafter the level appears in the glass and is rising.* 5]
[*I conclude that the control system is not at fault,* 1]

#### 6.5.5b Check for Consistency

Our basis is consistency. Systematically scan over the different elements for consistency in Section 6.2. 1) Label facts, opinions and remove the opinion from opinionated facts. 2) Check for cause → effect consistency. 3) Check that definitions did not change in different parts of the argument; that symbols are not defined in one way and used in another; that a variable is not treated as a constant; that a principle is not applied beyond its range of applicability, that an assumption is not made and then violated, that coupled variables are not treated as independent.

**Activity 6-15:** Adiabaticity
An adiabatic change is one in which no heat enters or leaves the system. A primitive Joule experiment in which a lead shot is shaken in a cardboard tube is approximately adiabatic. The lead shot gets hotter. Does it then have more heat in it? If so, how does the heat get there? Is the experiment adiabatic after all? Check for consistency.

*Example 6-26:* for **Case #3** and Michelle's reasoning:

[*With the control system on manual,* 2].
[*When the set point was increased manually,* 3]
[*the valve stem on the steam moves up,* 4],
is consistent with *behavior of a control system.*
Inference: [*if the set point is kept constant,* 6], [*the valve-stem position is constant,* 7], [*the steam flow is constant,* 8].

### 6.5.5c
### Which Evidence is Pertinent?

The overall context is to find the cause of the cycling level in the bottoms of the column. Therefore any evidence gathered should relate to what does and does not affect the level of liquid in the bottoms of the column. Here we focus on the hypotheses-symptom-action chart from the Trouble-Shooter's Worksheet. Alternatively we consider the symptom ← cause data given in Chapter 3, subject to the concerns expressed in Section 6.2.2b. Some might prefer to use a diagram to illustrate the symptom with all the related possible causes/hypotheses.

*Example 6-27:* for **Case #3**, Figure 6-5 is an example of symptom ← cause diagram.

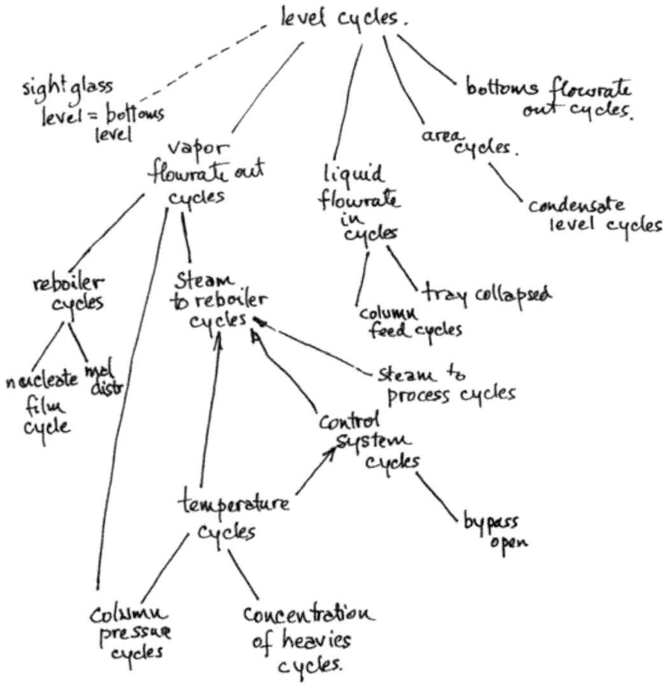


**Figure 6-5** Symptom ← cause chart for **Case #3**.

6.5.5d
### Diagram the Argument

Scriven and Halpern (1996) take different approaches in diagramming an argument. Scriven places the evidence at the top of the page and moves down the page to a conclusion. Halpern places the conclusion at the top and uses different evidence to "support" the conclusion. Here are some suggestions:

- Use the numbers for the evidence and conclusions that were used in Section 6.5.5a.
- Connect evidence to conclusions with an arrow with separate arrows for each different form of evidence.
- Rate the quality of the support that the evidence lends to the conclusion: weak, moderate, strong and write this rating on the arrow connecting the evidence to the conclusion.
- Include assumptions. I use a dotted line if the assumption is inferred.
- Include the counter-arguments with a wriggly line.

*Example 6-28:* **Case #3** Figure 6-6 shows a diagram of the arguments related to Michelle's thinking during the example scenario.

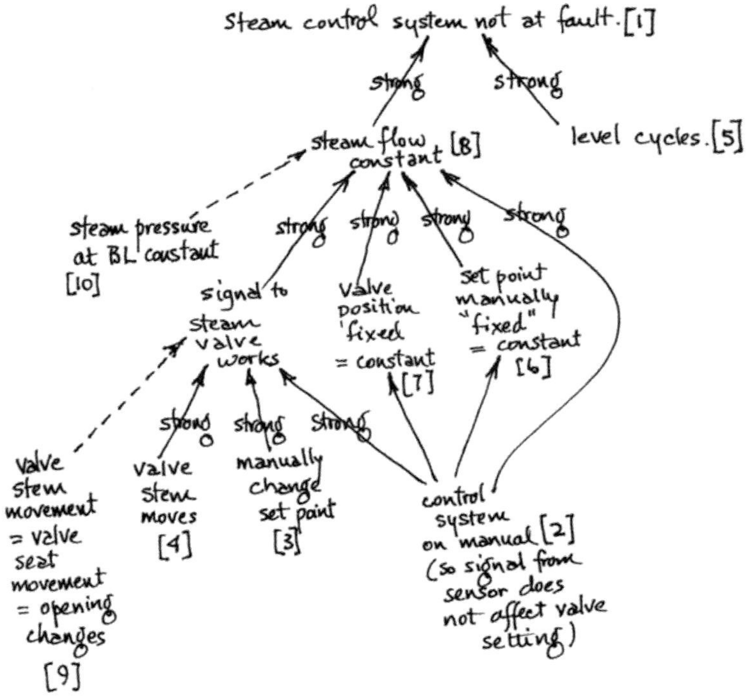


**Figure 6-6** Diagram of the arguments.

### 6.5.6 Step 6: Formulate the Assumptions

An assumption is a statement for which no proof or evidence is offered. Consider each bit of evidence and the resulting conclusion and identify the inferred assumptions. This should include the assumptions made by a) you, b) the designer, c) the process operator, d) the laboratory analysis team, e) the instrument technicians, f) the maintenance personnel and g) vendor. Show the assumptions on the diagram.

*Example 6-29:* **Case #3:** Michelle has assumed:

- valve-stem movement = valve opening; the seat is connected to the valve stem [9]
- the steam pressure from the utilities plant (and at the Battery Limits) is constant [10]

These are shown in Figure 6-6 as dotted lines

### 6.5.7 Step 7: Assess the Quality of the Reasoning

Consider the reasoning related to: i) the symptoms; ii) the symptom ← cause relationships; iii) the data gathered to test the hypothesis or probable causes; iv) the assumptions and v) the conclusion.

Focus on the assumptions and on consistency.

Use a chart of the arguments to guide the analysis. Consider each of the components in turn.

i) **The symptoms.** Were the symptoms correctly identified in Step 1? In the **Case #3**, we have no reason now to alter the symptom. There is an assumption related to the symptom, and shown on Figure 6-5 as a dotted line: cycling in the level in the sight glass = cycling in the level in the bottoms =? cycling in the level in the reboiler. This was noted but not addressed.

ii) **The symptom ← cause relationships.** Revisit Chapter 3 and see if any pertinent causes have been omitted. Also check that the diagram is consistent, based on the principles of Section 6.2.2. Recall that symptom ← cause evidence is where many mistakes are made because, although the cause → symptom connection is valid, the reverse (symptom ← cause) is not always true since multiple causes can display the same symptoms.

*Example 6-30:* **Case #3:** reconsider Section 3.3, reboilers; Section 3.4 distillation and Section 3.5 steam traps.

Section 3.3 reboilers: general:

*"Cycling (30 s–several minutes duration) steam flow, cycling pressure on the process side and, for columns, cycling $\Delta p$ and cycling level in bottoms":* instrument fault/ condensate in instrument sensing lines/ surging/ [foaming]* in kettle and thermosy-

phon/ liquid maldistribution/ steam-trap problems, see Section 3.5, with orifice $\Delta p$ across trap < design/ temperature sensor at the feed zone in a distillation column/ collapsed tray in a distillation column.

*[foaming]**: surfactants present/ surface tension positive system/ operating too close to the critical temperature and pressure of the species/ dirt and corrosion solids.

Section 3.3 reboiler: thermosyphon:

*"Cycling (30 s–several minutes duration) steam flow, cycling pressure on the process side and, for columns, cycling $\Delta p$ and cycling level in bottoms":* in addition to general, all natural circulation systems are prone to surging/ feed contains high w/w% of high boilers/ vaporization-induced fouling/ constriction in the vapor line to the distillation column. For horizontal thermosyphon: maldistribution of fluid temperature and liquid.

Section 3.4 distillation: Only lists cycling temperatures:

*"Cycling of column temperatures: "* controller fault/ collapsed tray/[damaged trays]*/ [jet flooding]*/ [foaming]*/ [downcomer flooding]*/ [dry trays]* *[jet flooding]*:* excess loading/ fouled trays/ plugged holes in tray/ restricted transfer area/ poor vapor distribution/ wrong introduction of feed fluid/ [foaming]*/ feed temperature too low/ high boilup/ entrainment of liquid because of excessive vapor velocity through the trays/water in a hydrocarbon column; *[downcomer flooding]**: excessive liquid load/ restrictions/ inward leaking of vapor into downcomer/ wrong feed introduction/ poor design of downcomers on bottom trays/ unsealed downcomers/ [foaming]*; *[foaming]**: surfactants present/ surface tension positive system/ operating too close to the critical temperature and pressure of the species/ dirt and corrosion solids. *[Dry trays]*:* flooded above/ insufficient reflux/ low feedrate/ high boilup / feed temperature too high. *[Damaged trays]*:* leak of water into high molar mass process fluid/ large slugs of water from leaking condensers or steam reboilers/ startup with level in bottoms > design/ attempt to overcome flooding by pumping out bottoms at high rate/ too rapid a depressurization of column/ unexpected change in phase.

Section 3.5 steam traps.

No specific entries for "cycling"

*"No condensate discharge":* bypass open or leaking/ scale in the orifice/ plugged strainer/ inlet pressure too high/ for inverted bucket trap, bucket vent clogged, incorrect $\Delta p$ across the orifice.

Comment:

The focus in Figure 6-5 was on the causes related primarily with the control system. The bold items listed above do not seem to be explicitly included in the chart. Scrutiny of these suggests that no major issues have been left out considering Michelle's purpose in creating this chart.

iii) **The data gathered to test the hypothesis or probable causes**. Check the evidence from the questions and tests that were gathered. The general options and the biases and interpretation issues were considered in Section 6.1 and analyzed in Step 5, Section 6.5.5.

*Example 6-31:* **Case #3**

From Figure 6-6 we note that Michelle focused on evidence [2], [3] and [4]. She inferred [6] and [7]. She should have explicitly checked this by setting the flow = constant; noting if the valve stem was constant and then checked for cycling [5]. Because Michelle did not do this explicitly, conclusion [8] is not proven. What Michelle should have concluded from evidence [3] and [4] is that the signal to the steam valve works.

If she had shown that the steam flow is constant and yet the level cycles [5], then she could conclude that the steam control system is not at fault [1].

iv) **The assumptions**. Focus on the assumptions and on consistency.

*Example 6-32:* **Case #3**

The two assumptions Michelle noted were [9] and [10]. Assumption [9] would have been easy to check.

v) **The conclusion**. Strong evidence should support the conclusion; or a large collection of weak to moderate evidence. There should be no strong counterarguments nor strong assumptions.

### 6.5.8
### Step 8: Assess the Strengths of the Counterarguments

Use *What if?* counterexamples to help clarify the reasoning. These can be the reverse of the assumptions.

### 6.5.9
### Step 9: Evaluate the Consequences and Implications

Once we have accepted a valid conclusion, explore the *So Whats?* In most troubleshooting situations this means identify another feasible cause or, once the cause has been identified, to suggest remedial or corrective action.

### 6.5.10
### Activity 6-14

Critique the arguments described in one of the Cases presented in Chapter 4.

## 6.6 Feedback and Self-Assessment

**Reflections:**

From this chapter, what have you discovered about trouble shooting? about planning and performing tests? about asking questions? about reasoning and drawing conclusions?

________________________________________

________________________________________

________________________________________

________________________________________

| Rate yourself on the following | already do this | might work for me | not for me |
|---|---|---|---|
| **Diagnostic actions**<br>– systematic and structured pattern to obtain information | ☐ | ☐ | ☐ |
| – use criteria to select tests | ☐ | ☐ | ☐ |
| – unique style and potential biases | ☐ | ☐ | ☐ |
| **Consistency**<br>– separate fact versus opinion | ☐ | ☐ | ☐ |
| – cause–symptom data | ☐ | ☐ | ☐ |
| – symptom–cause relationships | ☐ | ☐ | ☐ |
| – identify root cause versus “symptom” | ☐ | ☐ | ☐ |
| – rules for English, mathematics, science fundamentals | ☐ | ☐ | ☐ |
| **Classification**<br>– use single basis/criteria per level | ☐ | ☐ | ☐ |
| – apply classification principles to identify symptoms | ☐ | ☐ | ☐ |
| **Patterns**<br>– methods to identify patterns | ☐ | ☐ | ☐ |
| **Reasoning**<br>– systematically use a process similar to the 9-step | ☐ | ☐ | ☐ |
| – draw/construct symptom ← cause diagrams | ☐ | ☐ | ☐ |
| – diagram an argument and use this to critique | ☐ | ☐ | ☐ |
| – well-developed methods to identify assumptions | ☐ | ☐ | ☐ |

## 6.7 Summary

A variety of critical-thinking skills are used when we trouble shoot. The first critical-thinking skill relates to how we *gather information* about the process. As we select diagnostic actions to gather information and evidence about the system we usually start with general background information (although if we are familiar with the process we know this already). Then we build up our mental image of the system and context by determining what happened recently, reminding ourselves of the details of what should be happening and then exploring what really is happening.

In *testing hypotheses/causes* we should start simply and test the most-likely cause early. Above all we should refrain from the expensive tactic of shutting down, opening and inspecting. Much crucial information can be gained from the instruments, simple calculations and sound reasoning. A range of biases and mistakes made in gathering and interpreting data are listed: confirmation bias, over-interpretation, under-interpretation and mis-interpretation, availability bias, premature closure, anchoring. Jungian typology (or MBTI) dimension P-J provides a useful indicator of one characteristic of your personal style for gathering and interpreting data.

A second critical-thinking skill relates to the importance of being *consistent* in our use of words, and our use of knowledge about processes and process equipment. Specifically the focus was on identifying fact versus opinion, gathering accurate cause–symptom data and astutely reversing the connection to link symptom–cause. Our usage should be consistent with the rules of English, mathematics, the fundamentals of science and engineering and practical experience.

A third critical-thinking skill is *classification*; the process of dividing large sets of information into meaningful parts. In making the division we should use a single basis/criteria per level, use no single entries and avoid faulty coordination or subordination. This was illustrated for the task of classifying the starting information into "symptoms" and "triggering" events.

A fourth critical-thinking skill is to be able to identify *patterns*.

A final critical-thinking skill discussed here was *reasoning*. A systematic 9-step process is suggested and illustrated in the context of part of Michelle's reasoning in **Case #3**.

## 6.8 Exercises

1. Harry lives on the eleventh floor of an apartment building. Each morning he rides the automatic elevator down to the first floor and goes to his specialized job in the cramped quarters assembling components in the airplane fuselage. Each evening he rides the elevator up to the seventh floor and then walks up to his own floor. Why?
2. Mr Tom Jones goes to the doctor's office twice a week to pick up his pills. This particular Tuesday his wife decides to go with him so that she can continue on and do the shopping afterwards. Tom parks the car in front of the doctor's office, leaves his wife in the car and goes in to get the pills. While he is talking with the doctor, he hears a terrific crash, rushes to the window and upon seeing his car completely demolished exclaims "My wife's been killed!" whereupon the doctor reaches into the desk drawer, pull out a gun and shoots Tom. Why?

3. Check out the mathematics in the following:

$$n^2 = n(\underbrace{1+1+1+1+\ \ldots\ldots\ +1}_{n})$$

$$n^2 = (\underbrace{n+n+n+n+\ \ldots\ldots\ +n}_{n})$$

$$\frac{dn^2}{dn} = \frac{d}{dn}(\underbrace{n+n+n+n+n\ \ldots\ldots\ +n}_{n})$$

$$2n = (\underbrace{1+1+1+1+\ldots\ldots+1}_{n})$$

$$2n = n$$

$$2 = 1$$

# 7
# Polishing Your Skills: Interpersonal Skills and Factors Affecting Personal Performance

In any problem, people are involved. You as trouble shooter are the major person we might focus on. Yet there are others: the operators, the maintenance team, other team members, the designers, your supervisors and your colleagues. Interpersonal skills are needed as you work with others. These include skill in communication, listening, building and maintaining trust and building on personal uniqueness.

Furthermore, some factors affect performance: our performance and that of others. These factors include pride and willingness to risk admitting a mistake, stress and distress, and the environment.

Consider interpersonal skills and then factors that affect performance.

## 7.1
## Interpersonal Skills

Included here are skills in communication, listening, applying the fundamentals, building trust and accounting for personal uniqueness.

### 7.1.1
### Communication

Communication is speaking and writing that a) correctly identifies multiple *audiences*, answers their needs and questions; b) has *content* that includes evidence to support conclusions, c) is well *organized* with summary and advanced organizers, d) uses a *style* that is coherent and interesting, defines jargon or unfamiliar words, and e) includes a *format* that is grammatically correct and follows the expected format and style. These five elements should characterize every communication. In addition, in TS situations requests for changes in operation and for tests should be in writing and should consider the safety of the operator.

**Activity 7-1:** Communication
Critique the following request from the point of view of *audience, content, organization, style and form*. Then, if pertinent, rewrite the communication.

*Successful Trouble Shooting for Process Engineers.* Don Woods

ISBN: 3-527-31163-7

a. A verbal request from engineer Jose to process operator, Marco. The context is that Jose is trouble shooting the process and believes that the cycling observed in the controller on the reflux relates to the changes in composition in the overhead. Jose is looking for a time variation.

*"Take about a dozen samples of the overhead and send them to the lab for a full analysis. Tell them it's urgent."*

b. In March, hourly data have been collected on the day shift over three weeks so that you can do a detailed analysis of the process operation. In April the process will be shut down for routine maintenance. Andre realized it was important to ensure that the meter readings collected over the test period were accurate. The last week of March Andre wrote a memo, not an e-mail, to the Pulp Mill engineer:

*"Would you please request that the instrument shop determine the calibrations on the six flowmeters during the April shutdown. Thanks."*

c. "Record the temperatures at the top of the distillation column once every shift." Fact. There is only a temperature indicator at the top of a 30-tray, self-standing tower. Question: how many operators do you think actually climb the tower in the rain to read the instrument?

### 7.1.2 Listening

Listening includes focusing attention on the talker, avoiding distracting behaviors, showing respect and frequently acknowledging through appropriate body language and "ahums" and reflecting statements. The process can be modeled as Sensing, Interpreting, Evaluating and Responding or SIER. That is, we sense the message; we internally interpret what is being communicated; we evaluate the message in the context of the situation, our feelings, needs and goals; and we select how to respond.

Here's what we know about listening:

- sensing the message is complex because about 55% of the message is communicated by body language, 38% by tone and 7% by the words;
- listening is about four times slower than thinking;
- about 80% of our waking hours are spent in verbal communication; with about $^1/_2$ of that spent listening;
- untrained listeners understand and retain between 25–50% of a conversation;
- only about 5% self-assess themselves as being highly skilled listeners.

In TS, we usually need to gather key information from others. We ask them questions; we listen to their answers. Trouble shooters need to encourage the person to communicate clearly, to listen carefully and to interpret the answer correctly. Three listening skills to aid this process are: attending, following/tracking and comprehension checking/reflecting. Here are the details.

*Attending:* posture is inclined forward and open, facing squarely approximately 1 m apart; no distracting behavior and eye contact is called "soft focus" (contrasted with looking away or staring).

*Tracking/following*: provides minimal encouragement (for example, *"Tell me more", "sure.." "Oh..", " Then .."*) and infrequent questions (for example, prefer "*What?*" questions to "*Why?*") and attentive silence.

*Reflecting* is responding with a concise restatement of the content and feelings expressed in the listener's own words. That is, include the content and feelings of what was said, express it in the listener's own words, do not add new ideas, do not leave out ideas. Some example approaches include saying *"As I understand it.."* or *"Are you saying that .."* Reflecting is usually used when someone is very emotional, or when you see differences developing between you and the other person, when there is disagreement, when the talker seems to be confused or when the talker needs encouragement that his/her contribution is valuable.

**Activity 7-2:** Listening

Assess the quality of the listening by the engineer in the following situation. Note the five strengths and the two areas to work on.

*Engineer Ahmed goes out to the control room for* **Case #7**, *the case of the reluctant crystallizer. Upon entering the control room Engineer Ahmed says, "Hi Phil. I'm new here. I understand from Frank that you're a great plant operator. I hear that we've got trouble on this VC. Let's see, that gauge is vibrating all over the place, so is that one. Hey, this is a zoo. Don't worry I'll solve it soon."*

**Activity 7-3:** Reflecting

For **Case #7**, operator Phil said "When the liquid level started to drop I went up and listened to the booster ejector and it sounded fine. The only thing I noticed was that the pressure gauge on the bay water line to the booster condenser was fluctuating wildly."

Which of the following might you use to show Phil that you are listening to him?

a. "Ahum"
b. "Ok, please continue."
c. "As I understand it, the booster ejector sounded as you expected when the liquid level started to drop and the pressure gauge on the bay water line *really* fluctuated. Is that correct?"
d. "Why is that noteworthy, Phil?"
e. "I'm listening"
f. Other

### 7.1.3 Fundamentals of Interaction

The fundamentals of interaction are summed up in the seven RIGHTS and the four destroyers. Consider each in turn.

- Claim and honor the seven fundamental rights of individuals, RIGHTS (Woods, 1994)
  R the right to Respect
  I the right to Inform or to have an opinion and express it
  G the right to have Goals and needs and express these.
  H the right to Have feelings and express them.
  T the right to have had Trouble and make mistakes and be forgiven.
  S the right to Select your response to other's expectations.
  and the right to Claim these rights and honor these in others.

- Avoid the four destroyers of relationships (Woods, 1994):
  Contempt
  Criticism
  Withdrawal and stonewalling
  Defensiveness.

I remember these by recalling the phrase "Did the *contempt*uous *crit*ter sit on *de fence* or the *stone wall*?"

**Activity 7-4:** Fundamental rights and destroyers
Analyze the conversation and identify claiming of rights, honoring rights in others and evidence of the four destroyers.

Two engineers Tonya and Marcos are trouble shooting **Case #8:** *the depropanizer: the temperatures go crazy.* Let's listen in on their conversation. The parts of the conversation are coded to guide discussion.

Tanya: "OK, the six hypotheses are (i) 1. tray collapsed in the stripping section, 2. too much bottoms fed to the debutanizer, 3. too much overheads in the feed, 4. feed valve FV1 stuck, 5. pump F-26 not working, 6. not enough feed to the column (ii). What do you think? (iii)

Marcos: "It's got to be a collapsed tray. (iv) I encountered something like that on the S256 plantlast year. Same evidence. It's got to be a tray. (v)"

Tanya: "Hey! (vi) You've done it. (vii) You've zeroed in on one hypothesis when we need to keep an open mind and do some simple checks first. (viii)"

Marcos: "Don't get huffy about it. (ix) I'm just trying to get this solved fast. (x) My experience tells me it's the tray! So what great insight are you bringing to this problem? (xi) Ms. Smarty. (xii) Besides your hypothesis #6 is a symptom, and not a root cause. (xiii)"

Tanya: "Now you resort to criticizing my hypotheses. (xiv) That's not fair (xv)"

### 7.1.4 Trust

Build and maintain trust. Trust glues relationships together. Trust is based on integrity, competency and benevolence (not doing anything that will hurt the other purpose).

- We build trust by such acts of integrity as:
  - keeping commitments to yourself and others.
  - clarifying expectations that you have of yourself and of others.
  - showing personal integrity, honesty and loyalty to others.
  - promptly and sincerely apologizing when you know you are wrong.
  - honoring the fundamental RIGHTS listed above and avoiding the destroyers.
  - taking time to see things from the perspectives of others.
  - accepting others "warts and all."

  and by such benevolent acts as:
  - not saying ill of the person behind his/her back or when they are not present.
- We destroy trust by
  - the reverse of the Builders of trust listed above, and
  - not meeting commitments.
  - selectively listening, reading and using material out of context.
  - not accepting the experience of others as being valid.
  - asking others to give up their fundamental RIGHTS.

  and such non-benevolent acts as:
  - making changes that affect others without consultation.
  - playing the broken record until you've eventually worn them out.
  - subtly making changes in the context/issues/wording gradually so that they are unaware of what is happening until it is too late. They were side-swiped.

**Activity 7-5:** Trust

Someone requests "Would you please be chair of the upcoming conference. It won't take much work and you are the ideal person to do it". Your situation is that you have been chair of a similar conference before; this would take the equivalent of at least 2 months of concerted effort. You have promised your family to spend more time with them. You are just barely managing to meet your commitments now. The conference will draw many from abroad and being chair would bring you a lot of personal satisfaction as well as increase your visibility and reputation. How do you respond?

**Activity 7-6:** Self-assess trust

Complete the inventory about trust given in Worksheet 7-1.

**Worksheet 7-1:** Trust.

Trust is having the confidence that you can mutually reveal aspects of yourself and your work without fear of reprisals, embarrassment or publicity.

Trust works both ways: you trust them and they trust you. Trust is not developed overnight, trust takes time to develop. Trust can be destroyed by one incorrect act.

Check your current status

| **Building your trustworthiness getting them to trust you** | **already do this** | **needs some work** | **need lots of work** | **unsure if this is for me** |
|---|---|---|---|---|
| 1. Do what you say you will do. | ☐ | ☐ | ☐ | ☐ |
| 2. Be willing to self-disclose: don't hide your shortcomings; share yourself-honestly. | ☐ | ☐ | ☐ | ☐ |
| 3. Listen carefully to others and reflect to validate your interpretation. | ☐ | ☐ | ☐ | ☐ |
| 4. Understand what really matters to others; do your best to look out for their best interests. | ☐ | ☐ | ☐ | ☐ |
| 5. Ask for feedback. | ☐ | ☐ | ☐ | ☐ |
| 6. Don't push others to trust you more than you trust them. | ☐ | ☐ | ☐ | ☐ |
| 7. Don't confuse "Being a buddy" with trustworthiness. | ☐ | ☐ | ☐ | ☐ |
| 8. Tell the truth. | ☐ | ☐ | ☐ | ☐ |
| 9. Keep confidences. | ☐ | ☐ | ☐ | ☐ |
| 10. Honor and claim the 7 RIGHTS. | ☐ | ☐ | ☐ | ☐ |
| 11. Don't embarrass them. | ☐ | ☐ | ☐ | ☐ |

| **Checking your trustworthiness do they trust you?** | **always** | **most times** | **some-times** | **don't think applies** |
|---|---|---|---|---|
| 1. Do they disclose confidential information trusting that you will keep it confidential? | ☐ | ☐ | ☐ | ☐ |
| 2. Do they assign you challenging tasks to do without frequently checking up on you? | ☐ | ☐ | ☐ | ☐ |
| 3. Do they honor your RIGHTS? | ☐ | ☐ | ☐ | ☐ |
| 4. Do they seem to look out for your best interests? | ☐ | ☐ | ☐ | ☐ |
| 5. Honest and forthright. | ☐ | ☐ | ☐ | ☐ |
| 6. Do not leave you feeling that they haven't told you everything about the situation; they seem to be holding back. | ☐ | ☐ | ☐ | ☐ |

### 7.1.5
### Building on Another's Personal Uniqueness

- The Unique You and the Unique Them. Each of us has our biases, prejudices and preferences or style. A variety of questionnaires and inventories can be used to help understand preferred styles for managing conflict, making decisions, applying creativity, differing style of conversing, validating ideas, gathering and using data, accounting for facts versus feelings and considering details versus the big picture. Inventories of Johnson and Johnson (1986), Kirton (KAI: 1976), and Jung (MBTI: 1984) are examples of such inventories. Your style is unique; it will differ from others. Accept, respect and improve the quality of your interaction through these differences. Do not let these differences lead to conflict. Some applications of this are given in Section 6.1.3.3.

**Activity 7-7:** The unique you
You are on a trouble-shooting team with the following people whose personal style are given in the following table. Write in your scores.

| | Johnson style for conflict | | | | | Jungian | | | | Kirton |
|---|---|---|---|---|---|---|---|---|---|---|
| | Withd | Accom | Force | Comp | Negotiate | IE | TF | PJ | SN | |
| You | | | | | | | | | | |
| Marie | –1 | 2 | –3.5 | 3.4 | 9 | I | T | J | S | 82 |
| Phil | –6.1 | 1.3 | –5.8 | 0.6 | 7.5 | E | F | P | S | 98 |
| Jean | –1.1 | 3.7 | –3.1 | 2.4 | 6.7 | E | T | J | S | 87 |
| Terry | –4.1 | 3.5 | –6.4 | 1.2 | 7.8 | E | F | P | N | 83 |

1. Where are your blind spots? Describe this in actions. For example, if you are a dominant S then your blind spot might be *seeing the big picture, focusing too much on the details.*
2. Does the team have any blind spots?
3. With whom might you have minimum differences? What are those differences?
4. With whom will you have maximum differences? What are those differences?
5. How can you build on this to trouble shoot efficiently and well.

We should also recognize the personal tendency or bias to prefer to report interpretation and inferences, instead of "just the facts". This is discussed in Section 7.2.5.

## 7.2 Factors that Affect Personal Performance

If an operator makes a mistake, the type of mistake is likely to be:

90% *no action* taken (when some kind of action was needed),

5% took corrective action but moved the correct and appropriate variable in the *wrong direction*. Thus, he/she knew that the temperature should be changed but increased it instead of decreased it.

5% took corrective action on the *wrong variable*. Thus, he/she should have changed the temperature but, instead, changed the composition to the reactor.

Why do operators tend to take no action when a clearly defined action is needed? Such factors as inability to admit error, stress, alienation and lack of motivation, tendencies to infer and an "I know best" attitude all affect anyone's ability to perform a task. These factors can affect the operators and people with whom we must interact, and these factors can affect us.

### 7.2.1 Pride and Unwillingness to Admit Error

Each person has his/her own pride and self-esteem to keep intact. We do not want to admit:

- that we are wrong;
- that we made a mistake;
- that we are guilty of wrongdoing.

People want to behave so that they "look good" in the eyes of those who matter... their spouse, their colleagues, their supervisors, themselves.

Not all evidence may be presented because some mistakes may have been made or we are embarrassed and do not want to show we have made a mistake.

*Example 7-1:*
Marg, the design engineer, is now part of the startup team. She is convinced that she designed the reactor correctly. It may be difficult for her to work with data that suggests the design is wrong.

*Example 7-2:*
An operator, while making a preliminary inspection, was to observe and not touch. However, he opens and closes a valve. However, the valve does not go as far shut as he thinks it should.

*Example 7-3:*
For a working process, an operator accidentally drops a mercury thermometer into the receiving and blend tank. The mercury will be dispersed throughout the feed

but "the concentration will be very low" and the "glass should be all broken up anyway." In a couple of weeks the catalyst activity seems to fall off.

*Example 7-4:*
The new regulation for energy conservation requires that all bypass valves on steam traps be closed. Bill strongly disagrees with this policy and cracks open the bypass valve. He thinks the process works better that way. Later it is found that the temperature control on the reboiler seems to be inadequate.

*Example 7-5:*
Engineers may visit the plant site, but they may not touch or adjust things. Peter, on a recent visit to the site, notices a small leak around a valve stem. Believing that sometimes a leak can be stopped by turning the stem slightly and then returning it to its former position, he tries this. However, the valve jams after he turns it from half open to quarter open. He cannot get it to return to half open. He leaves and returns to his office. A few minutes later he is called out to the plant because there is trouble on the plant.

### 7.2.2
### Stress: Low and High Stress Errors

We encounter stress as an accumulation of events from home, work and play. The work environment also provides daily stress: through the amount and type of interruptions, the noise and cleanliness of the environment and the complexity of the tasks being done. The amount of stress one has experienced affects our performance. If there is not much stress or there is too much stress in our life, we are prone to make mistakes.

Powers and Lapp (1983) and Kletz (1986) summarize the probability of different types of operator error to occur. For tasks involving sensing – interpreting – acting, *if* the person is well trained, and motivated and with no stress then,

- he/she will make about 1 error in 1000 trials if feedback is given to the person after they have made the action;
- he/she will make about 1 error in 100 trials when there is no feedback to the person for the action taken.

*if* the person is under distress – not because of high stress levels cumulating throughout the year – but because of the situation, then

- he/she will make about 1 error in 10 trials. This might happen in a busy operating center where other alarms are sounding, the telephone is ringing and people are asking for information about a part of the process.
- he/she will make about 1 error in 2 trials if, for example, many complex actions are required and the implications if an error is made are frightening.

The distress comes from poor training, confusion because of poor training, conflicting data; from the need for fast action; or from a large penalty if a mistake is made, or from extensive confusing and contradictory types of demands.

The training should be competency based and oriented to develop skill, and not just to develop knowledge. Training is particularly important when new technology is introduced, for infrequent events and for new people.

When we TS, it is wise to monitor our own stress and to be sensitive to the stress experienced by others.

**Activity 7-8:** Stress and control

One common cause of stress is that we worry about – and get angry over – things over which we have no control. Research evidence suggests that we have control over only 1 out of 10 things we might worry over.

Trouble shooting can be particularly stressful. You are under a lot of pressure to find the root cause and correct it safely and quickly. You are trouble-shooting case #12 given in the list below. Which of the following do you have control over?

**Worksheet 7-2:** The problem given in **Case #12** arises at 8:30 am. The following background and resources are available. Identify which ones you think are directly under your control. If you think you have control over the item, circle Y; if you do not have control over it, circle N.

| | | |
|---|---|---|
| 1. The safety officer, Hack, is overbearing, not liked and gets carried away about simple things. | Y | N |
| 2. The laboratory can analyze liquid samples with their equipment but gaseous samples cannot be analyzed because their instrument is broken. | Y | N |
| 3. The lab schedule is busy with top priority analyses. Samples could not be analyzed until after 3:00 pm. | Y | N |
| 4. The union prevents you from analyzing any samples; if you do, there will be a strike. | Y | N |
| 5. The upstream styrene plant is operating at 80% capacity. | Y | N |
| 6. The upstream ethylene cracking furnace is operating at 95% capacity. | Y | N |
| 7. The upstream propylene plant is shut down. | Y | N |
| 8. The operator on the ethylene plant is cooperative. | Y | N |
| 9. Samples can be taken at any of the sewer gates within the Battery Limits and at A, B, and C. | Y | N |
| 10. No blueprints are available for the sewer system. | Y | N |
| 11. Your performance review to establish your salary is being done next Thursday at 4:30 pm. | Y | N |
| 12. You have to prepare your "record of progress" record for the performance review on Thursday. | Y | N |

**Activity 7-9:** Stress and self-talk

A common contributor to your stress is your self-talk. Too many people say to themselves *"You are stupid" "You can't do this"*. Make a list of any negative self-talk comments you say to yourself-during a week. Create an action plan to minimize your negative self-talk and to maximize your positive self-talk.

**Case #12** Safety on the ethylene plant (courtesy of John Gates, B. Eng. 1968, McMaster University)

The part of the ethylene plant that relates to this problem concerns the drying section to remove moisture from the feed gas and the distillation train to separate the gas stream into the desired component section. Drying section: Three alumina dryers are installed. One dryer is regenerated while two are hooked in series on stream. For example, dryer V106 is being regenerated with V107 and 108 removing the moisture in the process stream to less than 4 ppm. Then V107 will be regenerated with V108 and 106 in series and so on. The cycle lasts 12 hours. The dried gas goes to a knockout pot (to remove any entrained material) and then is chilled, in exchangers E107 and 108, before entering the separation towers.

During regeneration of the dryers, "fuel gas", heated with 2.8 MPa steam, flows through the dryer in the direction reverse to normal flow. Once it is through the dryer the regeneration effluent gas is cooled and returned to the fuel-gas system. During regeneration the dryer temperature rises to 190 °C and is maintained at this temperature for one hour. Then the fuel gas bypasses the heater and is sent directly to the dryer to cool it. For this plant, the "fuel gas" or regeneration gas is the non-condensable overhead from column T 101, the demethanizer.

The dryers are all appropriately manifolded and valved so that any dryer can be regenerated, bypassed or used. The regenerating dryer is separated from the line dryers by gate valves.

The separation is performed in a train of three distillation columns operating at about 3.2 MPa. These are a demethanizer, de-ethanizer and C2 splitter, T101, T102 and T103, respectively.

The process is illustrated in Figure 7-1. The overhead from Tower T101 is condensed with ethylene as refrigerant. The overhead from Tower T102 is condensed with propane as refrigerant. The overhead from Tower T103 is condensed with propylene as the refrigerant.

The safety inspector telephones to say that this morning's gas samples from the sewer drop boxes within our battery limits (for ethylene distillation units) are explosive. These sewers serve other plants: styrene, ethylene furnaces and propylene plants. The plot plan of the unit is shown in Figure 7-2. Clear up this matter immediately; this hazard cannot be permitted!

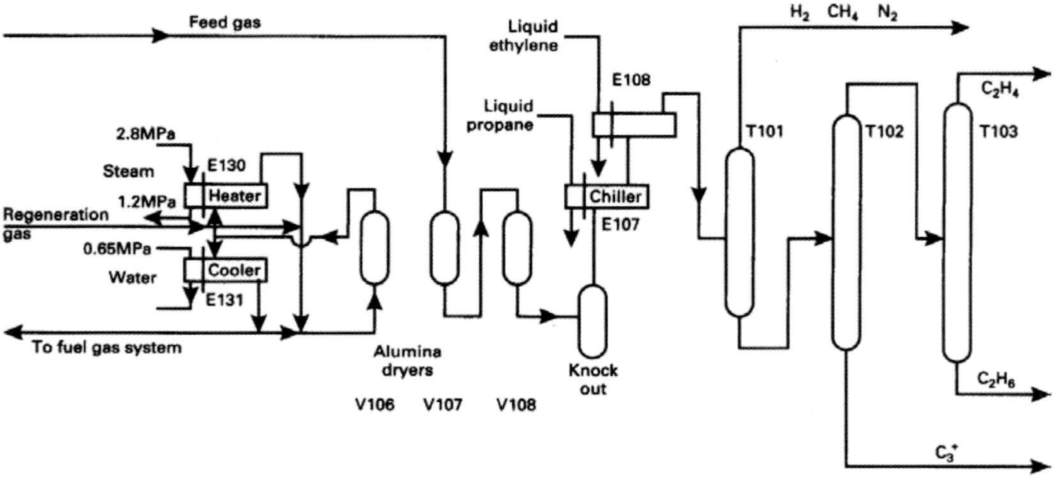


Figure 7-1 The ethylene plant for Case #12.

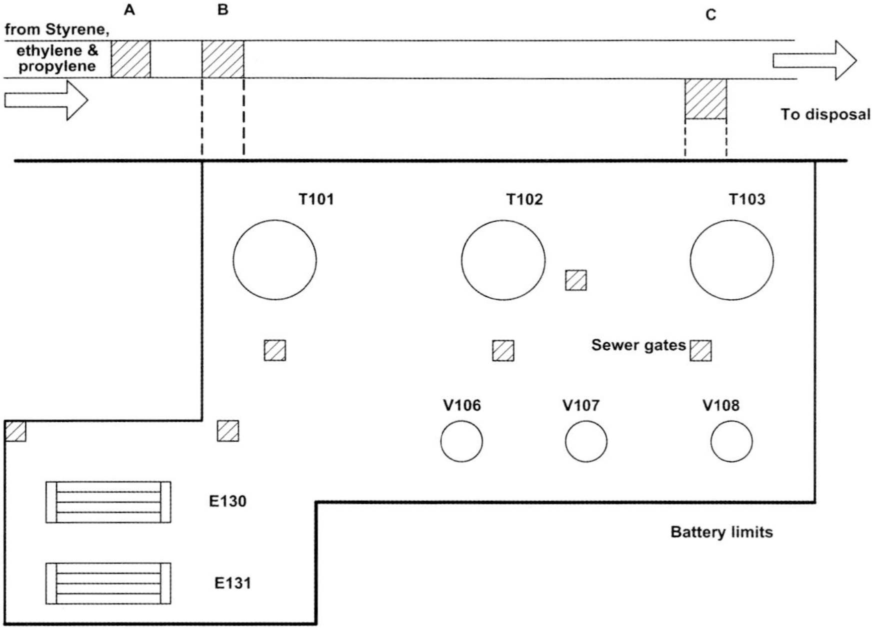


Figure 7-2 Plot plan of the ethylene plant for Case #12.

### 7.2.3
### Alienation and Lack of Motivation

The working environment should motivate employees. Employees should have a clear idea of the expectations, and rewards should be forthcoming for achieving those expectations. People are demoralized if people considered to be "stars" are always given the challenging work, if information sessions about the company are not given and if the promotion and salary adjustments are unclear or seem to be based on who you know and not on performance. People are alienated when a) tasks lack imagination or are to be performed under strained conditions, b) where the working conditions are sloppy maintenance, extremes in temperature, odors and dust and hazards, c) where the emphasis is on machines ... and not on people, and d) when management decisions lack consistency, predictability and transparency.

### 7.2.4
### "I Know Best!" Attitude

In other situations, sometimes, people deliberately fail to follow instructions because "they know better". An operator opens the bypass valve on the steam trap or overrides a control system because "those silly systems don't work! I know best." This attitude is often related to increased alienation and lack of motivation.

### 7.2.5
### Tendency to Interpret

Instead of saying "the gauge reads ..." we tend to say "the temperature is ..." It is vital that a trouble shooter listens carefully to what is said and how he/she interprets the information.

Communicating just the facts is boring. We want to describe what we think is wrong.

*Example 7-6:*
The designer might say "*the rogue data point for the heat capacity of acetic acid vapor, shown in Figure 7-3, is probably caused because the instruments or the researcher made a mistake*" and proceed to design the preheater assuming the heat capacity at 200 °C is about 0.45 CHU/lb °C.

Comment: the rogue point is correct.

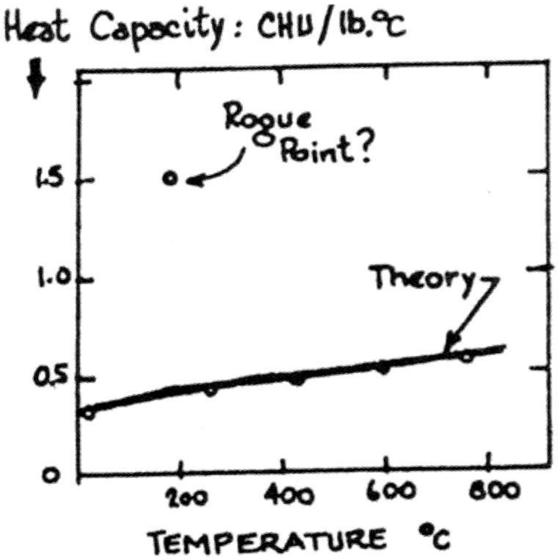


**Figure 7-3** Heat capacity of acetic acid vapor.

*Example 7-7:*
The operator might say that *the valve on the bypass is misbehaving again; see the instabilities in the flow control of the reactor feed.* Is it really the valve? Is there really an instability?

*Example 7-8:*
The engineer might suggest that the reason *the low yield for the reformer, compared with the expected yield, is caused by low catalyst activity.*

*Example 7-9:*
The engineer says that the *pressure in the line is 50 kPa; the temperature is 450 °C.* Comment: the engineer should have said the pressure gauge reads 50 kPa; the thermocouple reads 450 °C.

**Activity 7-10:** Triad talking
In groups of three, one is the "*talker*", one the "*fact-summarizer*" and the other is the "*inferrer*". The *talker* describes, for three minutes, his/her greatest frustration, favorite activity or hobby. The *fact-summarizer* takes one minute to summarize the facts given by the talker. No personal opinions or inferences should be given. The *inferrer* describes in one minute what inferences and messages came through from the ideas presented, by the body language and by what was not said. Rotate responsibilities so that each has a chance to play all three roles. This activity takes 15 minutes total. At the end, share your experiences.

**Activity 7-11:** Pair talking and "just the facts"
In pairs, with one being the *talker* and the other the *listener*. The *talker* reads over **Case #13** "The Lousy Control System" to himself/herself-for about five minutes. Then, referring only to the scenario for details the *talker* has three minutes to describe the facts in the Case to the *listener*. Do not simply read the scenario to the *listener*. You may show your *listener* the sketch of the process but you cannot show the written material. You have to describe the facts.

Change roles and repeat with **Case #14**, "The condenser that was just too big."

After each role, take two minutes and write out your reflections about the activity. Discuss with your partner.

**Case #13** The Lousy Control System (courtesy Esso Chemicals)
All the texts on process control say that controlling the overhead condensate exit temperature by varying the fan pitch is a slow and clumsy method of control. Yet, that is the method used on column T6. Today it is raining, the temperature of the condensate is 3 °C subcooled. Yesterday it was hot and sunny. The fans were running flat out but still the gas exit valve controlling the pressure was open half-way most of the day. All the uncondensed gas went to flare. The boss storms in "You've got trouble on this plant; too much stuff went to flare yesterday. What's wrong?" You're sure that the control system is lousy. The system is shown in Figure 7-4.

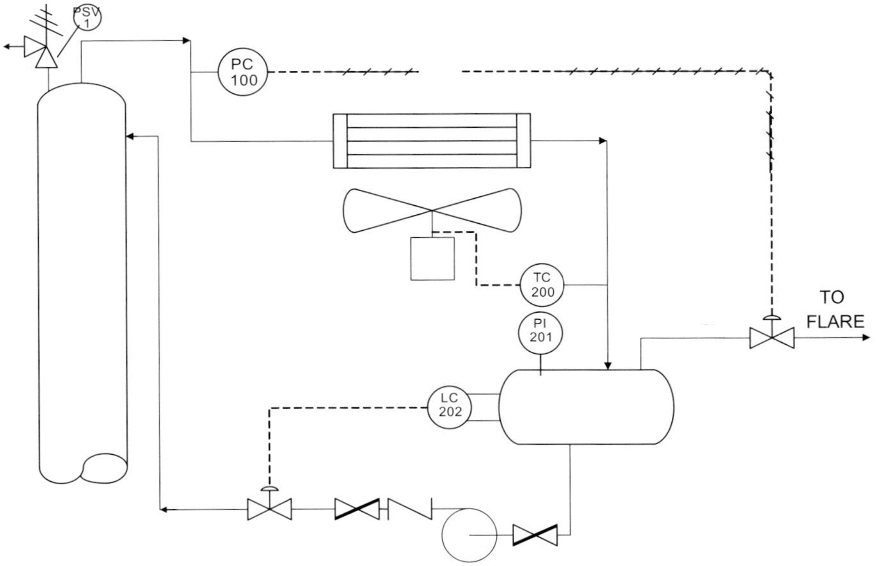


**Figure 7-4** The control system for Case #13.

**Case #14** The Condenser That was Just too Big

Fatty acids are solid at room temperature. Fatty acids are purified in a distillation column operating under a very high vacuum. The "coolant" in the overhead condenser cannot be water because the fatty acids would solidify in the condenser. Instead the coolant is boiling water with the "coolant" temperature controlled by the pressure at which the water boils. This plant is started up for the first time. A common fault in startup situations is that the condensers are over-designed for "clean conditions" because fouling factors have been used in the selection of the overall heat-transfer coefficient. This would mean that at startup the condensing fatty acids would be subcooled. When you go out on the plant one hour after startup, the worried operators say "Look at the overhead pressure gauge – it's much too high. It reads 2.7 kPa (20 mm Hg absolute) when it should be reading 0.4–0.7 kPa (3–5 mm Hg)." Just as you expected! the fatty acids are being subcooled in the condensers. Solid is starting to build up in the tubes. The excessive pressure drop across the tubes means that the vacuum system can't get rid of the air leaks fast enough and the pressure builds up. You pull out the drawings showing the configuration of the plant (Figure 7-5). This problem needs to be solved fast!

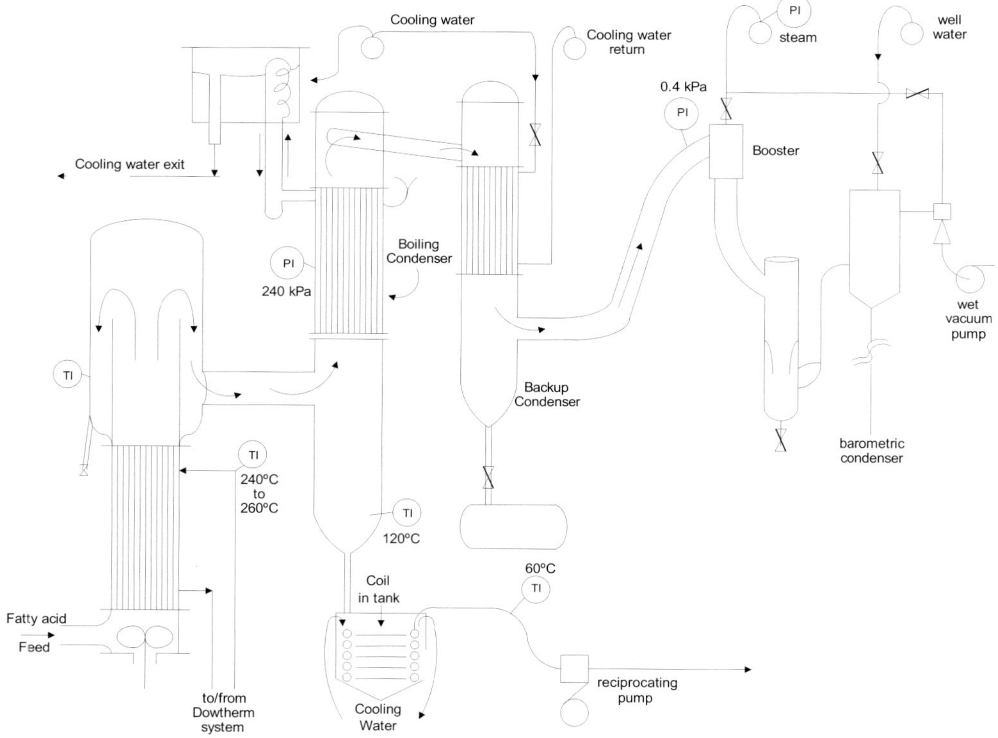


**Figure 7-5** Fatty-acid distillation column for Case #14.

## 7.3
## The Environment

The environment affects performance. If making a mistake is considered the worst possible thing to do; then no one will admit to making a mistake. If mistakes are accepted, understood and no serious repercussions occur, then people will be more ready to admit their mistakes.

**Activity 7-12:** Rate the environment
Use Worksheet 7-3 to help reflect on the environment and its impact this might have on the people with whom we interact.

## 7.4
## Summary

We interact with people when we trouble shoot. Our skill in trouble shooting often depends on our skill in communicating, listening, applying the basic fundamental principles of interpersonal relationships, building trust and understanding our own uniqueness and the uniqueness of others. These ideas were illustrated and activities provided to give you a chance to work with these skills.

We also need to be aware of the factors that affect our performance and the performance of others. These include unwillingness to admit having made a mistake, and the impact that stress has on our ability to work effectively. Sometimes people are frustrated and unmotivated. Sometimes people follow their own approach even though it may result in incorrect and even unsafe operation of the process. We also prefer to infer and interpret what we see or hear. Reporting the facts is usually boring. So we read a gauge and say "The pressure is 1.4 MPa" when in reality we should have said "The pressure gauge reads 1.4 MPa."

A questionnaire is included to give us a chance to evaluate the environment in which we trouble shoot.

## 7.5
## Exercises and Activities

1. Consider ten cases in Kletz's text "What went wrong?" From Kletz's description of the case, classify the cause of the equipment malfunction, human design mistake, human maintenance mistake, human operator mistake or other.
2. Typical feedback from industrial workshops on trouble shooting.

For **Activity 7-10:** *Fact-summarizing was either easy or difficult depending on the quality of the talker's presentation. Taking notes helped. Inferring was most enjoyable.* Compare your experience with this.

For **Activity 7-11**: *Giving facts was difficult. Take my time. Repeating was difficult. Write everything down.* Compare your experience with this.

3. Cases to consider. For Cases **#15, 16, 17** and **18** create hypotheses as to the cause.

**Worksheet 7-3:** Feedback about your environment.

To what extent do you agree with the following descriptors of your environment where you usually "trouble shoot".

**People are willing to admit error:** "The people that I work with are very unwilling to admit errors; they blame others, they pass the buck and, if necessary, would purposely mislead me rather than to admit error."

| Strongly Disagree | Moderately Disagree | Slightly Disagree | Slightly Agree | Moderately Agree | Strongly Agree |
|---|---|---|---|---|---|
| 1 | 2 | 3 | 4 | 5 | 6 |

**Encourage risk taking:** "Risking is rewarded. We are expected to take risks about 10 times a day. Risks should be wisely, not indiscriminately selected. But, nevertheless, we are not only encouraged but we are rewarded for risk taking."

| Strongly Disagree | Moderately Disagree | Slightly Disagree | Slightly Agree | Moderately Agree | Strongly Agree |
|---|---|---|---|---|---|
| 1 | 2 | 3 | 4 | 5 | 6 |

**General stress at work *they* are under:** "Their environment is very stressful. People have many deadlines and interruptions. The consequences of making mistakes is very high. The issues are complex. The environment changes often and includes a lot of uncertainty."

| Strongly Disagree | Moderately Disagree | Slightly Disagree | Slightly Agree | Moderately Agree | Strongly Agree |
|---|---|---|---|---|---|
| 1 | 2 | 3 | 4 | 5 | 6 |

**General stress at work *you* are under:** "My environment is not stressful. I do not have many deadlines or interruptions. The consequences of making mistakes are low. The issues are straightforward. The environment is safe, stable and secure."

| Strongly Disagree | Moderately Disagree | Slightly Disagree | Slightly Agree | Moderately Agree | Strongly Agree |
|---|---|---|---|---|---|
| 1 | 2 | 3 | 4 | 5 | 6 |

**People's listening and responding:** "The people are open, communicate well, can clearly identify *facts*, will offer opinion when asked for, and are very competent but are not aggressive "know-it-alls"".

| Strongly Disagree | Moderately Disagree | Slightly Disagree | Slightly Agree | Moderately Agree | Strongly Agree |
|---|---|---|---|---|---|
| 1 | 2 | 3 | 4 | 5 | 6 |

**Case #15** The flooded boot (problem supplied by W.K. Taylor, B. Eng. McMaster 1966)

As part of our energy conservation program we recycle condensate to our boiler house. This condensate comes from large, steam-driven turbines whose exhaust conditions are full vacuum. These turbines drive the feed gas, air, refrigeration and synthesis gas compressors on the reformer section of the ammonia plant. The exhaust steam goes to the shell side of a surface condenser. Chilled water is on the tube side. Vacuum is maintained by a two-stage steam ejector with inter and after condensers. These are located downstream of the surface condenser and pull a vacuum on the condenser itself.

Lately we have been increasing the load on the compressors and hence the steam consumption has increased. However, for these new conditions we can't seem to get rid of the condensate from the boot at the bottom of the surface condenser. We sometimes have to run the spare pump 122.JA (electric drive) because the turbine-driven pump 122.J cannot handle all the flow. "These high levels of condensate and the unusual operation with our spare pump are keeping my fertilizer production down. It's got to stop! My production rate is barely reaching 120 Mg/d and I think we can easily produce 132 Mg/d. That's $5000 a day. Fix this bottleneck!". The system is given in Figure 7-6.

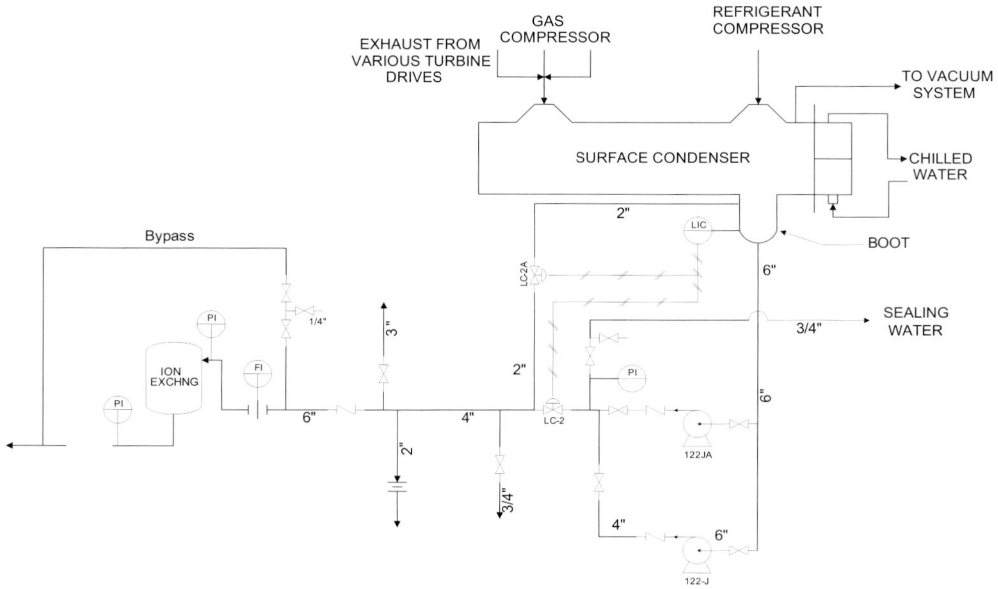


**Figure 7-6** The configuration of the flooded boot for Case #17.

**Case #16** The case of the dirty vacuum gas oil (problem supplied by R.J. Farrell, B. Eng. McMaster University, 1974)

Word spread quickly. "There's trouble on the vacuum tower!" Normally the side product, vacuum gas oil, is water clear. Today the product has been off-specification in color. Indeed "It looks brown" explained a frustrated supervisor. The feed to the vacuum tower is crude oil. The crude oil is heated by a series of three exchangers and the furnace. The first two exchangers extract heat from the vacuum gas oil (API 27.8), and from the mid-side-draw cut that has an API gravity of 25.4. The third exchanger extracts heat from the tower bottoms (of API gravity 11.5). The column has pump-around draw-off for the bottoms, to recycle to crude, to mid-side, and to vacuum oil-overhead condensates. The mid-side and vacuum gas oil are blended and cooled to yield a product of API 26.2. The blend is sampled after the cooler; the blend is off-specification. Fix the problem.

**Case #17** Is it hot or is it not? (case supplied by Jonathan Yip, B. Eng. 1998, McMaster)

The crude fatty-acid high-vacuum fractionation column is having trouble. The bottoms recirculation and transfer pumps need maintenance immediately. To cope with this the engineer placed the column on safe-park. The feed was stopped and the sump and overheads recycled back to the crude fatty-acid tank. The steam ejectors were turned off and all the valves were shut to keep the column under vacuum.

The pump was dismantled for repair and an hour into the repair job the plant operator noticed that the temperature gauge at the top of the column was showing an increase in temperature. Indeed, the temperature alarms started screaming.

"That's crazy!" said the operator. "There's nothing going on in the column. The column is isolated. The column is not running. That temperature should be decreasing instead of increasing. That temperature gauge and those alarms have been temperamental for a long time." He manually suppressed the nitrogen purge valve and cut off the alarms so that he could hear himself-think.

**Case #18** The streptomycin dilemma

In our process for the production of streptomycin we evaporate the eluate in tall rising film evaporators. These are fixed tube sheet with a shell expansion bellows to allow for the thermal expansion. Our plant operates on a single, eight-hour shift so that after each shift the evaporator is shut down. However, because we process pharmaceuticals we must clean and sterilize all units after each shift. The procedure is to open the end of the bundle, to brush the inside of all tubes, rinse, fill the tubes with water and then apply steam to the shell to boil and thereby sterilize the tubes in preparation for the next day's operation. The plant has operated without a hitch for the past four years. However, over the past months the tubes are starting to fail; leaks are developing at the tube/tube sheet and the tubes are buckling. The pattern seems to be random. Get this fixed.

# 8
# Prescription for Improvement: Put it all Together

This is the time to put all your skills together and work on some trouble-shooting cases. The purpose is to polish your skill. In this chapter we suggest several approaches you could take: work in triads or as an individual. Then an outline is given of the different cases included in this book. Then details are given for the cases. Enjoy.

## 8.1
## Approaches to Polish Your Skill

You can work in triads or as an individual.

### 8.1.1
### Triad Activity

One of the most effective ways to develop skill is to use a triad activity in which each person, in turn, plays the role of expert system, trouble shooter and observer. In a 2-hour period each can play each role once. Here are the three roles.

**a. The Expert System**

The *expert system* poses a trouble-shooting Case and then provides responses "from the system" to any request made by the trouble shooter.

In preparation for this role, 1) select a case; read it over carefully, 2) then track through all of the data requests given at the end of the case; 3) locate the answers and then read Appendix E for a debrief and answer. Understand the process extremely well. You might wish to complete a Trouble-Shooter's Worksheet (from Chapter 2) for yourself. Think about how "the fault" will affect all of the process variables. Try to anticipate the kinds of actions that the *trouble shooter* might take. What would the fault do to the system under those conditions? Give the results of experiments. Do not give explanations. Give correct information but do not be generous. If, for example, the fault occurs periodically, and you are asked to give the lab analysis for one sample taken, then assume Murphy's law applies and give them the result when the system was operating normally. Insist that they write out all

*Successful Trouble Shooting for Process Engineers.* Don Woods

ISBN: 3-527-31163-7

requests; write down the results opposite. Do not talk. . . . . just acknowledge that they are working on it by saying "Ahemmm, mmmmm,"

Insist on their instructions or requests be written precisely. If they write "Inspect the instrument" then respond "It's OK". If they ask what you did, then say "I went out and looked at it." Be tough. Do not offer more information than they asked for.

### b. The Trouble Shooter

You have a challenging role to play.

- you are to talk aloud so that the *Observer* can track what you are doing. Think about the process you will be using: are you searching for change? or for the basics?; are you clarifying the situation or testing an idea? You may feel frustrated because the *Expert System* is going to supply written responses to your requests for tasks to be done; the *Expert System* will not discuss things with you. He/she may say "Hmmm" or "mmmmmm" to you occasionally so that you do not feel as though you are talking to a wall, but the *Expert System* is there to provide an instant system written response to your requests.
- you are to display all the good problem-solving skills we have developed. Do this by verbally monitoring your progress, being active with pencil and paper to keep track of the route you are following. Indeed, you might wish to prepare a Trouble-Shooter's Worksheet. Recall, from Chapter 2 (and Chapters 5 and 6), the performance characteristics of successful trouble shooters. Try to display those performance characteristics.
- you are to write out your requests for information from the *Expert System*. These should be written out precisely and unambiguously.

### c. Observer for trouble shooting

Your worksheet is the Feedback form given in Chapter 2, Worksheet 2-2. As the *trouble shooter* is tackling the problem, your task is to assess how well the problem-solving components are handled. This is challenging because the skills are difficult to identify – let alone observe and assess. To some extent your role is similar to the listener role in TAPPS, described in Section 5. 1. The feedback form is made to help you look at the mental process used by the *trouble shooter*. Try to focus more on the "actions taken", and on the "talk-aloud description of the thought process". Look at the organizational pattern used; listen for the monitoring of the process. Consider, "Does he/she confuse activities unknowingly?"

Let the *Expert System* focus on "how well the trouble shooter wrote out the questions and tasks to be done".

**Activity 8-1**

Prepare for the activity ahead of time with each person in the triad selecting a Case and preparing for the role of *Expert System*. As *expert system*, make a copy of the Case to give to the *trouble shooter*. When it's convenient, the triad members meet for at least 2 hours.

1) The person with surname first in the alphabet starts as *trouble shooter.* Next in the alphabet, as *expert system* and last in the alphabet as *observer.* Seat yourselves approximately as illustrated in Figure 8-1, although the barrier between the *trouble shooter* and *expert system* is imaginary.

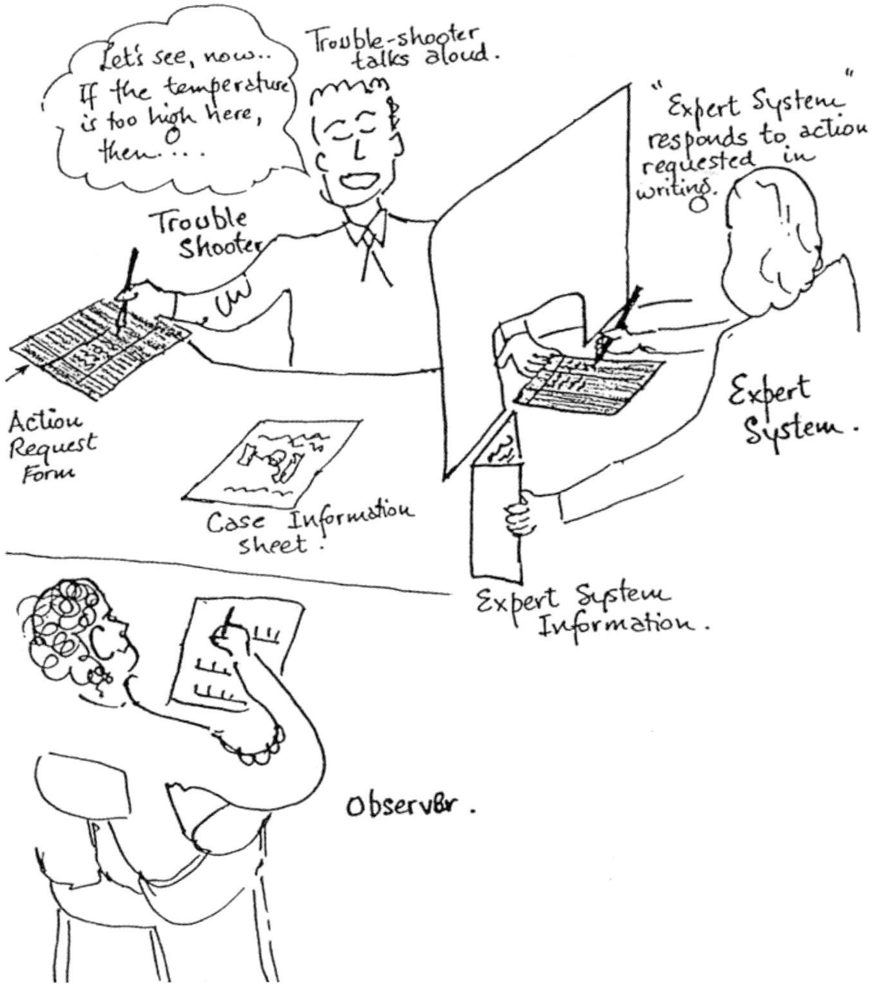


**Figure 8-1** The triad activity.

2) Refresh your memory as to how to play each role.
3) Set the timer for 20 minutes, start with the *expert system* handing the Case to the *trouble shooter.* The *trouble shooter* reads the case aloud and then, by talking aloud proceeds to "solve the case". He/she gathers information by writing out actions to be taken. These actions should be written one at a time with a

written response being given by the *expert system* to each task. The expert system responds in writing and so the role-playing activities continue. Focus on the process and not on rushing to trying to solve the problem in the available time. In the time available, the *trouble shooter* may only complete the Trouble-Shooter's Worksheet and never even ask a question of the *expert system*. That's OK.

4) When the 20 minutes has elapsed, the *expert system* reveals the root cause and possible solution. *This is not a discussion*. This is simply the *expert system* sharing the root cause about the case. Time = 2 minutes.
5) The *observer* completes the feedback form and gently shares his/her feedback with the *trouble shooter* about what he/she observed. 4 minutes.
6) The *trouble shooter* collects the evidence about the process: the case-problem statement, any worksheets, perhaps the Trouble-Shooter's Worksheet, and the action-request form with responses that were written between the *trouble shooter* and the *expert system*.
7) All three people write their reflections of what they learned from the activity. 5 minutes.
8) Discuss this briefly with the people in your triad. 3 minutes.
9) Rotate roles and repeat until everyone has had a chance to play all three roles and gather evidence about how they played the role of *trouble shooter*.

As an aside, although the primary goal of this activity is to improve your trouble-shooting skill through the role as *trouble shooter*, most participants report that they learned the most from playing the role of *expert system*. Interesting.

### 8.1.2
### Individual Activity

As an individual, read over the cases given in Section 8.2. The cases are graded according to degree of difficulty; some cases are written in a general context; they could occur in any process. Others are process specific. For the case you selected, you might start by completing the Trouble-Shooter's Worksheet given in Chapter 2. Then, scan the list of diagnostic actions given, prioritize, select your choices in sequence and obtain the answer *in sequence* from the code in Appendix D. For example, *in* **Case #8** *you might decide that your first activity is "put on safe-hold". The code is* 1880. *From Appendix D, the result of this activity is* 1880: *"Not needed at the beginning until after we have collected data. Indeed if put on safe-hold we won't be able to collect data to figure out what is wrong. $3000"* Keep a record of the codes you use. Continue until you believe you have identified the root cause or have corrected the fault. Check your answer with the fault encountered in industry, recorded in Appendix E. Reflect on the problem-solving process used and compare the sequence of actions with those reported for the case in Appendix E. Total up the cost. You might also give yourself-feedback using the form given in Worksheet 2-2.

## 8.2
## Cases to Help you Polish Your Skill

First, the cases are listed and guidelines provided for each. Then the cases are given together with a set of diagnostic tasks from which you can select. These tasks have been listed from the choices made by previous trouble shooters. I hope that I have included all the options you like.

### 8.2.1
### Guidelines for Selecting a Case

The actual process problems have been selected carefully from our files of several hundred. These represent varying levels of difficulty, different types of situations (from startup to usual operations) and different varieties of equipment. Table 8-1 summarizes these options. In selecting the sequence of cases to work on, I recommend that you start simply and build up your confidence and skill. The criteria to use in selecting the case are:

- Degree of difficulty. Although this is a subjective rating, I think it is useful. The basis for the ratings is given in Appendix E. Start looking at cases at Level 3 and 4. The lower ratings start with **Case #19**.
- Type of equipment involved. The key pieces of equipment (plus people) that are in the process are listed. For example, **Case #19** relates to filtration and pumps. If your *experience with process equipment* is being developed for this type of equipment, then you may wish to return to this case later. Alternatively, you can consult Chapter 3 and trouble shooting and suggestions for good practice early in your consideration of the case. As you scan the diagnostic actions listed for each case, you might find the *Vendor files* and *"call to the vendor"* activities helpful to do early.
- Type of process. Some cases apply to any process. Others include characteristics that are unique to the industry. As a start, you may want to select cases that are "general". Alternatively, some information unique to certain processes may be obtained by consulting the diagnostic actions: *MSDS, process description*, and *Handbook*.
- Continuity. Some cases all relate to similar processes. For example, four cases relate to the depropanizer-debutanizer system. Two, to the ethylene process; five, to the ammonia-reformer. The interconnections are given in column six in Table 8-1. Once you are on a roll with one type of process you might want to continue with problems for the same type of process.

### 8.2.2
### The Cases and Understanding the Choice of Diagnostic Actions for each Case

The cases are listed in Table 8-1. Each case has two parts: the case statement (and diagram) and the list of coded diagnostic actions from which you can select. Here are some more details about how I have designed the diagnostic actions (and the feedback provided for each from Appendix D).

**There are no diagnostic tests listed for the case that *tells you the fault*.** The diagnostic actions provide you with sufficient evidence that you can identify the fault. Identifying the fault, correcting the fault and preventing the fault from reoccurring are left to you. But you have to select the most pertinent diagnostic actions to give you the key evidence.

You are free to select any diagnostic action you wish from among those listed for each case. Every action you request incurs a cost. Unless otherwise specified, I have used \$600/h as a cost for loss in production. To this needs to be added the cost of the people and equipment needed. For example, the diagnostic action of determining "what's the weather today and in the past?" costs \$50 (representing about a 10-minute activity). The diagnostic action of "sampling and analysis" would cost \$3700 to \$15 000/sample depending on the type of analysis involved. (This represents a 6–24-h activity). For each case I tried to assign reasonable costs for each action.

The diagnostic actions fall into four classes. The first actions relate to safety! The second set of actions help you understand the background. The third set are factual gathering of information about the process as it operates now. The fourth set are various tests, calculations and activities to test hypotheses, and provide evidence as to the fault. As mentioned at the start of this section, the activities do not necessarily tell you the fault. Even the actions called *Take Corrective Action* may not correct the fault; they may, but they may not! Do not look for "the answer" to be among the options called *Take "corrective" action.*

Make your selections based on logic combined with safety and economics.

Here are more details of the four sets of activities that you may select.

1. **The first two activities relate to safety**

The first priority is safety: recognize hazards (via MSDS) and take action.

- **MSDS**

Usually all engineers working on any site should know this information already. If you don't for the particular case, MSDS information is available: this usually is expressed as the three National Fire Protection Association, NFPA, ratings for health, fire and spontaneous reaction/explosion for individual chemicals. The ratings range from 0 to 4 with 0 meaning negligible and 4 meaning extreme. Thus an NFPA rating of 0, 4, 0 would mean that the species is not an issue for health, it is extremely flammable and it is stable. I tried to include information about how one chemical might react with another when this is important.

- **Immediate action for safety and hazard elimination**. The first decision relates to safety. The initial evidence might suggest a health, a fire or an explosive hazard. Act now! There are four options:
  - *Continue with the trouble shooting without implementing activities related to safety:* there is no hazard.

- *Put on safe-park:* this keeps the process going but under conditions that are safe. This could mean isolating a distillation column and keeping it on total reflux; or reducing the throughput to conditions that previously did not pose a hazard.
- *Safety interlock shut down, SIS*: this should happen automatically if the control system has been designed with the four levels expected. However, sometimes this has to be actively initiated. This gives you a chance to reflect on the situation and decide if this should have happened.
- *SIS plus evacuation:* the SIS should happen automatically if the control system has been designed with the four levels expected. Now, because of the hazard posed we should add evacuation.

2. **The second set of activities: background that you probably know**

You probably know most of this information, except the weather and information related to maintenance. However, as you are developing your skill this section provides easy access to background information related to the case.

- **More about the process**. This is to provide some reassurance about what the process is about; this additional information might help. I have not included key information that is needed to solve the case in this activity.
- **IS and IS NOT**: this is based on given problem statement. This might help as feedback to you about how you have done this task.
- ***Why? Why? Why?***: may be helpful to put some of the cases in the larger context. You can do this activity on your own and use this question to give you feedback.
- **Weather** Today and past. All cases refer to weather conditions in Ontario, Canada, where there are four seasons with snow and freezing weather for December through March; hot humid summers June to August. Many of the cases are sensitive to the weather.
- **Maintenance: turnaround**. Three conditions might apply: new plant startup, startup of an existing plant that has just been through its annual turnaround and operation after some maintenance has been done.
  - That the problem is with a *first time or startup of a new process* is usually given in the statement of the problem. Therefore, no separate question is posed related to this.
  - The *Maintenance: turnaround* activity relates to startups after the annual turnaround. During the turnaround the minimum is usually
    - the inspection of most pieces of equipment,
    - the replacement of worn parts,
    - the installation of changes to the process, repiping, and changing the operation to implement ideas to optimize or improve operation,
    - the calibration of sensors,
    - cleaning exchangers, and
    - changing the catalyst.

It is important to know "*When and what done?*"

- **Maintenance: routine**. During routine maintenance something will have changed. It is important to know what and the extent. It could be that all the isolation valves were not correctly opened after the maintenance was completed; that the key was not correctly fitted into the drive shaft; that the pump was not primed. The answers I give to this question will not admit to such mistakes. However, this will open your mind to possible things to check and look at.
- **What should be happening** based on office records and files: the resources include the simulation/design computer background; records of the design; the collection of information from the equipment vendors and any internal reports on past tests, or trials done on the plant. Handbook data, mainly properties, for the conditions and species in the case are available. Trouble-shooting suggestions are also available as are simple calculations you might do *based on the given information in the case.*

**Design and simulation files** (allowances made for fouling, overdesign and uncertainties). Here I tried to list all the possible pieces of equipment involved in the case. All equipment is designed according to the Codes and Standards; the information given here are decisions within a designer's judgement.

**Vendor files**: practical information and some specifications from the suppliers of heat exchangers, steam ejectors, pumps and equipment that would be purchased from a vendor. The information may be slightly different from the information in the design files.

**Commissioning data, P&ID, internal reports**. Often new plants are constructed under contract. The contract may include penalty clauses stating a financial penalty that is paid if the process does not perform up to specification within a stated period of time. For insurance and governmental regulations, certain performance standards have to be met. Usually records are kept of the performance trials. However, don't look for much information here for the startup of a new plant.

**Handbook:** physical and thermal properties of the chemicals in the case: vapor pressure–temperatures, steam tables, and thermal properties may be given.

**Trouble-shooting files.** Here I refer to specific sections of Chapter 3 that pertain to the equipment in the case.

- **Calculations and estimations** that can be done in the office based on the information given in the case statement and the diagram (if available). The given information varies from Case to Case. Nevertheless, some simple checks and calculations can provide neat insight. These include:
  - Pressure profile: in most cases we can obtain the pressures on either side of a barrier and then state the direction of flow that would incur if there is a leak in the barrier.
  - Mass balance: nice idea but you usually can't do this at this early stage in the problem because the key data are not given in the case statement.
  - Energy balance: sink = source; the heat lost by one fluid = heat gained by the other. Often we are given enough information in the case statement to allow us to do this check.

- Thermodynamics: use the principles of thermodynamics to predict trends, in general.
- Rate: use ΔTs to estimate whether nucleate or film boiling might predominate. Similarly, we might be able to estimate rates of mass and heat transfer.
- Equipment performance: we *might* be able to estimate the performance of equipment; usually, however, we need more information than that given in the case statement.

3. **The third set are factual gathering of information about the process as it operates now.**

- **What is the current operation**

Unless you arranged over the phone to meet the operator at the piece of equipment, you will go to the control room first and check in. You might scan the data reported in the control room; discuss the operating activities with the operator and explore the operating procedures.

**Visit control room:** control-room data: values now and from past records.

**Process operators:** you can ask for information about what happened (from their perspective) and they might offer ideas as to what is the fault. The operators are usually a valuable source of information.

Operating procedures: knowledge of the usual operating procedure to be used for this condition might help.

- **Check with colleagues about hypotheses**

The purpose of this book is to develop your confidence and trouble-shooting skill. To provide feedback about reasonable hypotheses (that you will have generated by this time in your journey through the case), I give this diagnostic check, but only for the Cases up to **Case #29**. You may wish to create a *symptom* ← *cause* diagram similar to the one illustrated in Figure 6-5, in Section 6.5.5c but I give no feedback about this for the cases.

4. **The fourth set are various tests and activities selected to test hypotheses/or correct**

This is the central action for each case. You may choose the diagnostic actions in any order. For example, you might suspect that the sensors are incorrect. One test is to gather data that checks for consistency. In the format of this text, I rarely include a section called "consistency tests". I expect *you* to identify neat ways to check for consistency. Methods might include:

1) to compare two sensors at the same or close-by location,
2) to look for agreement between composition, temperature and pressure (say at the top or bottom of a distillation column),
3) to check for agreement between temperature and pressure on a pressure–enthalpy diagram for pure refrigerant,

4) to check that the conditions on a stream are the same at two locations. This latter type of information we often obtain by contacting the operators of the utilities or of the plants upstream or downstream from our process called **Contact with on-site specialists.**

The diagnostic activities from which you can select are:

- **visit the site**, go out on the plant, see, smell, feel and listen; and read gauges whose values are not shown in the control room,
- check that the diagram given with the case agrees with reality, **Check diagram and P&ID versus what's out on the plant**,
- do **on-site simple tests**, such as "turn and seal" test on valves; and check for *trends in the data.*
- **gather data needed to do key calculations**,
- check the sensors. This includes checking the **sensor's response to change**, **using temporary instruments**, and **calibrating the sensors**;
- consider the **control system**: put the system on manual, retune, call in instrument specialists.
- get **information from vendors**,
- gather **samples** and have these **analyzed,**
- do **more complicated tests**, such as gamma scans, tracer studies.
- **open and inspect**, and
- **take corrective action**.

Cautions must be given about the last two diagnostic options. **Open and inspect** is expensive and usually a last resort. Use this only when you have narrowed the problem down to a specific fault and piece of equipment. As mentioned at the beginning, the cause is not usually found in the list of **Take "corrective" action**. Sometimes this does corrects the fault. Sometimes it doesn't.

When you have completed your set of diagnostic actions, then you should have all the evidence needed for you to write down the fault, think about how to correct it and how to prevent its reoccurrence. Enjoy!

Now here, in Table 8-1, is the list of cases that I have selected to help you develop your skill in trouble shooting.

**Table 8-1**

| Case number and name | | Degree of difficulty | Equipment involved | Type of process | Continuity: related to Case # |
|---|---|---|---|---|---|
| 1 | Ammonia startup | 6 | reactor, startup heater and compressor | ammonia | #29, 15, 36, 42, 50–52 |
| 2 | Leak | 4 | pipe and storage tank | ammonia | #29, 15, 36, 42, 50–52 |
| 3 | Cycling column | 4 | distillation column plus auxiliaries | general | |

**Table 8-1** Continued.

| Case number and name | Degree of difficulty | Equipment involved | Type of process | Continuity: related to Case # |
|---|---|---|---|---|
| 4 Platformer fires | 4 | heat exchanger, platformer reactor | refinery | |
| 5 Acid pump | 5 | pump, storage tank | general | |
| 6 Air dryer | 8 | adsorption, regeneration | general | |
| 7 The reluctant crystallizer | 10 | vacuum, crystallizer | general | |
| 8 Depropanizer: the temperatures go crazy | 8 | distillation column plus auxiliaries | depropanizer | #32, 38, 41,43, 45, 48 |
| 9 Bleaching plant | 7 | conveying powder for adsorption in vacuum deodorizer | foods | |
| 10 Dry and not to dry | 8 | crystallizer, screen, hydrocyclone, centrifuge and steam dryer | foods | |
| 11 Lazy twin | 5 | pumps | general | |
| 12 Drop boxes | 3 | distillation, adsorption, regeneration dryers, exchangers, pumps | ethylene | Case #24 |
| 13 The lousy control system | 4 | distillation column, overhead condenser | general | #30 |
| 14 Condenser that was just too big | 3 | distillation column, vacuum, overhead condenser | fatty acids, food | |
| 15 The flooded boot | 9 | vacuum, condenser and pumps | general, reformer, ammonia | see Appendix C |
| 16 The case of the dirty vacuum gas oil | 5 | distillation plus auxiliaries | refinery | |
| 17 Is it hot or is it not? | 4 | vacuum distillation | foods, pharmaceutical | |
| 18 The streptomycin dilemma | 7 | evaporator | food, pharmaceutical | |
| 19 The belt filter | 2 | filter, screen, pump, sedimentation tank | wastewater | #37 |
| 20 The fussy flocculator pump | 3 | pump, storage tanks, flocculation | general, wastewater | |
| 21 The flashy flare | 3 | refinery, flare system, compressor | refinery | |
| 22 pH pump | 4 | pump, mixing | general, wastewater | |
| 23 The hot TDI | 4 | polymerizer, mixer, cooling system | polymer | |
| 24 Low production on the ethylene plant | 4 | distillation, adsorption, regeneration dryers, exchangers | ethylene | Related to #12 |

**Table 8-1** Continued.

| Case number and name | Degree of difficulty | Equipment involved | Type of process | Continuity: related to Case # |
|---|---|---|---|---|
| 25 The case of the delinquent exchangers | 4 | reformer reactor, furnace, exchangers, pump | refinery | |
| 26 The drooping temperature | 5 | furnace, pump | general | |
| 27 IPA column | 5 | distillation column plus auxiliary equipment | petrochemical, refinery | |
| 28 The boiler feed heater | 5 | shell and tube exchanger to heat boiler feedwater | general | |
| 29 Reluctant reactor | 5 | reactor, compressor, separator, condensers, refrigeration unit | ammonia | #36, 15, 42, 50–52 |
| 30 The case of the reluctant reflux | 6 | distillation column plus auxiliaries | general | #13 |
| 31 Ethylene product vaporizer | 6 | heat exchangers, boiling | ethylene. | |
| 32 What does the alarm mean? | 6 | sequence of distillation columns plus auxiliaries | petrochemical | #8, 32, 38, 41, 43, 45, 48 |
| 33 Chlorine feed regulation | 6 | slurry, pump, control system, storage tank | minerals processing | |
| 34 The cement plant conveyor | 6 | solids conveyer, bagging, dust filters, cyclones | general, ceramic | |
| 35 The cycling multiple effect evaporator | 6 | long tube evaporators, vacuum system | glycerine | |
| 36 The really hot case | 7 | reactor and heat exchanger, steam generation | reformer; ammonia | #29, 15, 42, 50–52 |
| 37 Mill clarifier | 7 | thickener, sludge pumps, control | pulp and paper | #19 |
| 38 More trouble on the deprop | 7 | sequence of distillation columns plus auxiliaries | petrochemical, refinery, | #8, 32, 41, 43, 45, 48 |
| 39 The case of the lumpy sunglass display | 7 | feed bin, molding machine, mold and mold design | polymer, injection molding of thermoplastics | |
| 40 The cool refrigerant | 7 | refrigeration system, compressor, turbine, control, knockout pot | general | |
| 41 Ever increasing column pressure | 7 | distillation column plus auxiliaries | general | #8, 32, 38, 43, 45, 48 |
| 42 The weak AN | 7 | storage tank, pump, reactor, cooling coil | ammonia | #29, 15, 36, 50–52 |

**Table 8-1** Continued.

| Case number and name | Degree of difficulty | Equipment involved | Type of process | Continuity: related to Case # |
|---|---|---|---|---|
| 43 High pressure in the debut! | 7 | sequence of distillation columns plus auxiliaries | petro-chemical, refinery | #8, 32, 38, 41,45, 48 |
| 44 Reactant storage | 8 | storage tank, heating coil, steam traps | general | |
| 45 The deprop bottoms and the ISO dilemma | 8 | sequence of distillation columns plus auxiliaries | petro-chemical, refinery | #8, 32, 38, 41 43, 48 |
| 46 Not so cool chiller | 8 | pumps, exchangers, refrigeration cycle | polymeriza-tion | |
| 47 The fluctuating production of acetic anhydride | 8 | feed vaporizer, vacuum pump, absorber, condensers, reactor | petro-chemical, acetic acid | |
| 48 The column that just wouldn't work | 8 | sequence of distillation columns plus auxiliaries | petro-chemical, refinery | #8, 32, 38, 41, 43, 45 |
| 49 The case of the faulty stretcher pedal | 8 | feed bin, molding machine, mold and mold design. | injection molding of thermo-plastics | |
| 50 The cleanup column | 9 | vacuum distillation column plus auxiliaries | general, ammonia | #29, 15, 36, 42, 50–52 |
| 51 More trouble on the cleanup column | 9 | vacuum distillation column plus auxiliaries | general, ammonia | |
| 52 Swinging loops | 9 | reactor, compressor, separator, condensers, refrigeration unit | ammonia | |

**Case #8: Depropanizer: the temperatures go crazy** [8, distillation column plus auxiliaries, depropanizer]
The problem statement is given in Chapter 2, Section 2.4.

- **MSDS, 1495**
- **Immediate action for safety and hazard elimination**, Put on safe-park, **1881**
  Safety interlock shut down, **1531**
  SIS plus evacuation, **732**
- **More about the process, 1010**
- **IS and IS NOT**, Where? upstream, **537**
  Where? downstream in the debutanizer, **979**
- ***Why? Why? Why?***, Best goal? To keep bottom level steady and stop tray temperature cycling, **1335**

- **Weather**, Today and past, **23**
- **Maintenance: turnaround**, When and what done?, **421**

**Maintenance: routine**, When and what done?, **2**

- **What should be happening**,
- **Design and simulation files** (allowances made for fouling, overdesign and uncertainties)
  Condenser, E-25, **591**
  Distillation column, C8, **515**
  Thermosyphon reboiler, E-27, **380**
  Feed pump, F-25; F-26, **16**
  Turbine drive, **1193**
  Motor drive, **1072**
  Reflux pump, F-27, **286**
  Feed preheater, E-24, **1989**
  Feed drum, V-29, **1478**
  Overhead drum, V-30, **1124**

**Vendor files**: Condenser, reboiler and preheater, **10**
  Steam traps, **481**
**Commissioning data, P&ID, internal reports**, **1245**
**Handbook**, Cox charts, **803**
**Trouble-shooting files**, **1430**

- **Calculations and estimations** (that can be done in the office before special tests are done and based on rules of thumb and information given in the case)
  Energy balance: sink = source, Estimate the steam flow to the reboiler based on the reflux rate and the fact that each kg steam boils 5 kg typical organic, **821**
- **What is current operation**

**Visit control room:** control-room data, Feed to the C8, depropanizer. FC/1, **563**
  Reflux flowrate, FIC/4, **155**
  Pressure drop $\Delta$p I/1, **78**
  Level bottoms LIC/2, **1436**
  Temperature bottoms TI/4, **1765**
  Temperature mid-column TIC/5, **1502**
  Temperature top, TI/3, **1947**
  Pressure on overhead drum, PIC/10, **1610**
**Process operators**, This shift, **334**
  Previous shift, **33**
Operating procedures,
  Column pressure, **659**
  Feed to the column, **1834**

- **Check with colleagues about hypotheses**, 2395

**Call to others on-site**,
Call operators of process supplying the feed, V29, about possible upsets and details of flowrate and composition of feed, **2371**
Call operators of downstream units receiving the propane, **2757**
Call operators of downstream unit receiving butane, **769**
**Visit site**, read present values, observe and sense.
Column pressure, PI-4, **303**
Look at the flare; same size as usual?, **799**
Pressure relief to flare PSV-1, **1177**
Temperature mid-column TI- 8, **1025**
Valve position for column feed, FV- 1, **1951**
Valve position for steam to preheater E-24, **208**
Valve-stem position on PV-10, **614**
Level in feed drum, V-29; LI- 1, **885**
Pressure on exit of pump F-26, PI-3, and compare with head-capacity curve at feed flowrate, **691**
Listen to the check valve on turbine-driven pump, downstream of PI-2, **1273**
Observe whether shaft is rotating for the feed pump F-26. Is the pump noisy?, **1906**
**Check diagram and P&ID versus what's really out on the plant, 2769**
**On-site simple tests:**
Shut isolation valves around turbine-driven pump F-25, **1585**
Put TIC/-5, FIC/-1 and -4 and LIC/-2 controllers on manual and try to steady out the column, **1737**
Shut exit valve on discharge of pump F-26 and read pressure, **1316**

- **Gather data for key calculations**,

Pressure profile,
Drum V-29 to pumps, F-25, 26, **1792**
$\Delta p$ across pump, converted to head, **1666**
Pump F-25–26 exit to feed location, **63**
Drum V-30, pump F-27 and reflux into column, **917**
Pump F-27, **1868**
Vapor from top of column to vapor space in V-30, **1171**
Thermosyphon reboiler process fluid side, **133**
Mass balance, over column, **442**
**Perform more complicated tests**, Gamma scan over the stripping section to locate collapsed tray, **255**

- **Take "corrective" action**,
Put column on "safe-park", **1390**

**Case #9: The bleaching plant** [7, conveying powders, adsorption, vacuum deodorizer, foods]

The problem statement is given in Chapter 5, Section 5.3.2.

- **MSDS, 1480**
- **Immediate action for safety and hazard elimination,**
  Put on safe-park, **1196**
  Safety interlock shut down, **1345**
  SIS plus evacuation, **2400**
- **IS and IS NOT**, What, **1416**
  When, **1896**
  Who, **2838**
  Where, **2245**
- ***Why? Why? Why?***, Best goal? to suck Fullers earth into deodorizer, **1008**
- **Weather**, Today and past, **2977**
- **What should be happening**

**Design and simulation files** (allowances made for fouling, overdesign and uncertainties)

For the diameter and length of conveying line, the design flowrate of Fuller's earth and the design conveying velocity, what vacuum is needed in the bleacher?, **2690**
Hopper design, **2562**
Piping for pneumatic conveying, **347**
Vacuum system, **59**

**Handbook**, Physical properties of Fuller's earth, **938**
**Trouble-shooting files**, **535**

- **Calculations and estimations**

Pressure profile,

Is the pressure difference between atmospheric and vacuum ($\Delta p = 80$ kPa) sufficient to convey Fuller's earth?, **1368**

- **What is current operation**

**Visit control room:** control-room data, none of the data are shown in the control room; all are on the unit,

**Process operators**, When you say "nothing happened" and "you couldn't see powder dumping into the liquid", what did you hear and see? **1097**

Operating procedures, **1925**

- **Check with colleagues about hypotheses, 1762**

**Call to others on-site,**
Utilities: any upsets in the steam? Check that the steam pressure at the boiler house is approximately the steam pressure at the ejectors, **2431**
**Visit site,** read present values, observe and sense.
Pressure gauge, **193**
Check that there is Fuller's earth in the hopper, **2920**
Is the level of liquid in the bleacher so high that it covers the inlet line for the Fuller's earth into the bleacher, **2630**
**Check diagram and P&ID versus what's out on the plant, 1206**
**On-site simple tests:**
Check for leaks into the bleacher from around the agitator shaft, **2988**
Rap the pipe and the side of the hopper to try to dislodge any bridging that might be occurring in the humid weather, **2767**
Use "turn and seat" to check the valve on the conveying line and leave open, **2334**
**Sensors: check response to change,** Pressure gauge on the bleacher, **2150**
**Sensors: calibrate,** Replace/calibrate the pressure gauge, **240**
**Contact vendor supplier,** Did other customers receive Fullers earth similar to batch number 4853 that we received, and if so, have they had any comments or queries?, **366**
Are there any particular precautions we should be taking?, **831**
**Samples and measurements,** Sample Fuller's earth, **457**
**More ambitious tests**
Remove the Fuller's earth from the hopper, and crack open the valve. Listen for air flowing into the bleacher and observe vacuum gauge, **2456**
Insert porous tubes into the bed of Fuller's earth and position them such that compressed air is blown in to try to fluidize the powder near the inlet of the conveying line, **1856**
**Open and inspect,** Open conveying line and check for plugs, **1442**

- **Take "corrective" action,** Replace the valve, **2437**
  Relocate the inlet to the conveying line. Instead of using a pipe stuck into the bed, attach the inlet to the bottom of the conical hopper, **2750**

**Case #10: To dry or not to dry** [8, crystallizer, screen, hydrocyclone, centrifuge, steam dryer, foods]
The problem statement is given in Chapter 5, Section 5.4.2.

- **MSDS, 52**
- **Immediate action for safety and hazard elimination,**
  Put on safe-park, **308**
  Safety interlock shut down, **1331**
  SIS plus evacuation, **1103**
- **More about the process, 2234**

- **IS and IS NOT**, What, **2209**
  Where, **2279**
  When, **2642**
- ***Why? Why? Why?***, Best goal? To get the exit crystal moisture content < 1.5%, **2779**
- **Weather**, Today and past, **290**
- **Maintenance: turnaround**, When and what done?, **492**

**Maintenance: routine**, When and what done?, **270**

- **What should be happening**,

**Design and simulation files** (allowances made for fouling, overdesign and uncertainties)
  Crystallizer, **20**
  Hydrocyclone, **427**
  Pump, **45**
  Surge vessel installed between the screen and the centrifuge?, **4**
  Surge vessel installed between the centrifuge and the dryer?, **405**
**Vendor files**:
  Centrifuge, **816**
  Hydrocyclone, **510**
  Pump, **992**
  Rotary dryer, **1490**
  Feeder screw, **1224**
  Parabolic screen, **711**
**Commissioning data, P&ID, internal reports**, 878
**Trouble-shooting files**, 533

- **What is current operation**

**Visit control room:** control-room data, all data on the plant.
**Process operators**, This shift; elaborate on the changes you have made, **1411**
Operating procedures,
  What is the new cycle time on the centrifuge and how does this compare with before?, **1009**
  Cycle time for screening and washing for the screen, before and now **1041**

- **Check with colleagues about hypotheses, 1311**

**Call to others on-site**,
  Utilities: steam plant: any upsets, pressure delivered to our site, **2219**
  Stores: any available spare equipment: centrifuges, screens, **1804**
  Utilities: wash water for screen and centrifuge: change in quality, upsets, **2786**
**Visit site**, read present values, observe and sense.
  Records of analysis of moisture content in feed to dryer. Samples taken at the beginning and end of a shift, **71**

**Check diagram and P&ID versus what's out on the plant**, 233

**On-site simple tests:**

Reduce throughput for the unit and thus reduce feedrate to dryer, **360**

Steam condensate trap on dryer working?, **1971**

Estimate feedrate to the screen over the screening cycle and compare with feedrate before the change, **1514**

Observe feed condition from the screen to the centrifuge during the wash cycle of the screen, **1114**

Estimate feedrate from the screen to the centrifuge over the filter cycle and compare with the feedrate before the change, **1458**

Estimate the feedrate to the centrifuge during the screen wash cycle and compare with the feedrate before the change, **561**

Estimate the feedrate from the centrifuge to the dryer during the filter cycle and compare with the feedrate before the change, **909**

Estimate the feedrate to the dryer during the centrifuge wash cycle and compare with the feedrate before the change, **545**

Stop the centrifuge when the screen is in its wash cycle. Test the degree of compaction of the residual crystals under the peeler, **54**

**Gather data for key calculations**

Pressure profile,

Recycle through hydrocyclone, **1671**

Energy balance: sink = source, Measure amount of condensate by submerging exit line in bucket with preweighed amount of cold water. Does energy lost from condensation of steam = energy gained by evaporation?, **603**

Equipment performance,

Rating dryer, **1860**

Screen, **1566**

Centrifuge, **1977**

**Call to vendors, licensee**, Rotary dryer, **482**

**Samples and measurements,**

Sample the crystal discharge from the screen every 30 s for three cycles of screen operation; measure % screen overs as a % of feed. Compare with results for old operation, **1968**

Sample the discharge from dryer every 20 s for three cycles of screen operation; measure moisture content. Compare with data for previous operation, **1520**

Sample the liquid-fine underflow from the screen every 30 s for the screening cycle. Compare with data from previous operation, **38**

Draw a composite sample the wash water exit from the screen. Analyze for particle-size distribution. Compare with previous operation, **318**

Draw a composite sample of the wash water from the centrifuge. Analyze for particle-size distribution. Compare with previous operation, **1911**

Sample the feed to the dryer every 20 s for three cycles of screen operation; measure moisture content. Compare with previous operation, **196**

**Return to previous operating conditions**,
Sample the discharge from dryer every minute for three cycles of screen operation; measure moisture content, **1705**
Sample the feed to the dryer every minute for three cycles of screen operation; measure moisture content, **1032**
Sample the crystal discharge from the screen every 30 s for three cycles of screen operation; measure % screen overs as a% of feed, **558**
**Open and inspect**,
Centrifuge, **913**
Dryer, **689**
**Take "corrective" action**,
Revamp the condensate removal system: install a thermodynamic trap, slanted exhaust pipe to the trap and minimized the distance between the trap and heating tube, **795**
Operate the centrifuge on former operation, **1194**

**Case #11: The lazy twin** [5, pumps, general]
The problem statement is given in Chapter 6, Section 6.2.2.

- **MSDS**, **376**
- **Immediate action for safety and hazard elimination**,
  Put on safe-park, **42**
  Safety interlock shut down, **490**
  SIS plus evacuation, **984**
- **IS and IS NOT**, What, **337**
  When, **2368**
  Who, **2912**
  Where, **2728**
- **More about the process**, **1811**
- ***Why? Why? Why?***, Best goal? get the flow up to design rate when pump A is running, **1667**
- **Weather**, Today and past, **1415**
- **Maintenance: turnaround**, When and what done?, **1400**

**Maintenance: routine**, When and what done?, **1822**

- **What should be happening**,

**Design and simulation files** (allowances made for fouling, overdesign and uncertainties)
Centrifugal pump A, **1595**
Centrifugal pump B, **1510**
Heat exchanger, **166**
Check valves, **395**
**Vendor files**: Centrifugal pumps A and B, identical and from the same vendor, **930**
Heat exchanger, **540**

**Commissioning data, P&ID, internal reports, 551**
**Trouble-shooting files, 976**

- **Calculations and estimations.** None can be done based on the given information
- **What is current operation**

**Visit control room:** control-room data
What is the temperature after the heat exchanger when pump A is on-line?, **386**
What are the levels in the downstream equipment when pump A is on-line?, **470**
Is FRC/-100 local or remote?, **506**
"What is the pressure in the storage tank T-200?", **973**
**Process operators**, Current shift, **1470**
Operating procedures, About use of pumps, **1006**

- **Check with colleagues about hypotheses, 90**

**Call to others on-site**, Contact downstream cat-cracking unit. Everything working as expected? is the behavior consistent with the flowrate signalled from FRC/-100?, **648**

- **Visit site**, read present values, observe and sense.
  Do either pumps A or B sound as if they are cavitating?, **1998**
  Look and see that valves V200, V201, V202, V205 and V206 are all open, as expected, **518**
  Read the controller output to valve V100 when pump A is running, **2325**
  Read the controller output to valve V100 when pump B is running, **2441**
  Observe the valve-stem position on F100 when pump A is pumping, **2086**
  Observe the valve-stem position on F100 when pump B is pumping, **2845**
  Observe whether the arrow on each valve V200, V201, V202, V205, V206 is in the direction of flow through the valve, **2935**
  Observe whether the direction of flow through both check valves agrees with valve installation, **1587**
  Check the tab on the orifice plate FRC-100 that it is the correct diameter and facing the correct direction, **1160**
  Pump A running hot?, **1617**
- **On-site simple tests:**
  Shut exit valve on pump B; read pressure on gauge PI-220 when pump is running. Convert to head of fluid, **1030**
  Shut exit valve on pump A; read pressure on gauge PI-210 when pump running. Convert to head of fluid, **1465**
  Put pump A on-line and check that the motor started. Repeat for pump B, **602**
  Test valves V200, V201, V202, V205 and V206 with "turn and seal" to ensure they are working properly, **1461**

What is the pressure at P210 and at P220 when pump A is on and pump B is off? Express this is head (so I can compare with the head-capacity curve), **550**
What is the pressure at P210 and at P220 when pump A is off and pump B is on? Express this is head (so I can compare with the head-capacity curve), **170**
Stop pump B and close valve V205. Start pump A. What is the flow on FRC/-100?, **102**
Stop pump A and close valve V201. Start pump B. What is the flow on FRC/-100?, **1482**
**Check diagram and P&ID versus what's really out on the plant**, **2372**
**Gather data for key calculations**
Pressure profile,
From Tank T-200 through pump A to cat cracker, **1432**
From Tank T-200 through pump B to cat cracker, **7**
Energy balance: sink = source,
Read clamp-on ammeter, voltmeter and power-factor meters, assume density = design density, compare power for motor to drive pumps A and B, **140**
**Sensors: check response to change**
Change set point on FRC/-100; does the valve V100 respond?, **1081**
**Sensors: calibrate**, Flowmeter, FRC/-100, **1552**
**Open and inspect**,
Pump B and ask maintenance to inspect and look for reasons why the pump is not functioning well, **1886**
Pump A and ask maintenance to inspect and look for reasons why the pump is not functioning well, **213**
Stop the process. Open and inspect valves V201 and V202, **453**

- **Take "corrective" action**,
  Shut down the process. Replace valves V201 and V202, **276**
  Shut valve V205 whenever pump A is running, **2027**
  Stop the process. Replace check valves on the exit lines from both pumps, **678**

**Case #12: Drop boxes** [3, distillation, adsorption, regeneration dryers, evaporators, ethylene]
The problem statement is given in Chapter 7, Section 7.2.2.

- **MSDS**, **2667**
- **Immediate action for safety and hazard elimination**,
  Put on safe-park, **925**
  Safety interlock shut down, **1428**
  SIS plus evacuation, **2268**
- **More about the process**, About the dryers and dryer cycle, **1715**
  About the distillation, **264**
  About the sewer systems, **420**
  About the low-temperature condensation, **499**

- **IS and IS NOT**, What, **827**
  When, **2001**
  Where, **1571**
- ***Why? why? why?***, Goal: "clear up the hazard as stated by the safety inspector", **637**
- **Weather**, Today and past, **2494**
- **Maintenance: turnaround**, When and what done?, **2085**

**Maintenance: routine**, When and what done?, **2982**

- **What should be happening**,

**Design and simulation files** (allowances made for fouling, overdesign and uncertainties)

Exchanger E131: cool regeneration gas to the dryer during last portion of regeneration: water tube side; regeneration fuel-gas shell side, **1773**
Heater E130: heat regenerative fuel gas to dryer, **256**
Exchanger E107and E108: precool feed from 38 °C to –29 °C, **444**
Demethanizer overhead condenser: process fluid shell side; ethylene refrigeration on the tube side, **462**
Demethanizer reboiler: process stream shell side; propylene on tube side, **987**
Steam trap on regenerative gas heater, **2997**
Knockout pot, **2990**

**Vendor files**:

Heat exchanger, **2518**
Dryer system, **2022**
Steam trap, **2533**

**Commissioning data, P&ID, internal reports**, P&ID, **437**

Sewer plot plan indicating which streams go to each "sewer gate" and which go to drop box B and which to drop box C, **1101**

**Trouble-shooting files**, **2223**

- **Calculations and estimations.** no calculations can be done based on the limited information in the problem statement
- **What is current operation**

**Process operators**, Any changes in operation?, **2330**

- **Check with colleagues about hypotheses**, **1318**

**Call to others on-site**,

**Safety inspector** Where were the samples taken and what was the composition?, **1699**
Operators of upstream styrene, ethylene and propylene plants: any upsets, any calls from the safety inspector, **2882**

- **Visit site**, read present values, observe and sense,
  Look at the flare, **2674**

**Check diagram and P&ID versus what's really out on the plant**, **2164**

**On-site simple tests:**

Vent valves closed on process (or shell side) of exchangers E107, E108, condenser E114, E131?, **2477**

Vent valves closed on utility side (or tube side) of exchangers E107, E108, condenser E114, E131?, **2364**

Vent valves closed on shell and tube side of heater E130?, **2876**

Drain valve on knockout pot shut?, **2617**

For exchanger E131 cooler; block off the in and out water lines; open vent on tube side and note fluid leaking out, **1211**

For the vent valves on the process (or shell side) of exchangers E107, E108, condenser E114, E131, test by "turn and seal", **2433**

For the vent valves on the utility side (or tube side) of exchangers E107, E108, condenser E114, E131, test by "turn and seal", **2363**

Test the vent valves on shell and tube side of heater E130 by "turn and seal", **2812**

Test the drain valve on knockout pot by "turn and seal", **2557**

Retest drop box **B** and **C** for explosive mixture at half-hour intervals for 2 hours, **554**

Retest drop box **A** for explosive mixture at half-hour intervals for 2 hours, **119**

**Gather data for key calculations**

Pressure profile,

Exchanger process fluid pressure relative to utility pressure: direction of leak, **2067**

Exchanger utility pressure relative to atmospheric; direction of leak, **2472**

Exchanger process fluid relative to atmospheric; direction of leak, **2814**

Mass balance, Over demethanizer, **2698**

**Samples and measurements**, Sample cooling water leaving cooler E131, **1488**

Gas sample from drop boxes **B** and **C** using an evacuated bomb. Lab analysis for the overall concentration of light hydrocarbons and a breakdown of the hydrocarbon portion, **2434**

**Open and inspect**,

Cooler E131 and look for leaks in tubes, **2576**

- **Take "corrective" action**,
  Replace the vent valves on the shell side of E107, 108, 130, 131, **2923**
  Replace drain valve on KO pot, **168**

**Case #13: The Lousy Control System** [4, distillation, overhead condenser, general]
The problem statement is given in Chapter 7, Section 7.2.5, Activity 7-9.

- **MSDS**, **2266**
- **Immediate action for safety and hazard elimination**,
  Put on safe-park, **2439**
  Safety interlock shut down, **2932**
  SIS plus evacuation, **2510**

- **More about the process**, **497**
- **IS and IS NOT**,
  What, **2317**
  When, **1770**
  Where, **621**
- ***Why? why? why?***, Goal: "to prevent too much stuff from going to the flare", **232**
- **Maintenance at turnaround**, When and what, **1106**
- **Weather**, Today and past, **30**
- **What should be happening**,

**Design and simulation files** (allowances made for fouling, overdesign and uncertainties)
Fan: control system on the air flow to the condenser, **324**
Condenser, air-cooled, **739**
Distillation column, **1996**
Reflux drum, **2070**
Reflux pump, **2458**
**Vendor files**: Condenser, air cooled, **2662**
Reflux pump, **2910**
**Commissioning data, P&ID, internal reports**, **2543**
**Trouble-shooting files**, **2218**

- **Calculations and estimations**

Equipment, Rate condenser: heat load, **411**

- **What is current operation**

**Visit control room:** control-room data
Control-room values now and from past records, Temperature at the top of the column, **2502**
Temperature on liquid-gas exit from the condenser, **2685**
Pressure at the top of the tower, **2722**
Pressure on the overhead receiver, **778**
Level in the overhead receiver, **1521**
**Process operators**, This shift, **2179**
Previous shift, **2356**

- **Check with colleagues about hypotheses**, **143**

**Call to others on-site**, Operators of the upstream process, **2185**
Operators of the downstream process, **1613**
**Visit site**, read present values, observe and sense.
Hot exhaust air recirculation to the intake of the air-cooled condenser, **220**
Inspect the hydraulic configuration on the exit header from the condenser, **89**
Look at the flare, **456**

Sounds around the condenser, **325**

**Check diagram and P&ID versus what's really out on the plant, 1462**

**On-site simple tests:**

Temperatures and humidities of inlet and exit air for the air-cooled condenser, **1866**

Lower the elevation of the exit from the header to decrease the amount of flooding, **2879**

**Gather data for key calculations**

Pressure profile,

On process pipe from the top of the column to the overhead receiver, **2346**

Across condenser, **2062**

Equipment performance,

Rating of air-cooled condenser, **2844**

Fans, **1631**

Reflux pump, **1876**

Column, **1670**

**Sensors: check response to change**

Temperature sensor exit of condenser, **1087**

**Sensors: use of temporary instruments**,

Surface temperature near thermowell at exit of condenser measured by a contact pyrometer, **192**

**Sensors: calibrate**,

Temperature sensor on exit of condenser, **684**

**Control system**,

Put on manual; change the set point and note response of fan system, **487**

**Sampling and measurements**

Concentration: vent from the overhead receiver; concentration of heavies, **2587**

Concentration: vent from the overhead receiver; concentration heavies, ten samples taken at one-hour intervals, **2152**

**Open and inspect**,

Air-cooled condenser: visual inspection plus air and water pressure tests, **1254**

Fan and fan blades, **2293**

- **Take "corrective" action**,

  Operate on manual, **2793**

  Install hose and let spray of water fall over tubes, **367**

  Stop operation; open the headers and water-wash with high pressure hoses to remove scale, **1230**

**Case #14: The Condenser that was just too big** [3, distillation, vacuum, overhead condenser, fatty acids, food]
The problem statement is given in Chapter 7, Section 7.2.5, Activity 7-9.

- **MSDS**, **51**
- **More about the process**, **273**
- **Immediate action for safety and hazard elimination**, Put on safe-park, **400**
  Safety interlock shut down, **88**
  SIS plus evacuation, **1500**
- **Weather**, Today and past, **1994**
- **What should be happening**,

**Design and simulation files** (allowances made for fouling, overdesign and uncertainties)
Condenser, **1660**
Backup condenser, **2446**
Booster ejector, **3000**
Ejector, **2524**
Wet vacuum pump, **1781**
Barometric condenser, **21**
Coil cooler in tank, **966**
Reciprocating pump, **1271**
**Vendor files**: Reciprocating pump. F1400, **1528**
Ejectors and booster ejector, **1961**
Wet vacuum pump F1401, **2163**
**Commissioning data, P&ID, internal reports**, Any files?, **2004**
**Handbook**, Vapor pressure of fatty acids, **2490**
**Trouble-shooting files**, **2969**

- **Calculations and estimations**

Pressure profile, Estimate the pressure in C1400; vapor side of E1400; vapor side of E1401, **2511**
Mass balance, Mass balance on the overhead pumped by F1400 compared with expectations based on feed composition and rate, **2781**
Energy balance: Check coolant temperature for E1400, **2580**

- **What is current operation**

**Visit control room:** control-room data,
P1, **2273**
T4, **2377**
T3, **390**
**Process operators**, Flow pumped from F1400, **486**
Trends in pressures and temperatures, **188**
Liquid level in bottom of E1400 as feed to V1400, **871**
Anything strange happening?, **1317**
Was the vacuum pump started up according to operating procedures?, **1020**

**Visit site**, read present values, observe and sense.

Cooling-water flow into E1403, **1322**

Cooling-water flowrate and temperature to E1404, **1091**

Read P3 and look for oscillations in pressure, **1851**

Is the water level in the hot well above the exit from the downcomer from the barometric condenser? Water temperature?, **1594**

Note whether steam valves to booster and regular ejector are full or partially open, **2201**

Does the booster ejector sound as though it is "kicking out"?, **2094**

Is the level in the boiling condenser E1400 dropping? Test by trying to add more liquid via the top up funnel at the top of E1400, **2655**

Listen for cavitation in pump F1400, **326**

**On-site simple tests:** Visually and audibly check for any leaks of air into the system, **716**

**Sensors: check response to change**

T4: **933**

T3: **522**

**Sensors: use of temporary instruments**,

Put surface temperature sensor on suction line to F1400 and check value, **1155**

**Call to vendors, licensee**, Reciprocating pump: what could cause the knocking its head off?, **1142**

**Sensors: calibrate**, T4, **1920**

T3, **1614**

**Samples and measurements**,

Feed: usual composition? and does this contain more volatiles than expected?, **1719**

**Open and inspect**, Suction line from bottom of E1400 through to F1400, **2859**

- **Take "corrective" action**, Flow cold water over the outside casing of pump F 1400, **2939**

**Case #19: The case of the reluctant belt filter** (supplied by Mike Dudzic, B. Eng. 82, McMaster University) [2, filter, screen pump, deep thickeners, wastewater treatment]
The following diagram shows a section of the sludge concentration and dewatering section of our wastewater treatment process. Water containing sludge is concentrated in a deep settling tank and then pumped to a continuous-belt filter. Polymer is added just upstream of the filter to improve dewatering. The process is shown in Figure 8-2.

For the past three weeks the plant has been operating smoothly although lately the operator thinks that the flow to the filter has been a bit lower than normal. The operator is relatively new on this plant. He has been on the unit for three months. Today, the operator notes that the sludge has stopped flowing to the filter. The pump is running. If the fault is not corrected quickly, the sludge will build up in the settlers and shortly the whole process will be shut down.

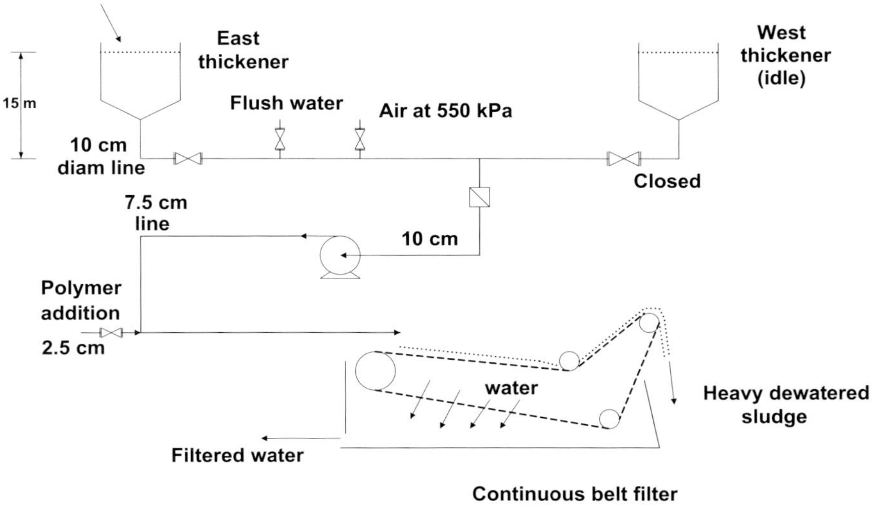


**Figure 8-2** The dewatering system for Case #19.

**Case #19: Reluctant belt filter**

- **Immediate action for safety and hazard elimination**,
  Put on safe-park, **8**
  Safety interlock shut down, **330**
  SIS plus evacuation, **268**
- **More about the process**, **2830**
- **IS and IS NOT**, What, **2608**
  When, **2161**
  Who, **1554**
  Where, **274**
- ***Why? why? why?***, Goal: "Get the sludge flowing to the filter", **596**
- **Weather**, Today and past, **494**
- **Maintenance: turnaround**, When and what done? **722**

**Maintenance: routine**, When and what done? **2660**

- **What should be happening**

**Design and simulation files** (allowances made for fouling, overdesign and uncertainties)
Thickener, **707**
Centrifugal pump for handling sludge, **765**
Strainer, **997**
Continuous-belt filter, **2143**
**Vendor files**, Centrifugal pump, **1518**
**Trouble-shooting files**, **223**

- **Calculations and estimations** that can be done in the office based on the information given in the problem statement

Pressure profile,

Calculate to see if 550 kPa g pressure is sufficient for air to actually backflow through the thickener, **1635**

- **What is current operation**

**Process operators**, When did the flow appear to decrease? **1330**

How long had the east thickener been idle before you started using it? **1757**

Have you ever seen anything like this before? and what did you do?, **2329**

Operating instructions, Standard procedure: if there is low sludge flow; flush the exit lines from the thickeners with high-pressure water and high-pressure air, **2052**

**Check with colleagues about hypotheses, 1805**

**Call to others on-site**, High pressure air; have there been any interruptions in service? what is the pressure of the air being delivered to site? **1258**

**Visit site**, read present values, observe and sense.

Is the shaft of the pump rotating? **322**

Is the block valve V101, to the idle thickener closed? **731**

Are the isolation block valves around the strainer and around the pump open? **294**

**Check diagram and P&ID versus what's out on the plant**, **2839**

**On-site simple tests:**

Flush the lines with high-pressure water and air for five minutes, **793**

Test by "opening and closing" the block valves V100, V101, **2672**

Test by "opening and closing" the block valves on the flush out lines V102, V103, **2216**

**Sensors: use of temporary instruments**,

Use a clamp-on ammeter to measure the amps to the pump motor and compare with design or usual value, **1855**

Use a clamp-on ammeter, portable voltmeter and power-factor meters and calculate the power drawn by the pump motor and compare with design or usual value, **1648**

**Call to vendors, suppliers or licensee**, Pump supplier. what might be going on here?, **1199**

**Open and inspect**,

The line between the thickener exit tee and the strainer; is it clear?, **1289**

Strainer, **1576**

Isolate, flush and when safe, open and inspect pump, **1923**

- **Take "corrective" action**,

Shut-off the polymer to the filter, **798**

Shut down. Block off and drain. Clean out the crud in the line from the bottom of the thickener to the screen, **249**

Replace the block valve V101 on the bottom from the idle thickener, **2300**

Install a recording ammeter on the pump motor, **2800**

**Case #20: The case of the fussy flocculator pump** (problem supplied by Jonathan Yip, B. Eng. McMaster University, 1997) [3, pump, storage tanks, flocculation, general] Pump 40P002 is a centrifugal pump that transfers wastewater from the buffer tank to the flocculation tank. The liquid overflows into the Dissolved Air Flotation, DAF, unit. Figure 8-3 shows the system. One day the operator noticed that the level in the flocculation tank was lower than normal and the resulting overflow to the DAF was less than expected. Pump 40 P002 just wasn't performing as it should! Get it fixed. The manual ball valve on the exit line is wide open.

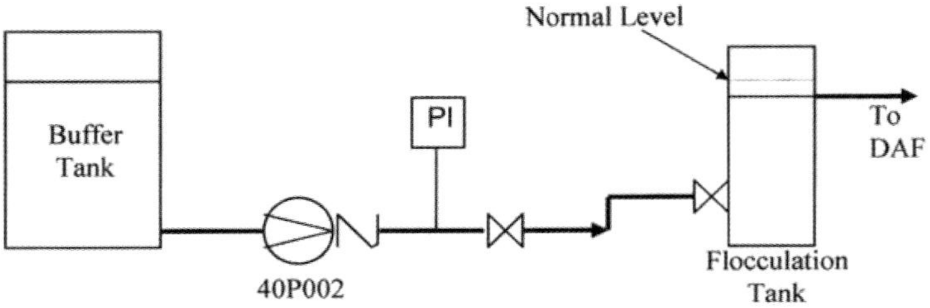


**Figure 8-3** The flocculation system for Case #20.

**Case #20: The case of the fussy flocculator pump**

- **Immediate action for safety and hazard elimination**, Put on safe-park, **416**
  Safety interlock shut down, **999**
  SIS plus evacuation, **1995**
- **IS and IS NOT**, What, **195**
  Where, **1892**
  When, **2744**
- ***Why? why? why?***, Goal: "Get the overflow to DAF to expected valve", **2596**
- **Weather**, Today and past, **2497**
- **Maintenance: turnaround**, When and what done?, **2987**

**Maintenance: routine**, When and what done?, **2624**

- **What should be happening**,

**Design and simulation files** (allowances made for fouling, overdesign and uncertainties), Centrifugal pump, **1297**
**Vendor files**: Centrifugal pump, **2525**
**Trouble-shooting files**, **2798**

- **What is current operation**

**Process operators**, Anything change in this portion of the plant?, **604**
Operating procedures, For pumping, **898**

- **Check with colleagues about hypotheses, 2176**

**Call to others on-site**, Operators of DAF: is the flow less than expected?, **779**
**Visit site**, read present values, observe and sense.
PI gauge, **2322**
Listen to the pump for sounds of cavitation, **2343**
Is drive motor on the pump running hot? Touch with gloved hand, **2420**
Is the pump running hot? Touch with gloved hand, **2411**
Does the handle on the ball valve, on the exit line, move easily and smoothly?, **2243**
Does the handle on the ball valve, on the entrance into the flocculation tank, move easily and smoothly?, **2460**
Current level in the buffer tank; usual value?, **2029**
**Check diagram and P&ID versus what's out on the plant**, **1342**
**On-site simple tests:** Close the discharge valve on the pump exit and read PI, **2764**
Backflush the pigtail on the pressure gauge with city water to clear any blockage, **1099**
Tachometer reading on drive shaft rpm for the pump, **2927**
Stethoscope on the check valve, **2877**
Measure motor amps with clamp-on ammeter, calculate power assuming volts, power factor and density, **2506**
**Gather data for key calculations**
Equipment performance, Pump, **641**
**Sensors: check response to change**
Pressure gauge PI response to partial closure of valve on exit line, **112**
**Sensors: calibrate**, Pressure gauge PI, **413**
**Samples and measurements**, Sample solution in buffer tank. Compare the density and composition with specifications, **676**
Sample solution in the flocculation tank. Compare the density and composition with specifications and with specifications for solution in buffer tank, **932**
**Open and inspect**,
Shut down system, isolate. Open and inspect ball valve, **953**
Shut down system, isolate. Open and inspect centrifugal pump: clearance between impeller and volute tongue; status of the wear rings; erosion of the impeller, key between the shaft and the impeller. Pluggage?, **1473**
Shut down system, isolate. Open and inspect line to the pressure gauge for dirt and pluggage causing sluggish response, **1630**
Shut down system, isolate. Open and inspect exit line from the valve to the flocculation tank, **2190**
Shut down system, isolate. Open and inspect check valve, **2894**

- **Take "corrective" action**,
  Change the pressure gauge, **2569**
  Replace the ball valve on pump exit, **2852**
  Replace the check valve, **2965**
  Replace the impeller in the pump, **1727**
  Realign drive motor and pump shaft, **1644**

**Case #21: The case of the flashy flare** (problem supplied by Mark Argentino, B. Eng. McMaster University, 1981) [3, refinery, flare system, compressor, refinery]
The flare system at your refinery works as follows. When there is a high-pressure build-up in a vessel, tower, or exchanger a relief valve opens and the high pressure vapors, and maybe some liquid, flow into a pipe network that eventually ties into a common header that flows into a large knockout drum where the liquid is removed and the vapors are drawn overhead. Other sources of vapor in the flare system could be off-specification products sent to flare, hazardous vapors educted from pumps, or any hydrocarbon or non-hydrocarbon sources in the refinery that are not of any product value but cannot be vented to atmosphere. The typical composition of the flare gases is as follows

| | | | |
|---|---|---|---|
| $CH_4$ | 20% | $C_4$ s | 3% |
| $C_2H_6$ | 15% | $C_5^+$ | 1% |
| $C_3H_8$ | 5% | $N_2$ | 50% |
| $C_3H_7$ | 5% | $H_2S$ | 1% |

The vapors then pass through a seal pot that is a vertical cylindrical drum with a smaller cylinder in the center. The vapors flow down through the central cylinder, out the bottom and into the annulus where they bubble through a liquid and are fed to the burner, flare, where they are burned. To visualize the operation of the seal, consider an analogous situation of a large drinking straw in a glass of water. When no air is blown down the straw, the water in the glass is at the same level as the water in the straw. As one blows down the straw the water level drops in the straw and rises in the water glass. This liquid differential in the system forms a certain liquid head that must be overcome for one to blow air out of the straw and bubble it through the water. As with the drinking-glass analogy, the seal pot has a certain level of liquid in it so that pressure must be applied on the inlet line (the *straw* in our analogy) for any vapors to pass through the seal and be burned. In this way it serves as a seal. The process is given in Figure 8-4.

Last turnaround a flare-gas compressor was added to the system. The compressor takes suction just downstream of the knockout drum. The function of the compressor is to recompress the flare gases from 106.5 kPa abs to 480 kPa abs so that the vapor can be recovered as fuel, which along with purchased natural gas, is used in the boilers and fired heaters. The compressor has a spill back or kick back valve and line, which is a small pipe that reroutes some of the compressed vapors in the discharge line, via a control valve on pressure control, back to the compressor suction, and hence to the flare line to keep the suction pressure at 106.5 kPa abs. The compressor will shut down on high suction pressure = 112 kPa abs. In the cold Canadian winter kerosene is used as the sealant liquid. Nevertheless, there are freeze ups in lines and equipment failures; hence frequent flaring when there is a high release of vapors to the flare line when the pressure builds up in the flare line. The pressure builds up in the flare line because the compressor will only pump a fixed maximum amount of vapors. If there is a higher flow of flare gases then the restriction of flow at the seal pot causes a pressure increase up to the blowpoint. The maximum blow pressure is 107.8 kPa. That is the level of liquid sealing the dip tube in the seal pot is equivalent to a 6.8 kPa differential.

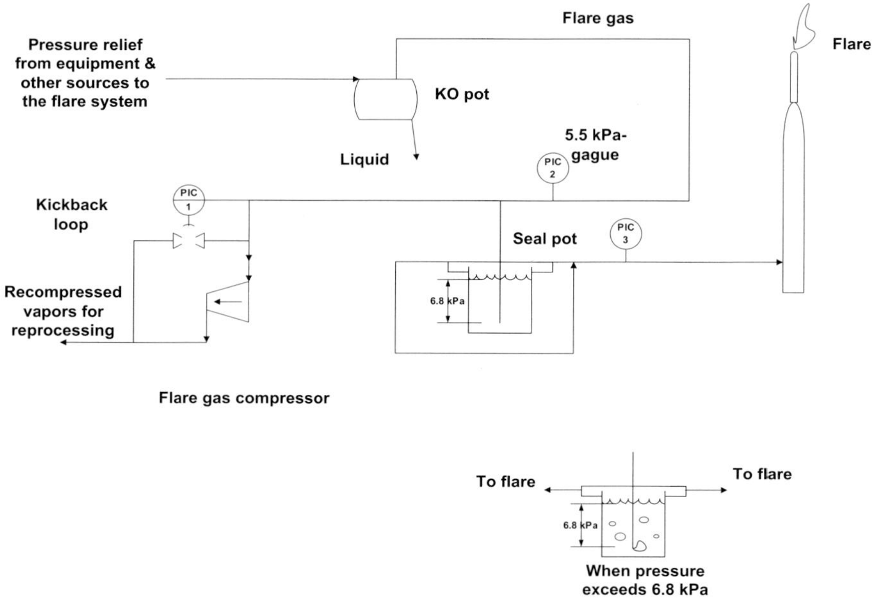


**Figure 8-4** The flare system for Case #21.

Today it is cold, dry and windy. The pressure gauge on the flare line reads 112 kPa abs; the compressor shuts down. The operator explains that "when the compressor shuts down, the accompanying surge in the flare gas flow is so great that the kerosene is all blown into the flare burner along with the usual flare vapors. The result is that the flare flashes and smokes like a giant fire." All the neighbors are phoning in with smoke complaints. With the flare seal blown, all the flare gas goes to the flare. When the compressor is down we are losing $1000/h in non-recovered vapors. Fix it.

**Case #21: Flashy flare**

- MSDS, 215,
- **Immediate action for safety and hazard elimination**,
  Put on safe-park, **84**
  Safety interlock shut down, **2341**,
  SIS plus evacuation, **2900**
- **IS and IS NOT**: (based on given problem statement),
  What, **2076**
  When, **2641**
  Who, **1992**
  Where, **1730**
- ***Why? why? why?***, Goal: "Prevent the compressor from shutting down", **597**
- **Weather**, Today and past, **1505**
- **Maintenance: turnaround**, When and what done?, **1395**

**Maintenance: routine**, When and what done?, **1451**

- **What should be happening**,

**Design and simulation files** (allowances made for fouling, overdesign and uncertainties)
Compressor, **75**
Control, **471**
Seal pot, **9**
Knockout drum, **500**
**Vendor files**: Compressor, **852**
**Commissioning data, P&ID, internal reports**, **576**
**Handbook**, Density of kerosene and color, **957**
**Trouble-shooting files**, **625**

- **Calculations and estimations** (that can be done in the office before special tests are done)

Pressure differential,
Height of kerosene in the seal pot = 6.8 kPa?, **1127**
Equipment performance, Compressor, **1497**

- **What is the current operation**

**Visit control room:** control-room data: values now and from past records,
Motors amps on the compressor, **1511**
**Process operators**, **1974**

- **Check with colleagues about hypotheses**, **2906**

**Call to others on-site**,
Call operators on other units to see if excessive pressure in the flare line might be originating on units, **1842**
**Visit site**, read present values, observe and sense.
Pressure gauge on flare line, **2382**
Pressure gauge by seal pot = pressure gauge on kick-back control = 112 kPa, **2610**
**Check diagram and P&ID versus what's out on the plant**, **70**
**On-site simple tests:** Atmospheric pressure, **2883**
Drain off the KO pot to verify that the liquid level is not > expected, **694**
**Sensors: check response to change**
Pressure gauge on flare line, **141**
**Sensors: use of temporary instruments**,
Surface sensor for temperature of suction gas to compressor, **410** Use clamp-on ammeter to measure amps when running under usual conditions, **2487**
**Sensors: calibrate**, Pressure gauge on flare line, **1034**
**Control system**, Control-valve stem moves when suction pressure < 106. 5 kPa abs?, **1284**
**Call to vendors, licensee**, Compressor, **742**

**Samples and measurements**, Flare gas: composition, **1240**
Liquid density of liquid from the knock-out drum, **1070**
Kerosene in seal pot, density, **1590**
Kerosene in seal pot, color, **1447**
**Open and inspect**,
Open seal pot and check for blockage, **2119**
Open and inspect KO pot for frozen ice in the demister, **729**

- **Take "corrective" action**,

Isolate the seal pot. Attach compressed air to the feed line to the seal pot and blow out line through the flare, **2465**

**Case #22: The pH control unit** (used courtesy of Scott Lynn, University of California, Berkeley, CA) [4, pumps, control, storage tank, acid-base wastewater treatment]
Concentrated hydrochloric acid is used to neutralize caustic wastes being fed to a newly built effluent treatment plant. The process is illustrated in Figure 8-5. The volumetric flowrate of wastes is approximately constant at 12.6 L/s but due to the nature of their source, the concentration varies from 1 to 10 g/L equivalent NaOH. The average is about 5 g/L. Control is usually good, but at times it becomes erratic and occasionally the acid flow stops altogether. Turning off and restarting the acid feed control system usually serves to get the acid flow going again, but this malfunction threatens to shut down the entire plant. Find the bug and get rid of it!

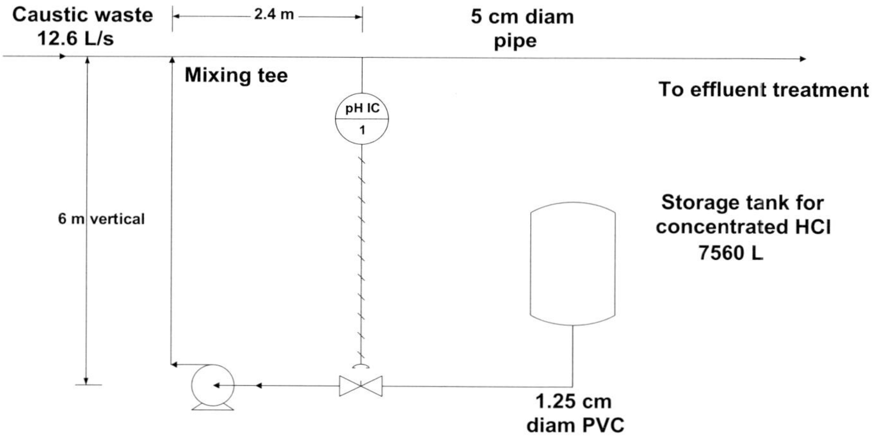


Figure 8-5 Feed system to the effluent treatment process for Case #22.

**Case #22: pH pump**

- **MSDS**, **2250**
- **Immediate action for safety and hazard elimination**
  Put on safe-park, **2167**
  Safety interlock shut down, **2035**
  SIS plus evacuation, **1740**
- **IS and IS NOT**,
  What, **2940**
  When, **2700**
  Who, **2081**
  Where, **2358**
- ***Why? why? why?***, Goal: "To prevent erratic control of pH", **2579**
- **Weather**, Today and past, **1953**
- **Maintenance: turnaround**, When and what done?, **2985**

**Maintenance: routine**, When and what done?, **234**

- **What should be happening**

**Design and simulation files** (allowances made for fouling, overdesign and uncertainties)
Feed pump, **468**
**Vendor files**:
Feed pump, **592**
**Commissioning data, P&ID, internal reports, 998**
**Handbook**, Density of 30% HCl, **530**
**Trouble-shooting files, 1326**

- **Calculations and estimations** based on given information
  Control system, Feed-forward control, **2909**
  Valve stiction and hysteresis, **426**
  Fluid dynamics, Calculate the residence time between the acid injection and pH sensor, **1425**
  Calculate if the flow is turbulent in the caustic waste line near the mixing tee, **1191**
  Velocity of waste in the line, **1403**

- **What is happening**

**Visit control room:**
**Process operators**, This shift, **2509**
Did the motor overload and trip off, **2677**

- **Check with colleagues about hypotheses, 97**

**Call to others on-site**,
Effluent treatment: variability in flowrate or pH, **1290**
**Visit site**, read present values, observe and sense.

Control valve: does the control-valve stem move in response to a signal from the controller, **2002**
Check that the block valves on the control valve are open, **1068**
Check that the valve on the bypass around the control valve is shut; if not shut it, **1878**
Is there a vent on the acid storage tank, **1835**
Note the characteristics of the response: cycling? amplitude and frequency?, **1682**
Sounds near the pump, **1150**

**Check diagram and P&ID versus what's out on the plant, 2129**

**On-site simple tests:**
Check valve for stiction: remove backlash by doing a bump of 2 to 3%; then move controller output slowly via a slow ramp or bumps of 0.1%; observe controller output: pressure to the actuator, the valve stem and the pH output. Repeat ramping up and ramping down, **2467**
Is period of oscillation in pH close to the time delay, **2799**
Manually start with a high acid flowrate demand gradually decrease the flow demand, **700**

**Gather data for key calculations**

Pressure profile,
For the feed acid from the storage tank to the injection location, **1003**

Energy balance:
Estimate power required for density of acid, **1499**

Equipment performance,
Rating pump, **2850**

**Sensors: check response to change**
pH, **184**

**Sensors: use temporary instruments,**
Use clamp-on ammeter to measure amps to pump, compare with expected, **896**

**Sensors: calibrate**, pH sensor, **850**

**Control system**, Put on manual, **498**

**Samples and measurements**, Sample acid and check that density is 1.48, **1075**

**Open and inspect**,
Pump discharge line from pump to injection point, **1700**
Pump suction line, **2100**

- **Take "corrective" action**,
  Replace pH sensor, **2500**
  Relocate acid injection line so that the acid enters the top of the caustic waste line, instead of the bottom, **2600**
  Install a check valve in the discharge of the acid pump, **378**
  To improve mixing, install a static mixer just after the mixing tee in the caustic waste line, **125**

Relocate the pH sensor to 3.2 m from the injection point to give a 2-s residence time, **1218**
Run cold water over the pump, **1300**

**Case #23: The hot TDI** (based on Barton and Rogers, 1997) [4, polymerizer, mixer, cooling system, polymer]
In the manufacture of polyurethane prepolymer, toluene diisocyanate (TDI), at room temperature, 25 °C, was charged to an 8-tonne reactor overnight. It was raining cats and dogs this morning. At the start of the 8 am shift, 1.7 Mg of polyol was added gradually over a 20-minute period as spelled out in the operating procedures. The mixer operated continuously. The temperature was monitored carefully. It rose to 127 °C as expected. After 35 minutes the operator in the control room noted that the reactor temperature read 170 °C or 43 °C hotter than expected. The red light indicated that the stirrer stopped. The operator couldn't get the stirrer restarted. Shortly thereafter the temperature sensor on the reactor read 200 °C.

**Case #23: The hot TDI**

- **MSDS**, **2252**
- **Immediate action for safety and hazard elimination**,
  Put on safe-park, **2168**
  Safety interlock shut down, **2034**
  SIS plus evacuation, **1739**
- **IS and IS NOT**,
  What, **2535**
  When, **2755**
  Who, **2422**
  Where, **14**
- ***Why? why? why?***, Goal: "get the reactor temperature to 127 °C and the mixer going", **1596**
- **Weather**, Today and past, **1954**
- **Maintenance: turnaround**, When and what done?, **2986**

**Maintenance: routine**, When and what done?, **2231**

- **What should be happening**,

**Design and simulation files** (allowances made for fouling, overdesign and uncertainties)
Custom designed reactor with coolant coils and stirrer, **2468**
**Commissioning data, P&ID, internal reports**, P&ID, **991**
**Handbook**, TDI, **531**
**Trouble-shooting files**, **1325**

- **What is happening**

**Visit control room:** control-room sensors and historical data,
TI in reactor and TRC in reactor, **1195**

Pressure in reactor, **2289**
Pressure relief on top, **2893**
Indication light that shows stirrer is turning; green means yes; red means stopped, **291**

**Process operators**, Was the TDI charged correctly?, **1201**
Is the feed cooling water cold?, **1422**
Is the cooling-water flowing to the cooling coils?, **1618**
Is there a power failure that might cause the mixer to fail?, **1986**
Please describe the procedure you used to add the polyol, **2188**

- **Check with colleagues about hypotheses**, 843

**Call to others on-site**,
Purchasing: did you change suppliers of the TDI or of polyol?, **1743**
Utilities: any upsets or changes in the cooling water supplied to our site, **2274**

**Visit site**, read present values, observe and sense.
Anything obvious interfering with the mixer shaft preventing it from turning?, **2791**
TRC valve position for the cooling-water valve controlling water to the jacket, **2824**
Emergency block valve on vent (that bypasses the PCV in case the PCV fails to open under SIS), **723**
Signal to the TRC control valve for the cooling water, **1261**

**Check diagram and P&ID versus what's out on the plant**,
No diagram supplied; only verbal description, **1837**

**On-site simple tests:** Glove test on temperature of reactor, **2996**

**Sensors: check response to change**
Temperature on reactor, **306**
Temperature on cooling water, **49**

**Control system**,
Put on manual to control temperature, **651**

**Sensors: calibrate**,
Temperature sensors on reactor, **977**
Temperature sensors on cooling water, **2736**

**Samples and measurements**,
Sample the polyol. Is it within specs?, **2522**
Sample the TDI. Is it within specs?, **709**

**Open and inspect**,
Isolate, drain, vent, make safe to enter. See if the cooling coils are fouled, **543**
Isolate, drain, vent, make safe to enter. See if there is anything preventing the mixer from turning, **1769**

- **Take "corrective" action**,
  Replace motor on mixer with motor with double the kW, **1503**

**Case #24: Low production on the ethylene plant** (courtesy of John Gates, B. Eng. 1968, McMaster University) [4, distillation, adsorption, regeneration dryers, exchangers, ethylene]

The part of the ethylene plant that relates to this problem concerns the drying section to remove moisture from the feed gas and the distillation train to separate the gas stream into the desired component section. Drying section: Three alumina dryers are installed. One dryer is regenerated while two are hooked in series on stream. For example, dryer V106 is being regenerated with V107 and 108 removing the moisture in the process stream to less than 4 ppm. Then V107 will be regenerated with V108 and 106 in series and so on. The cycle lasts 12 hours. The dried gas goes to a knockout pot (to remove any entrained material) and then is chilled, in exchangers E107 and 108, before entering the separation towers.

During regeneration of the dryers, "fuel gas", heated with 2.8 MPa steam, flows through the dryer in the direction reverse to normal flow. Once it is through the dryer the regeneration effluent gas is cooled and returned to the fuel-gas system. During regeneration the dryer temperature rises to 190 °C and is maintained at this temperature for one hour. Then the fuel gas bypasses the heater and is sent directly to the dryer to cool it. For this plant, the "fuel gas"or "town gas" is purchased from an off-site, independent utility supply pipeline. This gas is primarily methane with some hydrogen, is supplied at a pressure of 1.1 MPa and with a specified moisture content of $< 6$ ppm. This company has recently been expanding its facilities and pipelines.

The dryers are all appropriately manifolded and valved so that any dryer can be regenerated, bypassed or used. The regenerating dryer is separated from the line dryers by a gate valve.

The separation is performed in a train of three distillation columns operating at about 3.2 MPa. These are a demethanizer, de-ethanizer and C2 splitter, T101, T102 and T103 respectively.

The process is illustrated in Figure 7-1 (**Case #12**). The overhead from Tower T101 is condensed with ethylene as refrigerant. The overhead from Tower T102 is condensed with propane as refrigerant. The overhead from Tower T103 is condensed with propylene as the refrigerant.

The current situation. At low flowrates of 70 Mg/d of ethylene, we encounter no difficulties with production. Recently, we have had excessive pressure drops across the third column when all conditions were the same except that the production rate had increased to 150 Mg/d. At these higher rates of production, the pressure drop was so large that we could not operate satisfactorily. The lost capacity is worth $20 000/day. We cannot shut the plant down because the rest of the site uses ethylene as a raw material and our current inventories are very low. The usual operating conditions are given in Figure 8-6. Get our inventories up so that the rest of the site can function properly.

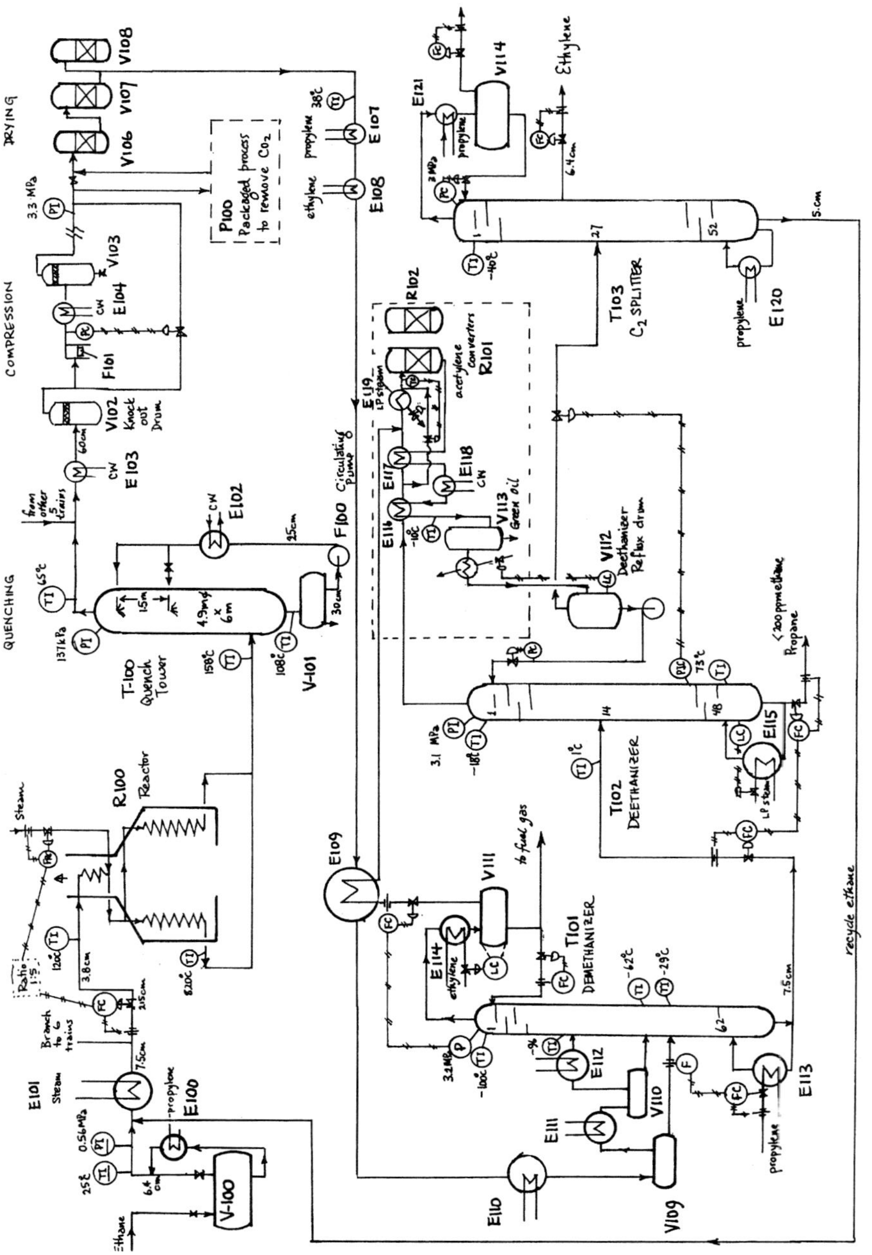


Figure 8-6 P&ID for the ethylene plant for Case #24.

**Case #24: Low ethylene production**

- **MSDS**, **320**
- **Immediate action for safety and hazard elimination**,
  Put on safe-park, **2948**
  Safety interlock shut down, **74**
  SIS plus evacuation, **98**
- **More about the process**,
  About the dryers and dryer cycle, **2888**
  About the low-temperature condensation, **2514**
- **IS and IS NOT**,
  What, **1800**
  When, **2200**
  Where, **2399**
- ***Why? why? why?***, Goal: "remove the high Δp across the trays in the C2 splitter, T103 at high throughput", **488**
- **Weather**, Today and past, **429**
- **Maintenance: turnaround**, When and what done?, **783**

**Maintenance: routine**, When and what done?, **1484**

- **What should be happening**

**Design and simulation files** (allowances made for fouling, overdesign and uncertainties)
  Exchanger E131: cool regeneration gas to the dryer during last portion of regeneration: water-tube side; regeneration fuel-gas shell side, **1773**
  Heater E130: heat regenerative fuel gas to dryer, **257**
  Exchanger E107and E108: precool feed from 38 °C to –29 °C, **451**
  Reboiler E115 on bottoms of De-ethanizer, **2130**
  Demethanizer overhead condenser: process fluid shell side; ethylene refrigeration on the tube side, **2885**
  Demethanizer reboiler: process stream shell side; propylene on tube side, **2999**
  Steam trap on regenerative gas heater, **2632**
  Distillation columns T101, 102 and 103, **2501**
  Knockout pot, **1777**
**Vendor files**:
  Heat exchanger, **1569**
  Dryer system, **280**
  Steam trap, **824**
  Utilities: town gas supplier, **501**
**Commissioning data, P&ID, internal reports**, **982**
**Handbook or Google**, Gas hydrates, **1493**
**Trouble-shooting files**, **2992**

- **Calculations and estimations** based on information given in the problem statement

Pressure profile,

Exchanger Process fluid pressure relative to utility pressure for E 131, 107, 108, 113, 114 to show direction of leak, **2726**
Exchanger Process fluid pressure relative to utility pressure for E 130 to show direction of leak, **159**
Exchanger utility pressure relative to atmospheric to show direction of leak, **466**
Exchanger process fluid relative to atmospheric, **2429**
Column T101 Δp estimate across trays at 150 Mg/d, **2936**
Column T102 Δp estimate across trays at 150 Mg/d, **2116**
Column T103 Δp estimate across trays at 150 Mg/d, **2112**

- **What is current operation**

**Visit control room:** read instruments and past records,

For 150 Mg/d; Temperatures: top and bottom for demethanizer T101, **1214**
For 150 Mg/d; Temperatures: top and bottom for de-ethanizer T102, **435**
For 150 Mg/d; Temperatures: top and bottom for C2 splitter T103, **1377**
For 150 Mg/d, measured Δp across demethanizer column T101, **2432**
For 150 Mg/d, measured Δp across de-ethanizer column T102, **2012**
For 150 Mg/d, measured Δp across C2 splitter T103, **2840**

**Process operators**, Any changes in operation other than increase in flowrate?, 2043

- **Check with colleagues about hypotheses, 162**

**Call to others on-site,**

Operators of upstream feed gas facilities: upsets? any moisture in feed? any changes in composition at the higher feedrates?, **559**
Operators of fuel-gas system; any changes in the fuel gas we are sending you from the regeneration process for our adsorbers?, **675**
Operators of the refrigeration units for the ethylene and propylene refrigeration loops: any changes or upsets?, **1129**
Utilities: any changes in the 2.8 MPa steam supplied to site, **1286**

- **Visit site**, read present values, observe and sense,
  Look at the flare, **1912**

**Check diagram and P&ID versus what's really out on the plant, 1767**

**On-site simple tests:**

Drain knockout pot after adsorbers, **2171**

**Sensors: check response to change**

Δp and pressure gauges on demethanizer T101, **2540**
Δp and pressure gauges on C2 splitter T103, **2232**

**Sensors: calibrate**,

Calibrate the pressure gauges at the top and bottom of the demethanizer column T101, **2834**

Calibrate the pressure gauges at the top and bottom of the de-ethanizer column T102, **2507**

Calibrate the pressure gauges at the top and bottom of the C2 splitter T103, **371**

**Call to vendors, licensee, supplier**,

Call utility supplying town gas about changes, **2032**

Call utility supplying town gas about specifications, **2050**

**Samples and measurements**,

Sample town gas entering battery limits. Analyze for water content. Samples every 30 min for 2 hours, **892**

Sample town gas leaving heater E130 before entering the dryer for regeneration. Analyze for water content. Samples every 30 min for 2 hours, **1000**

Sample town gas entering battery limits. Analyze for water content, **527**

Sample town gas leaving heater E130 before entering dryer for regeneration. Analyze for water content, **1874**

Sample process gas leaving KO pot as feed to tower T101. Analyze for water content, **2387**

Sample process gas entering dryers. Analyze for water content, **2880**

Sample process gas leaving the last dryer in the series. Analyze for water content, **2545**

**Perform more complicated tests**,

Gamma scan near the top of the demethanizer and of the C2 splitter to locate collapsed tray, **2790**

**Open and inspect**,

Shut down tower, vent to safety, open access hole near top and look for plugs in the downcomers or on the trays. Demethanizer column T101, **2370**

Shut down tower, vent to safety, open access hole near top and look for plugs in the downcomers or on the trays. C2 splitter T103, **2972**

Shut down column T102, isolate the reboiler E115 on the de-ethanizer, once conditions are safe, pull bundle, hydraulically pressure test to identify leaks in the tubes or between the tubesheet and the tubes, **1190**

Heater E130; isolate, once conditions are safe, pull bundle, hydraulically pressure test to identify leaks in the tubes or between the tubesheet and tubes, **348**

- **Take "corrective" action**,

  Replace the tube bundle on reboiler E115 on the De-ethanizer. T102, **818**

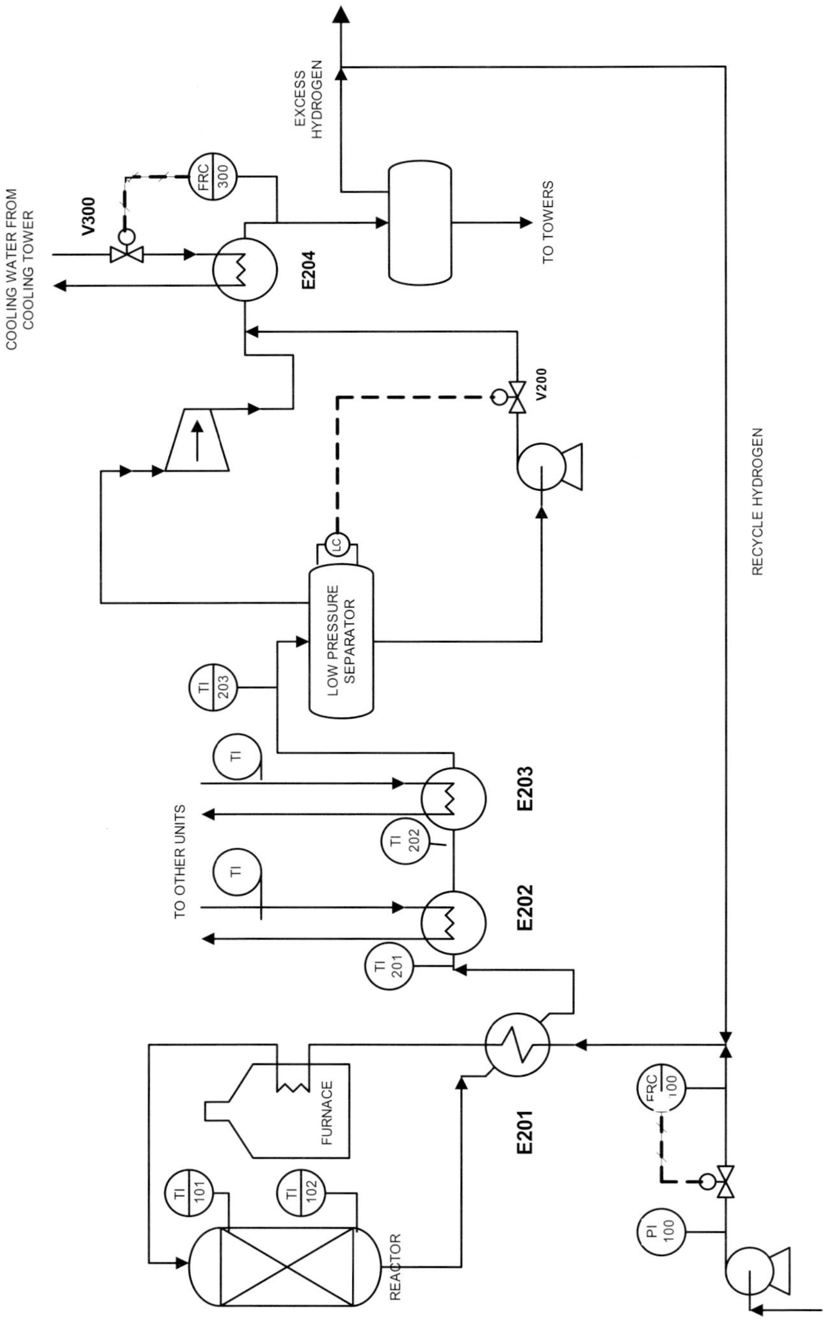


**Figure 8-7** The naphtha process for Case #25.

**Case #25: The case of the delinquent exchangers** [4, reformer, furnace, exchangers, pumps; refinery]

In the reforming process of naphtha, naphthenes are dehydrogenated (cyclohexane to benzene and hydrogen) and paraffins are isomerized (n-hexane to 2.methylpentane) and dehydrocyclized (n-hexane to benzene and hydrogen). In the reactor, a relatively high hydrogen:hydrocarbon feed ratio is maintained to minimize coking. The exit gas from the reformer is a very useful source of heat at a temperature of about 415 °C and is used to heat several streams for other units. Figure 8-7 shows the system. To keep up-to-date, we changed the catalyst so that the feed hydrogen: hydrocarbon flowrate could be reduced to about half its original value. Not only does this allow a reduction in the amount of hydrogen recycled, but this allows an increase in the naphtha flowrate to keep the space velocity in the current reactor the same as it was before we changed to the new catalyst.

We have just started up the unit after the new catalyst has been installed. Immediately the operators of the other units phone to say that their process streams are no longer getting the amount of heating (via E202 and E 203) they used to get from the reformer before the shutdown. The exchangers are delinquent! Fix the problem.

**Case #25: The case of the delinquent exchangers on the naphtha reformer**

- **MSDS**, **473**
- **Immediate action for safety and hazard elimination**,
  Put on safe-park, **243**
  Safety interlock shut down, **2774**
  SIS plus evacuation, **2966**
- **IS and IS NOT**,
  What, **2527**
  When, **27**
  Who, **496**
  Where, **564**
- ***Why? why? why?*** Goal: "to provide the usual amount of heat to the process streams of other units" **738**
- **Weather**, Today and past, **832**
- **Maintenance: turnaround**, When and what done? **951**

**Maintenance: routine**, When and what done? **1351**

- **What should be happening**

**Design and simulation files** (allowances made for fouling, overdesign and uncertainties)

Exchangers: E201, 202, 203, **1036**
Naphtha pump, **1474**
Furnace, **1235**
Reformer, **1526**

**Vendor files**:

Heat exchangers E201, 202, 203, **1636**

Pump, **1975**

**Commissioning data, P&ID, internal reports**, **2941**

**Handbook**, Approximate thermal properties of hydrogen and hydrocarbon vapors in the stream. Approx. Prandtl numbers, **2551**

**Trouble-shooting files**, **2998**

- **Calculations and estimations**

Equipment performance, Exchangers E201, 202, 203, **2451**

- **What is the current operation**

**Visit control room:** control-room data,

Into reformer, TI 203, **2111**

Ex reformer, TI 101, **2404**

Ex exchanger E203, TI 102, **356**

Naphtha flow, FRC 100, **65**

**Process operators**,

On other units: Has anything else changed for you because of the shutdown, **922**

On the reformer unit: anything changed with the new catalyst? **503**

Operating procedures, Please walk me through the startup procedure you used, **874**

- **Check with colleagues about hypotheses**, **2780**

**Visit site**, read present values, observe and sense.

TI 201, **686**

TI 202, **2418**

PI 100, **2122**

T on exit stream to other units from E202, **2816**

T on exit stream to other units from E203, **2503**

Visual check around exchanger E202 and E203, **2383**

Look at the flare, **2645**

**Check diagram and P&ID versus what's really out on the plant**, **2228**

**On-site simple tests:**

Vent exchanger E202 to check if inert gas trapped in shell side, use gas sniffer designed to test for expected gas, **2059**

Vent exchanger E203 to check if inert gas trapped in shell side, use gas sniffer designed to test for expected gas, **2212**

Vent exchanger E201 to check if inert gas trapped in shell side, use gas sniffer designed to test for expected gas, **1751**

**Gather data for key calculations**

Pressure profile,

Reformer exit to low-pressure separator, **2951**

Exchanger: tube side, **2016**

Naphtha from pump to reformer, **2316**

Mass balance,

Space velocity and mass flowrate through the reformer and the downstream exchangers, **2294**

Energy balance, Overall energy balance on the system, **1086**

Equipment performance,

Check heat exchanged on exchanger E202 based on new conditions, **1807**

Check heat exchanged on exchanger E203 based on new conditions, **1550**

**Sensors: check response to change**

TI 202, **1901**

TI 203, **11**

**Sensors: use of temporary instruments,**

Surface or laser thermometer on the line between E202 and E203, **313**

**Sensors: calibrate,**

TI 202, **600**

TI 201, **668**

T on exit stream to other units from E202, **1276**

T on exit stream to other units from E203, **1489**

**Control system,**

Controls related to the reformer; put on manual and check that working correctly, **851**

FRC 100: put on manual and check that it is working well, **701**

**Call to vendors, licensee,**

Phone call to catalyst vendor: any unexpected behavior difference between new catalyst and old catalyst given the conditions we are using, **392**

**Samples and measurements,**

Reformer exit gas and check for hydrogen content, **1956**

**Open and inspect,**

For exchanger E202: pull the bundle and measure baffle spacing; check sealing strips, check that baffles are not loose. Look for fouling; check that vent works. Look for condensed liquid, **1555**

For exchanger E201: pull the bundle and measure baffle spacing; check sealing strips, check that baffles are not loose. Look for fouling; check that vent works, **2866**

For exchanger E203: pull the bundle and measure baffle spacing; check sealing strips, check that baffles are not loose. Look for fouling; check that vent works, **2586**

- **Take "corrective" action,**
  Open exchanger E202 and decrease the baffle spacing to increased the vapor flow across the bundle by 20%, **2711**

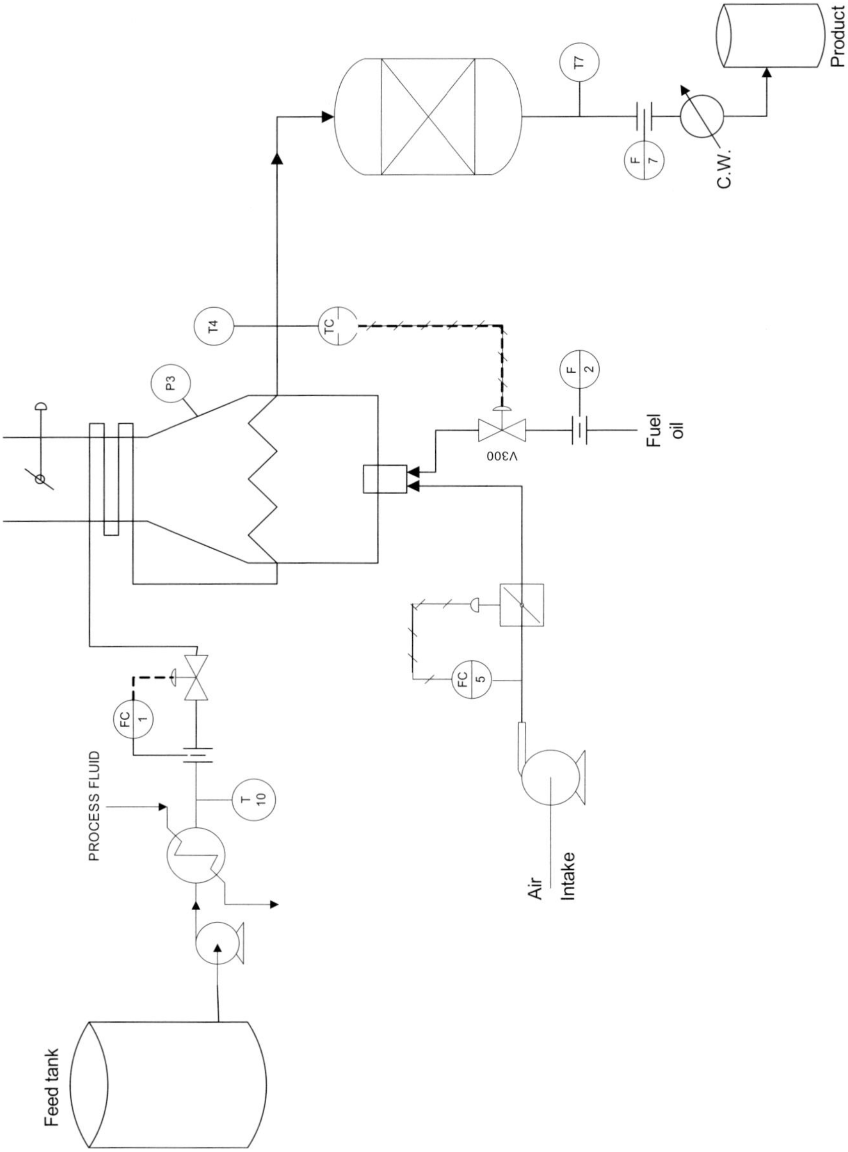


**Figure 8-8** The fired heater process for Case #26.

**Case #26: The drooping temperatures** (used with permission from T. E. Marlin)
[5, furnace, pump; general]
The process shown in Figure 8-8 consists of a fired heater, which raises the temperature of a hydrocarbon stream via convective and radiative heat transfer, and a packed-bed reactor. The process has been working well for over a year. Recently, the market for the product is growing, and the plant would like to maximize the production rate. Therefore, the operators have been slowly increasing the feedrate.

You happened to be in the control room one morning to collect some data when an operator asks for your assistance. She shows you the trend of selected variables seen in Figure 8-9. She is quite concerned; you better solve this problem fast!

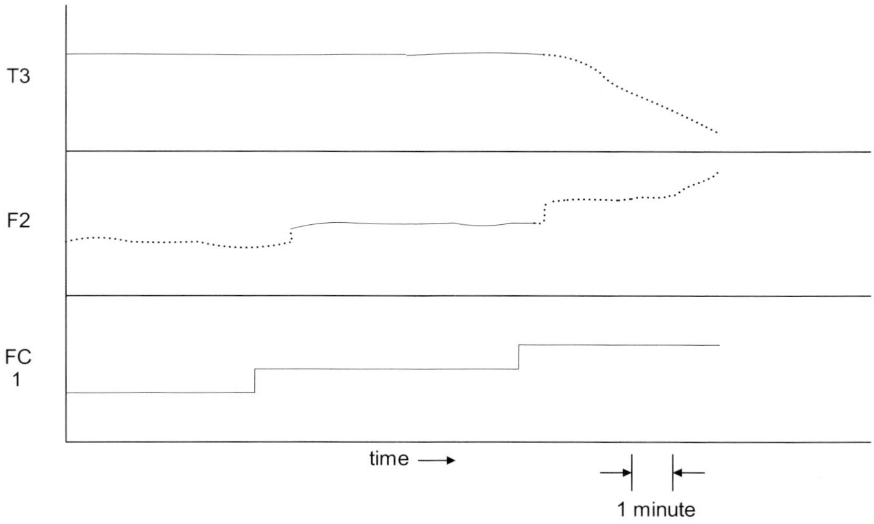


**Figure 8-9** Trend plot of some process variables for Case #26.

**Case #26: The case of the drooping temperature**

- **MSDS, 283**
- **Immediate action for safety and hazard elimination,**
  Put on safe-park, **340**
  Safety interlock shut down, **15**
  SIS plus evacuation, **476**
- **IS and IS NOT,**
  What, **1487**
  When, **2995**
  Who, **2636**
  Where, **218**
- ***Why? Why? Why?*** Goal: to operate plant safely and efficiently, **1444**
- **Weather**, Today and past, **2626**

- **Maintenance: turnaround**, When and what done? **1389**

**Maintenance: routine**, When and what done? **1130**

- **What should be happening**

**Design and simulation files** (allowances made for fouling, overdesign and uncertainties)
Pump, **1999**
Blower, **1782**
Heat exchanger, furnace/ fired heater, **1532**
**Vendor files**:
Blower, **1005**
Pump, **1381**
**Commissioning data, P&ID, internal reports**,
Packed-bed reactor: data about the capacity: can it handle the increased feedrate? **1401**
**Trouble-shooting files**, **1729**

- **What is the current operation**

**Visit control room:** control-room data,
Flow: Process liquid in. FC-1, **955**
Flow: process liquid ex reactor F-7, **2851**
Flow: Fuel gas F-2, **2542**
Flow: Air to furnace FC-5, **1225**
Temperature: exit process liquid TC-3, **1092**
**Process operators**, This shift, **1452**
Previous shift, **1255**
Operating procedures, Please guide me through the procedures you use, **1894**

- **Check with colleagues about hypotheses**, **1111**

**Visit site**, read present values, observe and sense.
Temperature: process liquid out. T4, **1623**
Pressure: on furnace P3, **2191**
Stack gas: Look at quality of flue gas out of stack, **2101**
Pump: process liquid: sound like cavitation? **2046**
Valve: control: fuel oil, **2881**
Valve: control air to furnace, **381**
Temperature of the process fluid entering fired heater, T-10, **2428**
Block valves around the fan fully open? **2140**
Readings of the draft gauges upstream and downstream of the damper, **2878**
Position of the damper; compare with expected 1/3 closed, **1279**
Inspect orifice tabs the three flow meters to ensure that the plates were installed with sharp edge upstream, **1139**
**Check diagram and P&ID versus what's really out on the plant**, **171**

**On-site simple tests,**

Use laser temperature sensor or pyrometer to measure temperature of tubes in the radiant section, **443**

Use laser temperature sensor or pyrometer to measure temperature of the tubes in the convection section using the observation port in that section, **820**

Shut the valve on the process pump discharge and compare the measured pressure with the head-capacity curve for that pump, **690**

Change set point on air flow, FC-5, and note response, **553**

- **Gather data for key calculations**

Pressure profile,

On process pipe through the furnace, reactor and product tank, **1902**

On fuel-oil line bringing fuel into the furnace, **1601**

On the combustion air line from the intake to the burner, **1801**

Direction of leak of process fluid versus furnace, **661**

Direction of leak of air: into the furnace or out of furnace? check pressure P3, **897**

Mass balance, On process liquid, **2301**

Energy balance: On combustion: temperatures, **2003**

On combustion: excess air, **101**

Equipment performance,

Heat exchanger, **307**

Air blower, **150**

Process liquid pump, **801**

**Sensors: calibrate,**

Temperature: process liquid out. T4, **287**

Pressure: on furnace P3, **469**

**Control system,**

Temperature: Controller: Signal output from controller TC/3, **2574**

Put the process fluid temperature controller, TC-3, on manual control to control under the high flowrate conditions, **2740**

Air-flow controller: signal output from controller FC/5, **2077**

**Samples and measurements,**

Oxygen in flue gas, **929**

**Open and inspect,**

Furnace, **1431**

Blower, **1189**

- **Take "corrective" action,**

Put TC3 on manual. Reduce output from valve V300, **1938**

Install sensor in the flue gas to measure percentage oxygen, **2379**

**Case #27: The IPA column** [5, distillation column plus auxiliary equipment; petrochemical or refinery]

Recently, the water-cooled condenser on the IPA column was replaced by an air-cooled condenser. Variable-pitch fans underneath the horizontal bank of finned tubes direct air upwards across the tubes. Since the installation, the ratio of IPA out to IPA fed to the column is 0.70. Previously we could account of 0.995. Where is the IPA going? This costs us $4,000/h.

**Case #27: The IPA column: where is it going?**

- **MSDS**, **2529**
- **Immediate action for safety and hazard elimination**,
  Put on safe-park, **1**
  Safety interlock shut down, **271**
  SIS plus evacuation, **122**
- **More about the process**, **2134**
- **IS and IS NOT**:
  What, **398**
  When, **1758**
  Who, **1002**
  Where, **1708**
- ***Why? why? why?*** Goal: "to prevent the apparent loss of IPA.", **2162**
- **Weather**, Today and past, **447**
- **Maintenance: turnaround**, When and what done? **1259**

**Maintenance: routine**, When and what done? **1589**

- **What should be happening**

**Design and simulation files** (allowances made for fouling, overdesign and uncertainties)

Fan: control system on the air flow to the condenser, **284**
Condenser, air-cooled, **126**
Distillation column, **2000**
Lines insulated? **485**

**Vendor files**: Condenser, air cooled, **2181**

**Commissioning data, P&ID, internal reports**, Commissioning report of tests done before startup, **2993**

**Handbook**, Pertinent properties of IPA, **1166**

**Trouble-shooting files**, **1017**

- **Calculations and estimations**: more data are needed before calculations can be made.
- **What is the current operation**

**Visit control room:** control-room data: values now and from past records,

Temperature at the top of the column, **1501**
Temperature on liquid exit from the condenser, **1685**

Pressure at the top of the tower, **1722**
Pressure on the overhead receiver, **777**
Level in the overhead receiver, **521**
Reflux flowrate, **1385**
Feedrate change? **2394**
Bottoms feedrate change? **2846**

**Process operators**
This shift, **1179**
Previous shift, **1356**

- **Check with colleagues about hypotheses**, **2926**

**Call to others on-site**
Of upstream suppliers of feed to IPA unit, **86**
Of downstream vent scrubber unit, **1291**
Of downstream processors of IPA, **377**

**Visit site**, read present values, observe and sense.
Hot exhaust air recirculation to the intake of the air-cooled condenser, **18**
Sounds around the condenser, **338**
Look at the flare, **239**

**Check diagram and P&ID versus what's really out on the plant**, Same information as given in **more about the process**, **2134**

**On-site simple tests:**
Temperatures and humidities of inlet and exit air for the air-cooled condenser, **106**
Tape accessible flanges on the line to the top of the condenser and after the condenser. Test for leaks with soap solution. Test valve stems with soap solution, **1879**
Record the vertical dimensions around the exit of the condenser, **1135**

**Gather data for key calculations**
Pressure profile,
On process pipe from the top of the column to the overhead receiver, **926**
Across condenser, **1062**
Mass balance, On process liquid, IPA, **1508**
Energy balance: Over the condenser: heat extracted in air = heat of condensation for all IPA? **1026**
Equipment performance,
Rating of air-cooled condenser, **844**
Fans, **632**
Column, **670**

**Sensors: check response to change**
Temperature at top of column, **50**
Temperature of the IPA leaving the condenser, **200**

**Sensors: use of temporary instruments**,
Temperature and humidity of inlet and exhaust air for the condenser, **1275**

Air velocity to the face of the condenser for the maximum and minimum blade pitch, **1423**

**Sensors: calibrate,**

Temperature at top of the column, **351**

Temperature of IPA leaving condenser, **1808**

**Control system,**

For fan, put on manual and increase blade pitch to maximum, **1069**

Ask control specialist about the quality of this type of control, **311**

Use clamp-on ammeters, voltmeter and power-factor meters to obtain data to estimate power to the fan, **475**

**Call to vendors, licensee,** Air-cooled condenser. Describe symptoms to the vendor, 1517

**Samples and measurements**

Sample bottoms of column; measure concentration of IPA and compare with past data, **1658**

Sample feed to column; measure concentration of IPA and compare with past data, **1172**

Concentration: vent from the overhead receiver; concentration IPA, single sample, **222**

Concentration: vent from the overhead receiver; concentration IPA, ten samples taken at one-minute intervals, **152**

Concentration: liquid from bottoms; IPA concentration. Three samples, at 2-min intervals, **461**

Concentration: feed to the column; IPA concentration. Three samples, at 2-min intervals, **956**

**Open and inspect,**

Air-cooled condenser: visual inspection plus air and water pressure tests, **1652**

Fan and fan blades, **251**

- **Take "corrective" action,**

  Run cold water over the tubes for an 8-hour shift, **1916**

  Tip the tube bank so that it is no longer level "so that the condensate can run out easily", **495**

  Repipe exit line so that the syphon is removed. "The condensate flows directly from the level of the bottom tubes to the overhead receiver", **1943**

**Case #28: The boiler feed heater** (adapted from case of P. L. Silveston, University of Waterloo) [5, shell and tube exchanger, steam heated, condensate traps; general]
Waste flash steam from the ethyl acetate plant is saturated at slightly above atmospheric pressure. It is sent to the shell side of a shell and tube exchanger to preheat boiler feed water to 70 °C for a nearby boiler house. Condensate is withdrawn through a thermodynamic steam trap at the bottom of the shell. The water flows through 1.9-cm nominal tubes. There are 100 tubes. See Figure 8-10. "When the

system was put into operation 3 hours ago everything worked fine," says the supervisor. "Now, however, the exit boiler feed water is 42 °C instead of the design value. What do we do? This difficulty is costing us extra fuel to vaporize the water in the boiler. " Fix it.

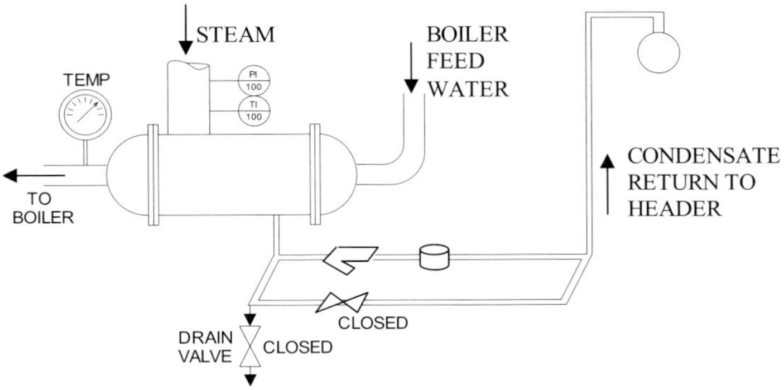


**Figure 8-10** The boiler feedwater heater for Case #28.

**Case #28: The boiler water preheater**

- **MSDS, 349**
- **Immediate action for safety and hazard elimination,**
  Put on safe-park, **6**
  Safety interlock shut down, **319**
  SIS plus evacuation, **266**
- **IS and IS NOT**: (based on given problem statement),
  What, **2861**
  When, **1320**
  Where, **2391**
- ***Why? why? why?***, Goal: "to heat the boiler water to 70 °C", **254**
- **Weather**, Today and past, **491**
- **What should be happening,**

**Design and simulation files** (allowances made for fouling, overdesign and uncertainties)

Exchanger, **2765**
Float steam trap, **995**
Piping: steam, **1990**
Piping: water, **1270**

**Handbook**, Approximate thermal properties of steam and ethyl acetate. Approx. Prandtl numbers. Saturation temperature for assumed 200 kPa g steam, **1504**

**Trouble-shooting files**, 224

- **Calculations and estimations** (that can be done in the office before special tests are done and based on information in the problem statement and rules of thumb)
  Flowrate, Estimate flow of boiler feed water based on tube area and estimated velocity in the tubes, **1237**

Energy,
  Difference in heat load, assuming inlet water temperature = 18 °C, **1788**
  Heat transfer rates at new and design conditions based on assumed steam pressure of 200 kPa g, **2199**
  Estimate steam condensation, **2699**

- **What is the current operation**

**Visit control room:** control-room data: values now and from past records. Data only available on site
**Process operators**, This shift, **1329**

- **Check with colleagues about hypotheses**, **2686**

**Visit site**, read present values, observe and sense.
  Inlet steam pressure, **323**
  Inlet steam temperature, **177**
**Check diagram and P&ID versus what's really out on the plant**, **2126**
**On-site simple tests:**
  Open bypass on the steam trap, **477**
  Open air vent on top of the exchanger; leave the vent open for 10 min. Read the temperature gauge on the exit boiler feed, **364**
  After the vent has been opened for 10 min, close the vent and read the boiler feed exit temperature after 3 h of operation, **876**
  At the end furthest from the steam inlet, remove a vertical section of the insulation about 10–30 cm wide. Tap the side with metal moving up the exchanger and listen for a change in sound indicating a liquid-vapor interface, **617**
**Collect data for key calculations**
Pressure profile,
  Exchanger: shell side, **1067**
  Exchanger: tube side, **66**
  Steam: from ethyl acetate plant to header to boiler feed heater, **465**
Mass balance,
  On steam-condensate; measure the condensate by measuring the volume collected over a timed period, **2005**
Energy balance: sink = source, Amount of steam = amount of heat picked up by water, **108**
Equipment performance,
  Rating exchanger, **144**
  Type of steam trap, **1632**
  Upstream Y strainer on condensate line upstream of the trap, **1930**

**Sensors: check response to change**
Steam temp TI 100, **3**
Exit water temperature, **210**
**Sensors: use of temporary instruments**, Water temp in, **502**
**Sensors: calibrate**, Steam temp TI 100, **368**
Exit water temp, **635**
**Call to vendors, licensee**, Heat exchanger, **1519**
Steam traps, **1533**
**Samples and measurements**,
Measure amount of condensate to get a measure of the steam. Compare the energy loss from the steam to the energy gained by boiler water, **388**
**Open and inspect**,
Exchanger: visually inspect. Water and air tests for leaks, **1577**
Exchanger: pull the bundle and measure baffle spacing; check sealing strips, check that baffles are not loose, **57**

- **Take "corrective" action**,
  Replace the float trap with an inverted bucket trap with an air vent, **1924**
  Operate with the bypass around the trap partially open, **480**
  Stop operation; pull the bundle and clean tubeside and shell side, **1944**

**Case #29: The Reluctant reactor** (courtesy of W. K. Taylor, B. Eng. McMaster, 1966)
[5, reactor, compressor, separator; ammonia]
Ammonia is produced on two interconnected reactor loops as given in Figure 8-11. Feed gas consists of hydrogen and nitrogen in the proper 3:1 ratio with about 1% methane as an inert. In this ammonia-synthesis reaction about 10% conversion occurs per pass through the reactor. Feed gas is compressed to 34.5 MPa abs and fed to a common header that feeds two reactor loops. Liquid product is condensed and removed from the system; gas is recycled back to the loop via the recycle stage compression. The reactor operates at 500 °C. There is an internal gas-gas heat-exchanger within the reactor. The $\Delta T$ due to exothermic reaction is about 50 °C.

Each compressor is a multistage reciprocating constant speed machine rated at about 3000 kW. Bypass valve B is operated to control the $\Delta p$ across the recycle stage that must not exceed 3.5 MPa. Opening valve B lowers the $\Delta p$ and the flow of recycle gas to the loop. The recycle flow is about five times the flow of fresh feed.

Bypass valve A is operated to trim the loop: closing this valve forces more gas over to the reactor; opening the valve causes gas to bypass the loop. Valve A is used to control the reaction temperature. If too much gas is fed to the reactor and the catalyst is inactive, the high flow might extinguish the reaction. Similarly if the flow to the reactor is too low, the reaction will go further because of the longer reaction time; the reactor will overheat because there is not enough flow to carry away the heat of reaction. Normally valve A is open slightly during plant operation.

Methane is an inert coming in with the feed. The methane concentration is kept about 15% in the loop gas to the reactors by maintaining a small purge.

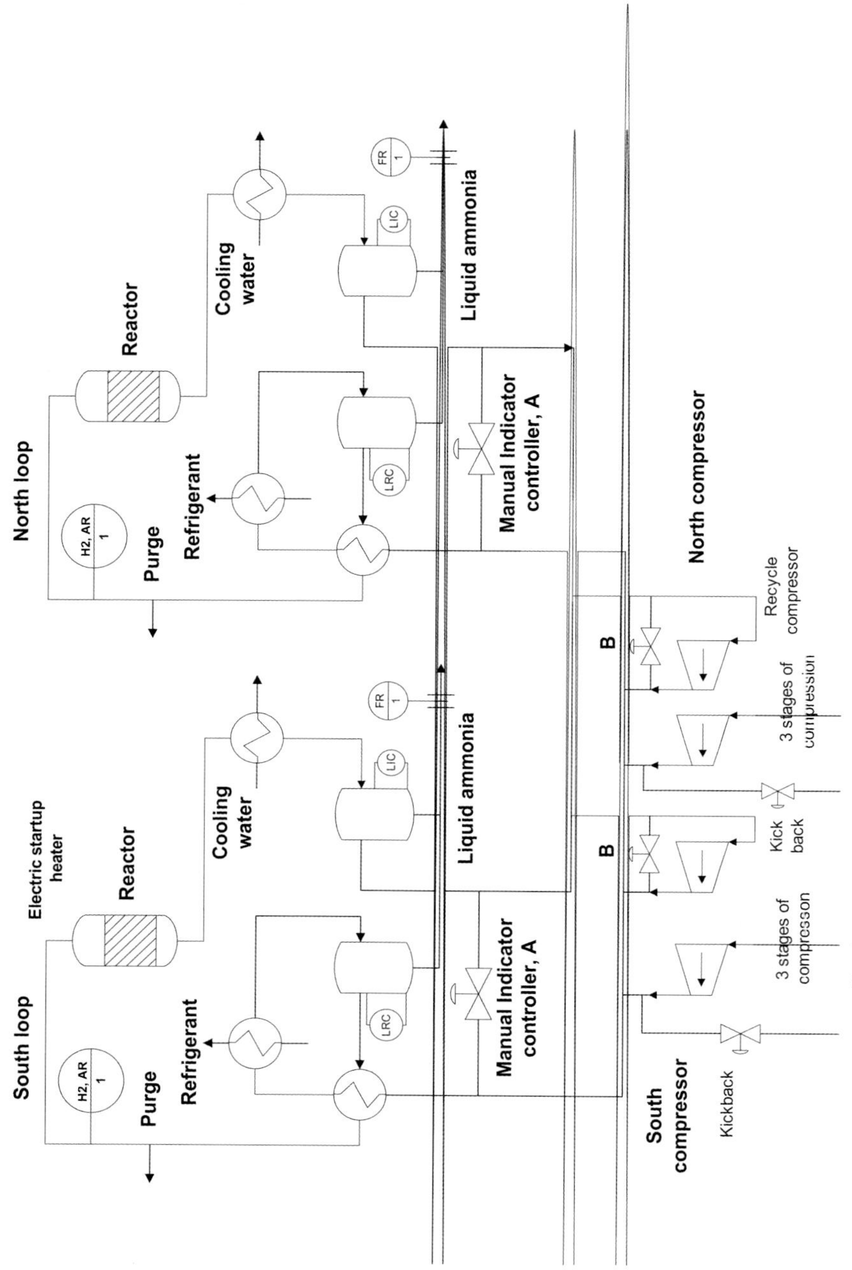


Figure 8-11 The Ammonia synthesis reactors.

The pressures, levels in the separators and the temperature profile in the reactor are shown in the control room.

The design provides operating flexibility. If one compressor breaks down the other machine can feed both loops thus keeping the reactors at operating temperatures while repairs are done. This avoids costly startup expense. Isolation block valves on the compressors are not shown on Figure 8-11. Furthermore, both loops are equalized in pressure thus evening out any slight variations introduced by the compressors.

*The problem:*
The plants are in the final phase of startup after a turnaround shutdown. The compressors are sending feed gas and recycled gas to the reactors. Startup pressure is 7 MPa. The electric cal rod heaters, used to heat up the reactor catalyst bed during startup, are on. Heat-up normally proceeds at 50 °C/h and at 8 am the reactors were up to 300 °C. By noon the reactors should be at 500 °C and in steady production and the plant in normal operation with all the gas vents closed. It is now 3 pm and the loops have only heated up 25 °C to 325 °C and have had no increase in temperature over the past hour. “Get these plants producing. This is costing us $20 000/h. “

**Case #29: The reluctant reactor**

- **MSDS, 352**
- **Immediate action for safety and hazard elimination,**
  Put on safe-park, **48**
  Safety interlock shut down, **2482**
  SIS plus evacuation, **2018**
- **More about the process, 483**
- **IS and IS NOT**: (based on given problem statement),
  What, **1319**
  When, **996**
  Who, **17**
  Where, **1978**
- ***Why? why? why?***, Goal: “to heat up the reactors to 500 °C”, **446**
- **Weather**, Today and past, **2991**
- **Maintenance: turnaround**, When and what done?, **2512**
- **What should be happening**

**Design and simulation files** (allowances made for fouling, overdesign and uncertainties)

Reactor, **2452**
Internal heat exchanger, **2010**
Reciprocating compressor, **1832**
Refrigeration system, **1507**
Condensers: refrigerant, **1733**
Condensers: water, **1960**
Gas-liquid separators, **1600**
Valves A and B, **1011**

**Vendor files**:

Reciprocating compressors, **178**

Refrigeration system, **484**

**Commissioning data, P&ID, internal reports**, **452**

**Handbook**, Thermal properties of hydrogen, nitrogen and ammonia, **385**

**Trouble-shooting files**, **751**

- **Calculations and estimations** (that can be done in the office before special tests are done)

Mass balance, Purge rate and methane buildup, **624**

Energy balance:

Energy transfer from Cal Rod heaters for startup (from experience). **532**

- **What is the current operation**

**Visit control room:** control-room data: values now and from past records,

Pressure at exit of compressors, in the loop, **1459**

Hydrogen concentration in the North loop, **1412**

Hydrogen concentration in the South loop, **1117**

Temperature in the catalyst bed, North reactor, **1048**

Temperature in the catalyst bed, South reactor, **1546**

Temperature leaving catalyst bed, North reactor, **1815**

Temperature leaving catalyst bed, South reactor, **1946**

$\Delta p$ across the reactor, North loop, **1664**

$\Delta p$ across the reactor, South loop, **2415**

Flow of liquid ammonia from North loop, **2897**

Flow of liquid ammonia from South loop, **2959**

**Process operators**, Anything surprising other than the temperatures?, **2558**

Refrigeration condensers colder than usual?, **2226**

Cooling-water condensers colder than usual?, **1720**

Operating procedures, Please tell me about the startup procedures you used, **1969**

- **Check with colleagues about hypotheses**,

**Visit site**, read present values, observe and sense.

Amps to cal rod heaters, **1564**

Check that sample valves are closed, **1789**

**Check diagram and the P&ID versus what's actually out on the plant**, **1216**

**On-site simple tests:** Increase the pressure in the loop to about 10 MPa, **1362**

**Control system**,

Over ride indicator control on board; go to site and manually open valve A fully. North loop. Check temperature after one hour, **1073**

Over ride indicator control on board; go to site and manually open valve A fully. South loop. Check temperature after one hour, **912**

**Sensors: calibrate**, Temperature sensors, reactor, North loop, **577**

Temperature sensors reactor, South loop, **2836**

**Samples and measurements,**

Cooling water, North loop condenser. Analyze for ammonia, **2601**
Feed gas. Analyze for hydrogen concentration, **2327**
Continuity check on cal rod for reactor in North loop, **333**
Continuity check on cal rod for reactor in North loop, **73**

**Open and inspect,**

Open and inspect the reactor and especially the top with the cal rod heaters. North loop, **277**

Open and inspect the reactor and especially the top with the cal rod heaters. North loop, **1281**

- **Take "corrective" action,**

  Open the kickback to reduce the pressure in the loop, **1085**
  Open the purge line to the maximum, **660**
  Shut down one compressor and provide loop gas to both North and South loops from one compressor. Allow one hour of operation, **640**

**Case #30: The case of the reluctant reflux** (courtesy Esso Chemicals) [6, distillation column plus auxiliaries; general]

The 20-tray column, operating at 520 kPa g, has just started up for the first time. The control configuration is illustrated below. The pump can deliver only 2/3 of the design value of the reflux, and although the reflux valve is fully open, the level in the accumulator continues to increase. Do something quickly before the reflux drum is full and sort out the problem. The financial penalty is high because of the clauses in the construction contract for producing specification product within the commissioning period, and because of insurance issues and government regulations. Figure 8-12 illustrates the system.

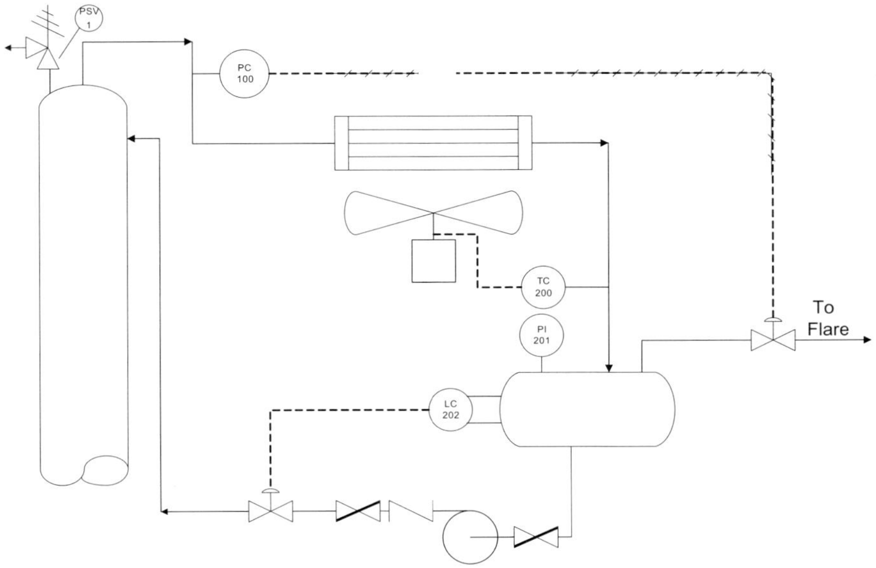


**Figure 8-12** The overhead system for the column in Case #30.

**Case #30: Reluctant reflux**

- **MSDS**, 272
- **Immediate action for safety and hazard elimination**,
  Put on safe-park, **450**
  Safety interlock shut down, **972**
  SIS plus evacuation, **780**
- **IS and IS NOT**: (based on given problem statement), What, **2930**
  When, **2621**
  Where, **2013**
- **Weather**, Today and past, **1530**
- **Maintenance: turnaround**, When and what done?, **1988**
- **What should be happening**

**Design and simulation files** (allowances made for fouling, overdesign and uncertainties)
  Reflux pump, **1058**
  Condenser, **1450**
  Reflux drum, **1364**
**Vendor files**: Reflux pump, **1137**
**Commissioning data, P&ID, internal reports**, 1823
**Trouble-shooting files**, 1828

- **Calculations and estimations** (that can be done in the office before special tests are done)

Pressure profile,
  Recheck NPSH supplied, **993**
  From reflux drum to column via reflux pump, **825**
Equipment performance, reflux pump, **26**
  Design of pressure-control system, **288**
  Design of level-control system, **138**

- **What is the current operation**

**Visit control room:** control-room data: values now and from past records,
  Overhead temperature, **460**
  Overhead composition from routine lab analyses, **752**
  Level in reflux drum, **1864**
**Process operators**, This shift. What's happening? **1606**
**Visit site**, read present values, observe and sense.
  Control valve on the reflux pump discharge line: valve position and other features of valve; for example, put in backwards, **1513**
  Note direction of rotation of the shaft and confirm that this corresponds with the arrow on the casing, **1985**
  Ask maintenance if the impeller might have been put in backwards, **2484**
  Difference in pressure between PC 100 and PI 200 and compare with usual $\Delta p$ across condenser, **2071**

Level indication on controller 200, **2320**
Listen for sounds of cavitation around the reflux pump, **2213**

**Check diagram and P&ID versus what's out on the plant, 236**

**On-site simple tests:**
Shut the valve on the discharge side of the reflux pump and read the pressure; compare with head-capacity curve, **2499**
Listen with a stethoscope to check valve. Note if the direction through the valve is the same as the flow direction, **2960**
Turn the valve stems on suction and discharge block valves on the pump, **2664**

**Sensors: check response to change**
LC202, **2582**
PI 201, **2186**
TC 200, **2409**

**Sensors: calibrate**,
LC 200 on reflux drum, **681**
PI 200; pressure on the reflux drum, **610**

**Control system**,
Put pressure-control system on manual, **1875**
Put level-control system on manual, **1675**
Put temperature control on manual, **1169**

**Samples and measurements**,
Feed to column and analyze for the amount of light components, **961**
Overhead vapor from column and analyze for heavies, **900**

**Open and inspect**,
Control valve, **595**
Reflux pump, **29**
Suction line from reflux drum to tower, **292**
Nozzle where line attaches to column, **407**
Line from pump to column, **1375**

- **Take "corrective" action**,
  Install larger-size impeller in the pump, **1775**
  Replace motor on air-cooled condenser, **2339**
  Replace the level sensor on the reflux drum, **2733**
  Reduce feed to the column, **2611**
  Replace the check valve on the discharge line of the pump, **2550**
  Install vent break line on condenser, **1266**

**Case #31: Ethylene product vaporizer** (courtesy of C. J. King, Chemical Engineering Dept., University of California, Berkeley) [6, heat exchangers, boiling, steam; general]
Our New Jersey petrochemical complex includes an ethylene plant that supplies 6.8 Mg/h of ethylene through a pipeline to various consumers. It is important that we maintain a steady flow of ethylene to our users, and, as a result, our plant contains a large storage sphere of liquid ethylene. The ethylene must be a vapor, however,

when it enters the pipeline and must be at a temperature close to ground temperature (slightly more than 1.6 °C) so as to avoid thermal stresses on the pipeline. For these reasons we have installed an ethylene vaporizer between the sphere and the pipeline. The ethylene is vaporized by condensing n-butane, which in turn is vaporized by steam. The cascade vaporization system is required so as to avoid undue thermal stresses across the heat-exchange surfaces.

Under normal operation, a small amount of ethylene, about 1.6 Mg/h is sent through the vaporizer, but the vaporizer frequently is called upon to provide more, or all, of the total ethylene supply. Before the vaporizer, the liquid ethylene is pumped up to 4.4 MPa gauge and metered through a flow-control valve; the ethylene pressure in the vaporizer is roughly the pipeline pressure of 3. 9 MPa gauge.

The butane pressure controller set point, PIC, can respond to anywhere from 0.58 to 0. 97 MPa. A set point of 0.8 MPa has been used successfully at all ethylene flow rates during the past year, although the outlet ethylene temperature has been slow to recover following a change in the ethylene flow rate.

In the past few months we have found it necessary to increase the PIC set point. Even so, we found yesterday when the ethylene unit came down that the vaporizer could not handle the full ethylene flow without tripping the low-temperature shutoff switch at the pipeline entry, which is set at 1.6 °C. This situation will cost us $6000/h plus inestimable customer good will if we stop the flow, or else it may well necessitate expensive and time-consuming pipeline repair if we continue. Figure 8-13 illustrates the layout.

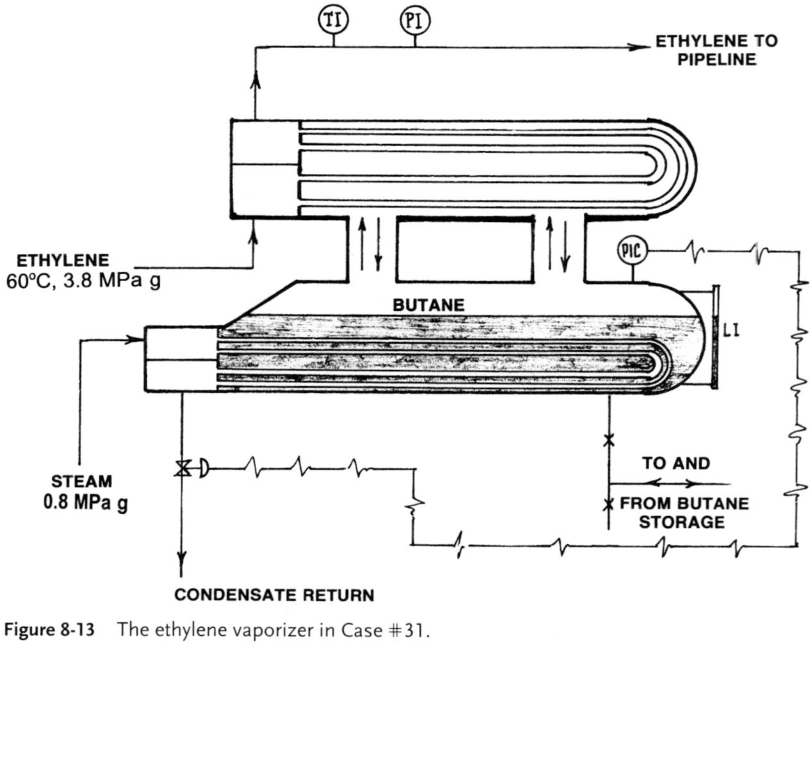


**Figure 8-13** The ethylene vaporizer in Case #31.

**Case #31: Ethylene product vaporizer**

- **MSDS**, **34**
- **Immediate action for safety and hazard elimination**,
  Put on safe-park, **2612**
  Safety interlock shut down, **2714**
  SIS plus evacuation, **2065**
- **More about the process**, **2980**
- **IS and IS NOT**: (based on given problem statement),
  What, **1108**
  When, **566**
  Where, **809**
- **Weather**, Today and past, **950**
- **Maintenance: turnaround**, When and what done?, **1359**

**Maintenance: routine**, When and what done?, **1063**

- **What should be happening**

**Design and simulation files** (allowances made for fouling, overdesign and uncertainties)
Butane vaporizer, **609**
Ethylene vaporizer, **120**
**Vendor files**: Butane-steam system, **414**
**Commissioning data, P&ID, internal reports**, **293**
**Handbook**, Temperature-vapor pressure data for ethylene and butane, **989**
**Trouble-shooting files**, **967**

- **Calculations and estimations** (that can be done in the office before special tests are done)

Pressure profile,
If there is a leak in the ethylene-butane system, the direction of the leak would be, **672**
If there is a leak in the steam-butane system, the direction of the leak would be, **1365**
Rate,
Butane-steam system: Is the boiling nucleate or film boiling?, **1147**
Ethylene-butane system: Is the boiling nucleate or film boiling?, **1717**

- **What is the current operation**

**Visit control room:** control-room data: values now and from past records, Past records, **2427**
**Process operators**, Please tell me what happened before the system shut down, **2248**
Tell me a bit about operation over the past few months, **2832**
**Call to others on-site**,
Steam utilities; any changes?, **2952**

Waste water treatment, **2528**

**Visit site**, read present values, observe and sense.

Steam pressure, **2603**

Ethylene pressure in pipeline, **2807**

Ethylene temperature in pipeline, **2549**

PIC setting, **2109**

Level in the butane, **2407**

Valve position on condensate line and size of control valve relative to line size, **2498**

Steam goes into the top channel of the butane evaporator, **2040**

Ethylene feed goes into the bottom channel of the ethylene evaporator, **1987**

Inspect insulation and look for cracks or holes in the external surface, **1776**

**Check diagram and P&ID versus what's out on the plant, 331**

**On-site simple tests:**

Open vent on butane evaporator for several minutes; close and observe operation, **1549**

Open vent on ethylene evaporator for several minutes; close and observe operation, **1745**

Use stethoscope and listen to the sound of the fluid flowing through the condensate valve, **1138**

Increase the level of butane in the butane evaporator as shown on the level indicator, **1257**

Open the bypass on the condensate control valve, **656**

**Sensors: check response to change**

PIC, **699**

Level indicator on butane, **105**

PI on ethylene line, **353**

TI on ethylene line, **2390**

Temperature shut-off sensor switch, **2469**

**Sensors: use of temporary instruments**,

Use a contact pyrometer to measure temperature of the ethylene pipeline, **2919**

**Control system**,

Reduce ethylene flowrate. Put PIC on manual and note response when increase the PIC setting, **2572**

Retune the PIC control system, **2020**

**Sensors: calibrate**,

PIC, **2154**

PI ethylene line, **1553**

TI ethylene line, **1963**

Level indicator on butane, **1238**

Temperature shut-off sensor switch, **1104**

**Samples and measurements**,

Butane from the shell side of the evaporator and analyze for ethylene and water contamination, **881**

Condensate and analyze for butane contamination, **958**
Ethylene and analyze for butane contamination, **55**
Butane from storage and analyze for contamination; compare with specs, **252**

**More ambitious tests**,
Drain all the butane and replace with fresh butane, **1914**

**Open and inspect**,
Ethylene evaporator for fouling, **1812**
Ethylene evaporator, pressure test and look for leaks, **1541**
Butane evaporator, pressure test and look for leaks, **2869**
Butane evaporator for fouling, **2055**

- **Take "corrective" action**,
  Replace the pressure gauge PIC, **2048**
  Install new control valve on condensate line; the size is same as line size, **1153**

**Case #32: The alarming alarm** (courtesy of T. E. Marlin, Chemical Engineering, McMaster University) [6, sequence of distillation columns with auxiliaries; depropanizer-debutanizer]

The process is the depropanizer-debutanizer described in **Case #8**, Chapter 2. A P&ID is given in Figure 2.4 accompanying **Case #8**.

The operation of the upstream process is being modified to accommodate a new catalyst and modified feed composition. The upstream units have been on-line and operating smoothly for nearly a shift. Suddenly the high-pressure alarm on the debutanizer, column C-9, rudely disrupts the quiet. And you thought everything was going smoothly!

**Case #32: The alarming alarm**

- **MSDS**, **1495**
- **Immediate action for safety and hazard elimination**, Put on safe-park, **111**
  Safety interlock shut down, **440**
  SIS plus evacuation, **2934**
- **More about the process**, **1010**
- **IS and IS NOT**, What, **2681**
  When, **2520**
  Who, **316**
  Where, **1407**
- ***Why? Why? Why?***, Best goal? To find out why the high-pressure alarm is sounding, **2335**
- **Weather**, Today and past, **1060**
- **Maintenance: turnaround**, When and what done?, **1921**

**Maintenance: routine**, When and what done?, **2928**

- **What should be happening**,

**Design and simulation files** (allowances made for fouling, overdesign and uncertainties)

Condenser, E-28, **2006**
Distillation column, C9: **2367**
Thermosyphon reboiler, E-30, **1382**
Reflux pump, F-29, **1112**
Overhead drum, V-31, **1455**

**Vendor files**: Condenser, reboiler, **1448**
Steam traps, **1244**

**Commissioning data, P&ID, internal reports**, **1771**
**Handbook**, Cox charts, **1910**
**Trouble-shooting files**, **1653**

- **Calculations and estimations**

Energy balance: sink = source, Estimate the steam flow to the reboiler based on the reflux rate and the fact that each kg steam boils 5 kg typical organic, **2132**

- **What is current operation**

**Visit control room:** control-room data,
Feed to C9, debutanizer, FI/2, **2227**
Feed to the C8, depropanizer. FC/1, **2304**
Overhead liquid product butane, flowrate FIC/7, **2011**
Reflux flowrate, FIC/6, **2461**
Pressure drop $\Delta p$ I/2, **1724**
Level bottoms LIC/4, **1957**
Level in feed drum, V-31; LIC-5, **1506**
Temperature bottoms TI/12, **1870**
Temperature top, TI/11, **1152**
Analyzer A-1, **1357**
Pressure on overhead, PIC/19, **1052**

**Process operators**, This shift, **1408**

**Call to others on-site**, Utilities: any change in the cooling-tower operation that might affect our site, **1323**

**Visit site**, read present values, observe and sense.
Column pressure, PI-12, **1173**
Pressure relief to flare PSV-3, **717**
Temperature of the feed to the column TI-10, **619**
Cooling-water temperature in to the condenser, TI-13, **910**
Cooling-water temperature out of the condenser, TI-14, **802**
Valve position for column feed, FV-2, **258**
Valve position on PIC/19; downflow from the condenser to the reflux drum, **13**
Signal to PIC/19, **359**
Pressure on exit of reflux pump F-29, PI-20, and compare with head-capacity curve at feed flowrate, **135**
Listen to the reflux pump F29 for sounds of cavitation, **455**

Observe whether shaft is rotating for the reflux pump F-29. Check that the direction of rotation is consistent with arrow on the housing, **767**

**Check diagram and P&ID versus what's out on the plant, 2769**

**On-site simple tests:**

Open bleed valves to fuel on the shell of the condenser for 10 minutes, then shut, **606**

Tap the side of the condenser along a vertical line to listen for change in sound associated with liquid level in the condenser, **858**

**Gather data for key calculations**,

Pressure profile,

Drum V-31 to pumps, F-29, **2863**

$\Delta p$ across reflux pump, converted to head, **2984**

Drum V-31, pump F-29 and reflux into column, **2723**

Pump F-29, **2594**

Thermosyphon reboiler process fluid side, **2531**

**Sensors: check response to change**, Temperature sensor at the top of column, TI-11, **963**

Pressure sensor at top of column PI-12, **586**

Pressure sensor on overhead line for PIC/19; PT-19, **1307**

Pressure sensor on reflux pump exit line, PI-20, **1013**

**Sensors: use of temporary instruments**,

Measure the surface temperature on the outside of the condenser, **1252**

**Sensors: calibrate**,

Temperature sensor at the top of column, TI-11, **1802**

Pressure sensor at top of column PI-12, **1997**

Pressure sensor on overhead line for PIC/19; PT-19, **1785**

Pressure sensor on reflux pump exit line, PI-20, **2202**

**Control system**,

Put overhead pressure control, PIC/19; reflux control FIC-6 controllers on manual and try to steady out the column, **2159**

**Sample and analyze**,

Feed to the upstream depropanizer for the amount of C4 and C5s, **2103**

Concentration of effluent from upstream reactor where catalyst was changed. Analyze for C4 and C5 and compare with previous, **2491**

- **Open and inspect**,
  Condenser, **2901**
  Pressure control valve PV-19, **2752**
  Reflux pump, F-29, **2602**
  Line from the top of the column to the condenser, E-28, **2742**
  Sensor tap for pressure on the overhead line, PT-19, **2818**

- **Take "corrective" action**
  Replace the pressure control valve PV-19, **2260**
  Replace the pressure gauge PT-19, **2139**

Direct water from the fire hose onto the outside shell of the overhead condenser, **401**
Reduce the flowrate to C8 and thus reduce the flowrate to C9, **2054**

**Case #33: Chlorine feed regulation** (courtesy of Scott Lynn, Chemical Engineering Department, University of California, Berkeley) [6, slurry, pump, control system, storage tank; minerals processing]
A copper mine is treating its crushed ore with a dilute solution (5%) of sodium hypochlorite to improve the recovery of molybdenum disulfide by flotation. Sodium hydroxide solution of the appropriate strength is reacted with chlorine gas in a 5.5-cm diameter pipe that serves as the reactor. Flow is continuous and relatively constant at 3.15 L/s. The pipe carries the bleach solution to the slurry tank. An oxidation-potential probe at the pipe outlet is used to regulate the flow of chlorine. The system has just been installed, as shown in Figure 8-14 and serious trouble has been encountered with the chlorine feed regulation. The flow of gas can be readily controlled manually but fluctuates wildly when put on automatic. A time recording of the oxidation potential, OPRC, is shown in Figure 8-15. Please correct this problem quickly!

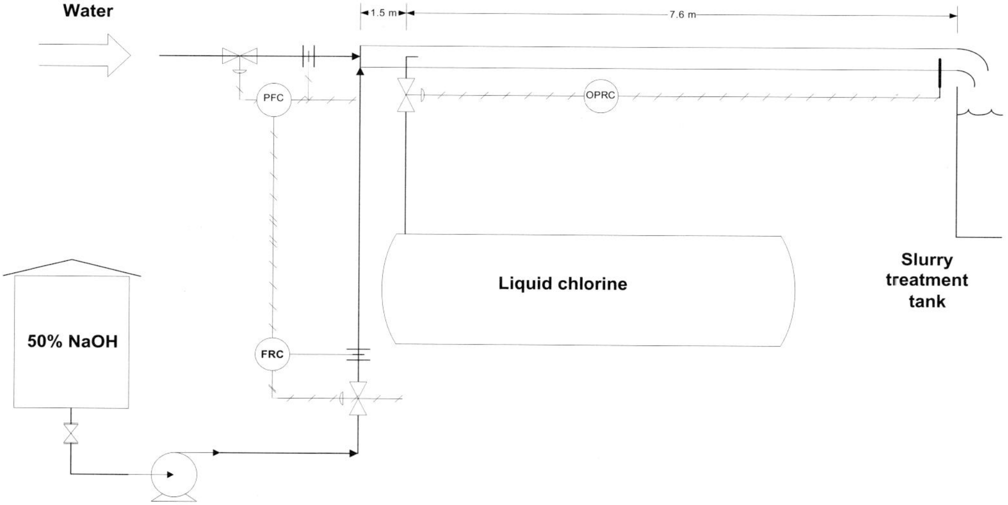


Figure 8-14 The chlorine feed system for Case #33.

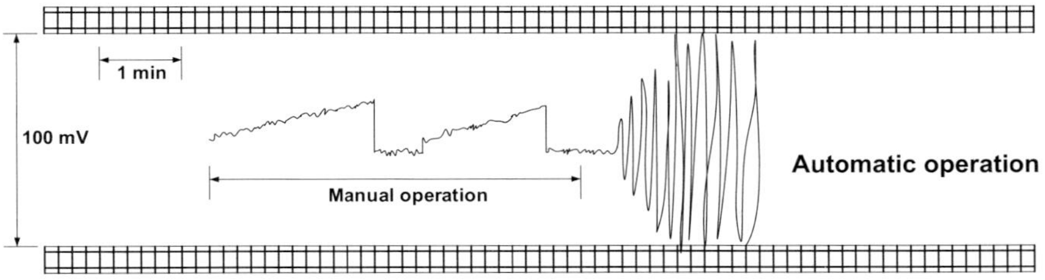


**Figure 8-15** Sample recording chart for the OPRC for Case #33.

**Case #33: Chlorine feed regulation**

- **MSDS, 110**
- **Immediate action for safety and hazard elimination,**
  Put on safe-park, **399**
  Safety interlock shut down, **846**
  SIS plus evacuation, **2899**
- **IS and IS NOT**: (based on given problem statement),
  What, **2588**
  When, **2758**
  Where, **1859**
- **Weather**, Today and past, **1642**
- **Maintenance: turnaround**, When and what done?, **1274**

**Maintenance: routine**, When and what done?, **1148**

- **What should be happening**

**Design and simulation files** (allowances made for fouling, overdesign and uncertainties)
Orifice plate on sodium hydroxide line, **1027**
Orifice plate in water feed line, **697**
Control valve on chlorine feed line, **542**
Control valve on caustic feed line, **145**
Control valve on water feed line, **2794**
Caustic feed pump, **2813**
Pipe-reactor, **2943**
Source of water, **2546**
Chlorine injection line, **2079**
**Commissioning data, P&ID, internal reports, 1621**
**Handbook**, Density and viscosity of 50% sodium hydroxide, **1915**
**Trouble-shooting files**, **1787**

- **Calculations and estimations** (that can be done in the office before special tests are done)
  Implications of cycling, observed frequency of fluctuation, **1167**

Observed amplitude of fluctuation, **1379**

Mass balance,

Flowrates of caustic and chlorine for reaction, **1096**

Fluid mechanics, mixing and residence time,

Estimate the Reynolds number in the "reactor", **1019**

Estimate velocity in reactor, **781**

Estimate residence time before chlorine addition, **981**

Estimate residence time between chlorine injection and OPRC sensor, **2889**

Rate,

Of reaction to form sodium hypochlorite, **2922**

Equipment performance,

Control system: check the degrees of freedom, **2682**

Control system: and stability, **225**

- **What is the current operation**

**Visit control room:** control-room data: values now and from past records,

Flowrate of water, **99**

Flowrate of caustic, **2848**

Check that the output from OPRC fluctuates rapidly and significantly, **2609**

**Process operators**, **345**

**Call to others on-site**, Operators of upstream units providing water: any variation in the composition, temperature or flowrate?, **176**

**Visit site**, read present values, observe and sense.

Valve movement of chlorine when operating on automatic, **1349**

Valve movement of caustic feed control valve, **1446**

Valve movement of the water, **1249**

Check that the bypass valves on all control valves are shut and that the block valves are fully open, **2299**

Sounds of cavitation in the caustic pump, **2099**

**Check diagram and P&ID versus what's out on the plant**, **2403**

**On-site simple tests:**

Check valve for stiction: remove backlash by doing a bump of 2 to 3%; then move controller output slowly via a slow ramp or bumps of 0.1%; observe controller output: pressure to the actuator, the valve stem and the pH output. Repeat ramping up and ramping down, **1561**

**Gather data for key calculations**

Temperature of the water such that the water may be flashing in the orifice meter sensor in the PFC loop, **1547**

Check that the oscillation is close to the time delay for the concentration to flow from the mixing point to the analyzer, **1931**

**Sensors: check response to change**

OPRC, **616**

**Sensors: use of temporary instruments**,

Use a contact pyrometer to measure the temperature variation in feed water, **945**

Use a contact pyrometer to measure the temperature of the chlorine storage tank to ensure it is < 50 °C, **593**
Use clamp-on ammeter to measure amps to pump, compare with expected, **62**

**Sensors: calibrate**, OPRC, **439**

**Control system**, Retune control system, **267**
Place controller on manual; observe the magnitude of the noise on the measurement, **2261**

**Call to vendors, licensee, or suppliers**, Suppliers of chlorine, **2457**
Suppliers of caustic, **2449**

**Samples and measurements,**
Caustic and check against the specs from the supplier, **2198**
Chlorine and check against the specs from the supplier, **2656**
Water and check for high levels of silt or humic acid from spring runoff, **1774**

**Open and inspect,**
Check that orifice plates are not installed backwards in the two locations, **1909**
Open reactor pipe and check that it is clear (free of obstructions) and that the chlorine injection line is centered, **1948**

- **Take "corrective" action,**
  Shutdown the process. Insert "static mixer" just downstream of chlorine-injection point, **1598**
  Replace the water controller with FFC (flow fraction control, a ratio) instead of PFC, **1209**
  Shutdown the process. Change control loop on water line to be feed-forward control, **663**
  Shut down the process. Add a support structure to the chlorine feed line to remove the vibration of the feed line, **899**
  Redesign the caustic-water mixing area to provide better mixing. Shut down the process and install the improved system, **749**
  Insert "static mixer" just downstream of caustic injection point, **295**

**Case #34: The cement plant conveyor** [6, solids conveyor, bagging, dust filters, blowers, fans, cyclone: ceramic]

This plant produces dry mixes of mainly cements and coarser aggregates. The raw materials are added – one batch at a time – to a mixing hopper, transported by a high-pressure, batchwise, conveying system to a solids blender or mixing vessel. The mixed material is conveyed from the mixer to a packaging hopper feeding a bagging machine. The finished, mixed product is packaged in 25-kg bags. The Quality Control Department checks regularly on the finished product to ensure that the material has adequate strength by forming casts and determining the break strength of the casts. Two or three days ago, the QC Department found that the strengths of the cast samples had decreased substantially. What's going on?

**Case #34: The cement plant conveyor**

- **Immediate action for safety and hazard elimination**,
  Put on safe-park, **262**
  Safety interlock shut down, **418**
  SIS plus evacuation, **919**
- **Weather**, Today and past, **5**
- **Maintenance: turnaround**, When and what done?, **516**

**Maintenance: routine**, When and what done?, **41**

- **What should be happening**,

**Design and simulation files** (allowances made for fouling, overdesign and uncertainties)
Specs for blend, **771**
Specs on mixing time, **1143**
**Handbook**, Pertinent properties of cement blends, **1917**
**Trouble-shooting files**, **1626**

- **Calculations and estimations** (that can be done in the office before special tests are done)

Equipment performance,
Dust collection, **212**
Fans, **441**
Conveying from mixer to packaging hopper, **25**
Mixer/ blender, **569**
Packaging, **1372**
Materials hoppers and star valve feeder, **1083**
Blower and batch-conveying system from the bins to the mixers, **1001**

- **What is the current operation**

**Visit control room:** control-room data: values now and from past records. No sensors or data in control room.
**Process operators**,
Mixing: What did you do differently three days ago?, **137**
Packaging: What did you do differently three days ago?, **728**
This shift on mixers, **538**
This shift on packaging, **504**
Previous shift on mixers, **941**
Previous shift on packaging, **1417**
**Check diagram and P&ID versus what's out on the plant**, **1260**
**Control system**,
For dust collector, **865**
For packaging, **585**
**Call to vendors, licensee**, Packaging unit, **336**
Solids mixing unit, **432**

Baghouse, **391**
Pressurized conveying system from the feed hopper to the solids mixer, **812**

**Samples and measurements**,
Monitor sampling of raw material, QC. Is the particle-size distribution within specs? QC perform tests, **1312**
Monitor sampling. after the mixer. Sample to QC: Measure strength of casts ; particle-size distribution; composition of components, **1078**
Monitor sampling. From the package Sample to QC: Measure strength of casts ; particle-size distribution; composition of components, **60**
Sample air leaving bag filter and measure concentration and size of particulates. Compare with previous, **2886**

**Open and inspect**,
Feed hopper, **508**
Solids mixer, **790**
Conveyor hopper, blower and conveying line, **115**
Dust collector on top of packaging hopper, **31**
Packaging unit, **382**

- **Take corrective action**,
  Bang side of the hopper to the bagging machine to break any bridging, **2075**
  Repipe the bagging machine so that the dusty air flows down through the bags instead of up through the bags. The goal is to prevent fluidization of the fines in some of the bags, **2496**

**Case #35: The cycling triple-effect evaporator** [6, long tube evaporators, vacuum system, steam; glycerine]

We are starting up a new plant, and it is like a zoo out there. We have three, multiple-effect evaporators to concentrate glycerine. However, we can't seem to get the system to behave. It will not steady out. The flowrates and pressures in the three evaporators all seem to cycle. Clear up the problem. All pressures are absolute pressures. Figure 8-16 shows the process. Figure 8-17 illustrates the ejector system used to pull the vacuum.

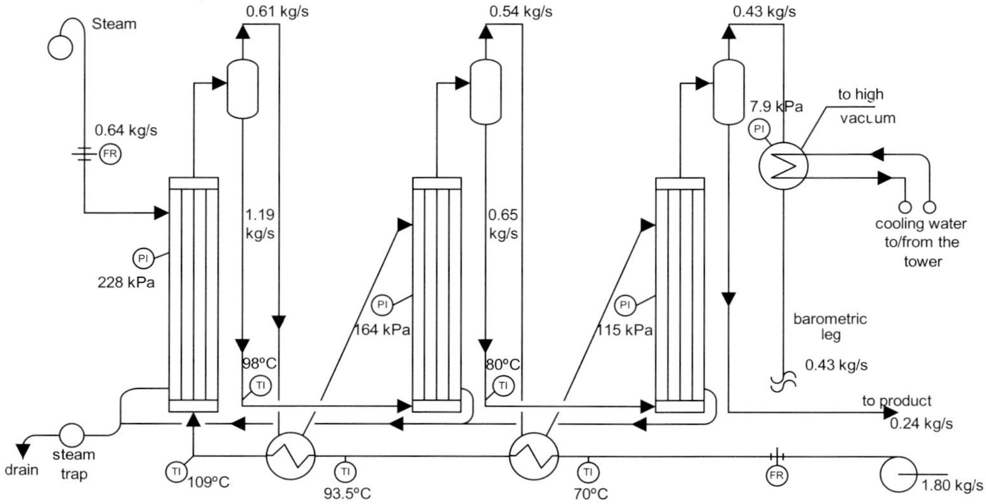


**Figure 8-16** Three multiple effect evaporators for Case #35.

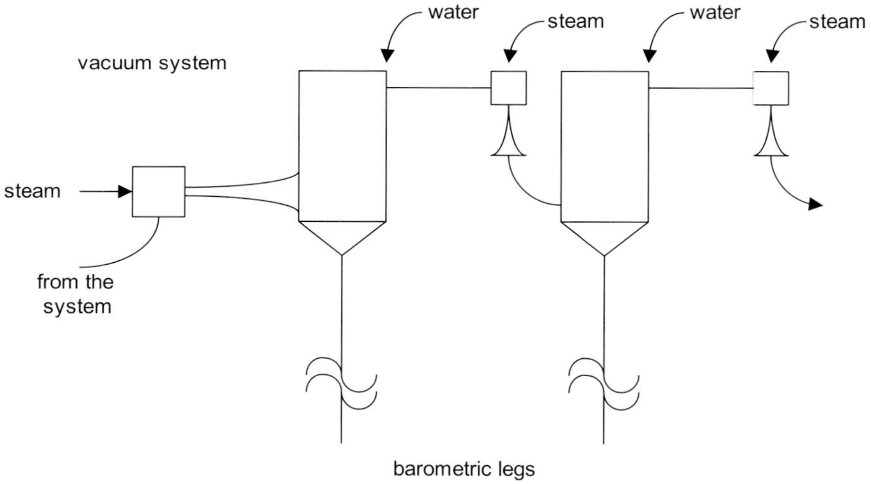


**Figure 8-17** The high vacuum system for the multiple effect evaporators for Case #35.

**Case #35: Cycling triple effect**

- **MSDS**, 1871
- **Immediate action for safety and hazard elimination**,
  Put on safe-park, **1611**
  Safety interlock shut down, **1744**
  SIS plus evacuation, **2331**
- **IS and IS NOT**: (based on given problem statement), What, **2128**
  When, **2095**
  Where, **2964**
- **Weather**, Today and past, **2668**
- **Maintenance: turnaround**, When and what done?, **2571**

**Maintenance: routine**, When and what done?, **2508**

- **What should be happening**

**Design and simulation files** (allowances made for fouling, overdesign and uncertainties)
Vacuum system, **2746**
Three evaporators, **2489**
Two preheaters, **2344**
Steam trap, **2147**
Feed pump, **2097**
**Vendor files**: Triple effect vacuum unit, **2041**
**Commissioning data, P&ID, internal reports**, 2255
**Handbook**, Steam tables, **2450**
**Trouble-shooting files**, 1818

- **Calculations and estimations** (that can be done in the office before special tests are done)

Pressure profile,
If there is a leak: stages 1, 2 and 3; which direction would it leak?, **1759**
If there is a leak in the preheaters, which direction would it leak?, **1853**
Mass balance,
Steam: condensate per stage, **1028**
Process liquid out from the third stage, **1635**
Energy balance: sink = source, estimate heat loads based on steady state information given, **1079**
Rate, boiling range: nucleate or film, **1128**
Equipment performance, three stages, **1178**

- **What is the current operation**

**Visit control room:** control-room data: values now and from past records. Sensors out on the plant.
**Process operators**, Please tell me about the startup, **1228**

**Contact on-site personnel**, Steam utilities: steam conditions to ejector system, **1278**

Cooling water to the barometric condensers; is the cooling water hotter than usual, **1328**

**Visit site**, read present values, observe and sense.

Glycerine feedrate, FR, **1378**

Pressure PI stage 1; 228 kPa, abs, **1427**

Pressure PI stage 2; 164 kPa, abs, **1476**

Pressure PI stage 3; 115 kPa abs, **814**

Pressure PI inlet to vacuum system. 7.9 kPa abs, **611**

Temperature TI inlet to preheater, **962**

Temperature TI exit stage 1.98 °C, **915**

Temperature TI exit stage 2.80 °C, **864**

Temperature TI into stage 1; 109 °C, **866**

Temperature TI exit preheater 1: 93.5 °C, **571**

What is the cycle time, **509**

Steam flowrate to stage 1, **44**

Steam pressure to the ejectors, **139**

**Check diagram and P&ID versus what's out on the plant**, **1304**

**On-site simple tests:**

Knock on the side of the vertical third effect and listen for a change in sound that corresponds to the cycle, **281**

Gloved hand to test the temperature of trap and upstream and downstream of the steam trap. Do this over the full cycle, **431**

**Sensors: use of temporary instruments**,

Use a stethoscope to listen to the steam going to each of the three ejectors in the vacuum system. Listen for "vibration", **472**

**Call to vendors, licensee**,

Triple-effect evaporator system, **983**

Steam-trap vendor, **770**

**Control system**, Put control system on manual, **860**

**Samples and measurements**,

Sample glycerine feed before pump and just as it enters the first stage; analyze for glycerine content and compare with specifications, **2265**

Immerse steam-trap discharge line in weighed amount of cold water and measure condensate flow over three successive one-minute periods, **2108**

**Open and inspect**,

Triple effects and pressure test for leaks, check for blockages, evidence of excessive foaming, and fouling of inside or outside of tubes, **2280**

Preheaters: and pressure test for leaks, check for blockages, and fouling of inside or outside of tubes, **2207**

Centrifugal feed pump, **2931**

Barometric condensers; look for blockage that could have flooded the condenser, **2678**

- **Take "corrective" action,**
  Raise the level of the overflow baffle on the hot well to give better seal of the downcomer, **2904**
  Retune the control system, **2136**
  Replace the valve on the steam line to the first ejector after the booster ejector, **1703**

**Case #36: The really hot case** (courtesy W. K. Taylor, B. Eng. 1966, McMaster University) [7, reactor, heat exchanger, steam drum; reformer, ammonia]
The heat from exit gas from the secondary reformer is used to generate steam according to the scheme shown in Figure 8-18. The temperatures and pressures for making 100 kg/s (as measured by F4) are as follows: the location, the design value and the current value for temperature and pressure, respectively.

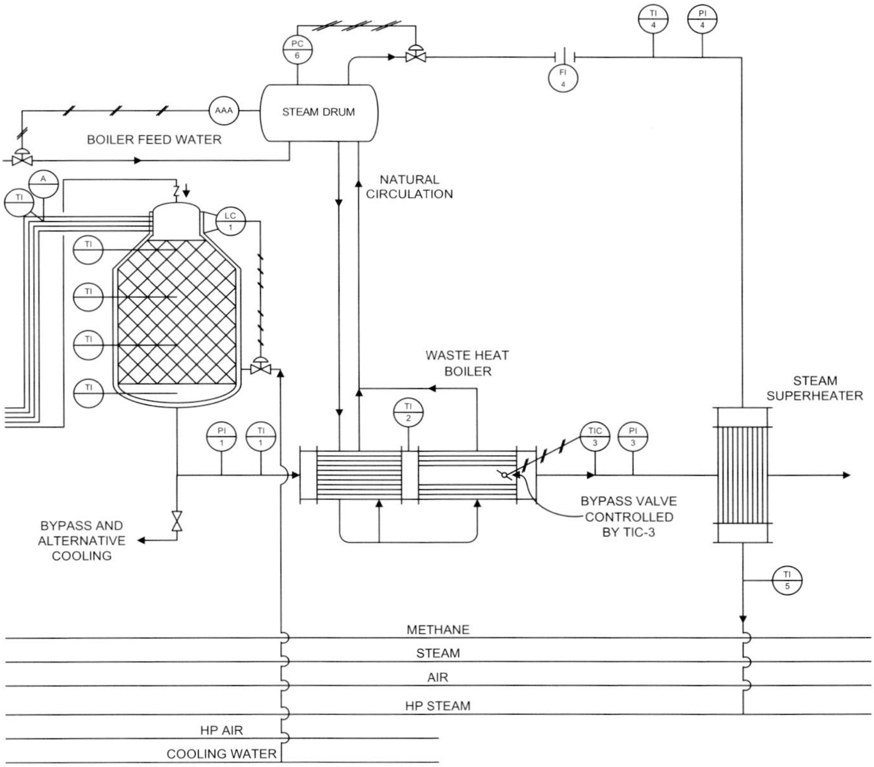


**Figure 8-18** Steam generation by the secondary reformer exit gas for Case #35.

| Location | Design T, °C; | Reads, °C | Design P, MPa | Reads, MPa |
|---|---|---|---|---|
| Reactor exit | T1 = 1000 | 1000 | P1 = 4.5 | 4.55 |
| Inside boiler | T2 = 750 | 820 | ? | ? |
| Boiler exit | T3 = 600 | 730 | P3 = ? | 4.5 |
| Superheat feed | T4 = 325 (sat) | 325 | P4 = 12.0 | 12.0 |
| Superheat exit | T5 = 353 | 443 | ? | ? |

We cannot seem to control the exit gas temperature TI-3; the TI-3 controller output to bypass is 100% closed. The steam is much too superheated and could damage the turbine where it is used. Correct the fault.

**Case #36: The really hot case**

- **MSDS**, **113**
- **Immediate action for safety and hazard elimination**,
  Put on safe-park, **116**
  Safety interlock shut down, **43**
  SIS plus evacuation, **1327**
- **More about the process**, **2981**
- **IS and IS NOT**: (based on given problem statement),

  What, **1018**
  When, **2408**
  Where, **2170**
- **Weather**, Today and past, **2015**
- **Maintenance: turnaround**, When and what done?, **2971**

**Maintenance: routine**, When and what done?, **2564**

- **What should be happening**

**Design and simulation files** (allowances made for fouling, overdesign and uncertainties)
Reformer, **2712**
Steam drum, **2853**
Waste-heat boiler, **2953**
Steam superheater, **2651**
**Vendor files**: Waste-heat boiler, **2253**
Steam superheater, **2453**
Reformer catalyst, **2056**
**Commissioning data, P&ID, internal reports**, **2017**
**Handbook**, Steam tables, **2435**
**Trouble-shooting files**, **555**

- **Calculations and estimations** (that can be done in the office before special tests are done)

Pressure profile,
  Direction of leak: waste-heat boiler, **902**
  Direction of leak: steam superheater, **520**
  $\Delta p$ for process gas through the boiler and compare with rule-of-thumb, **37**
Energy balance: sink = source,
  For waste-heat boiler: Heat loss in gas = steam generated, **428**
  For waste-heat boiler: heat load in first section vs second, **1742**
  For superheater: heat loss in gas = superheat, **1958**
Rate, Boiling regime (film or nucleate), **1778**
Equipment performance,
  Estimate performance of boiler based on given data, **1538**
  Estimate performance of superheater based on given data, **2967**

- **What is the current operation**

**Visit control room:** control-room data: values now and from past records,
  Temperature TI-1, **2526**
**Process operators**, What has been done? **2688**
**Call to others on-site**, Downstream users of superheated steam, **341**
  Upstream feed to the reformer: any change, **1449**
**Visit site**, read present values, observe and sense.
  FI-4: flow of steam, **2946**
  Output signal from TIC-3 to baffle, **2599**
**Check diagram and P&ID versus what's out on the plant**, **1134**
**On-site simple tests:**
  Are the thermocouple temperatures in the reformer bed consistent with the exit temperature TI-1?, **2120**
**Sensors: check response to change**
  Temperature TI-5; superheated steam exit superheater, **2352**
  Temperature TI-2; process gas inside boiler, **2047**
  Temperature TI-3; process gas boiler exit, **1677**
  Pressure PI-1, **1984**
  Pressure PI-3, **53**
  Pressure PI-4, **354**
  Flowmeter FI-4, **804**
**Sensors: use of temporary instruments**,
  Use contact or laser pyrometer to measure TI-5, **1413**
  Use contact or laser pyrometer to measure TI-1, **1861**
**Sensors: calibrate**,
  Temperature TI-5; superheated steam exit superheater, **854**
  Temperature TI-2; process gas inside boiler, **40**
  Temperature TI-3; process gas boiler exit, **459**
  Pressure PI-1, **2470**
  Pressure PI-3, **2350**
  Pressure PI-4, **2950**
  Flowmeter FI-4, **2634**

**Control system,**

Put TIC-3 on manual and open the bypass valve. Note changes is TIC-3 and TI-5, **2495**

Retune the TIC-3 control-baffle system, **2976**

Put PC-6 on manual. Open the valve and watch steam drum pressure and PI-4, **584**

**Call to vendors, licensee,**

Catalyst in the secondary reformer, **2360**

**Samples and measurements,**

Analyze the material found, **2083**

**Open and inspect,**

Orifice plate for FI-4 to check if the plate was in backwards. Tab is not clearly marked, **1556**

Isolate, drain the superheater. Pull the bundle and inspect for fouling; inspect the inside of the tubes for fouling, **1904**

Isolate, drain the superheater. Pressure test the tubes with water and look for leaks. Plug any leaking tubes, **627**

Isolate, drain the boiler. Pressure test the tubes with water and look for leaks. Plug any leaking tubes, **128**

Isolate and drain the boiler. Pull the bundle and check for fouling on the outside of the tubes, **2810**

Isolate and drain the boiler. Check the location of the bypass valve. Inspect the inside of the tubes for fouling, **2647**

- **Take "corrective" action,**

  Correct the linkage between the TIC-3 and baffle, **2149**

  Replace the sensor TIC-3, **1105**

**Case #37: The mill clarifier** (courtesy D. F. Fox, B. Eng. 1973, McMaster University) [7, thickener, sludge pumps, control; pulp and paper]

The mill clarifier is 45 m in diameter and designed for 90% reduction in suspended solids from an inflow of 0.53 $m^3/s$. The retention time is 2 hour. Parshall flumes measure the total inlet and outlet flows. There are 24-hour composite samplers on the influent and overflow effluent lines. The system is shown in Figure 8-19.

Our target is to have the clarifier effluent < 50 ppm. The operator cannot understand why the clarifier effluent has been around 200 ppm for the last 20 days. The sludge pump has been "wide open", yet all that occurs is flooding at the belt filter. The feed concentration is about 400 ppm. That's the problem. Get it fixed pronto.

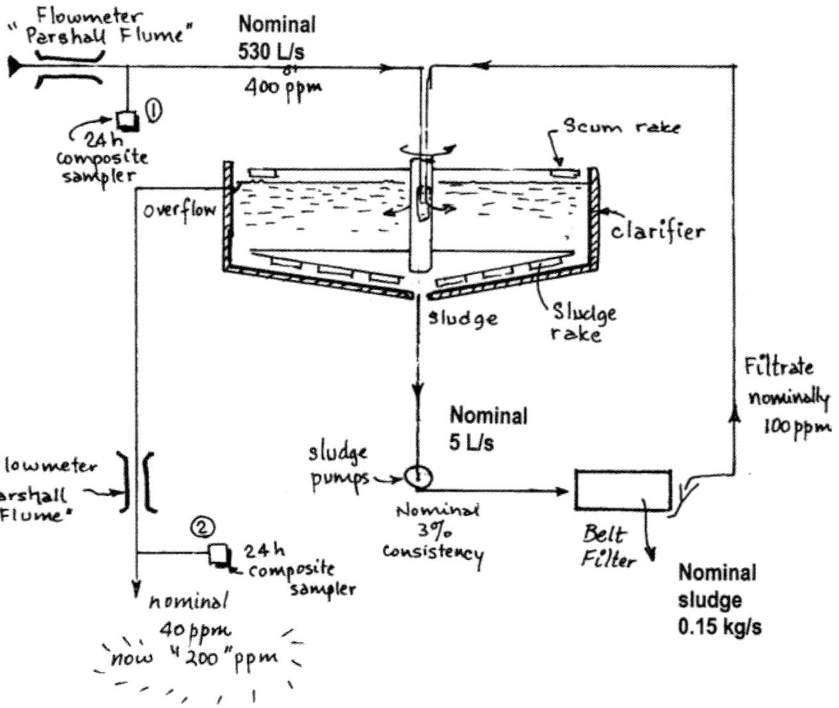


Figure 8-19 The mill clarifier for Case #37.

**Case #37: The mill clarifier**

- **Immediate action for safety and hazard elimination**,
  Put on safe-park, **438**
  Safety interlock shut down, **121**
  SIS plus evacuation, **2975**
- **IS and IS NOT**: (based on given problem statement),
  What, **2676**
  When, **2801**
  Where, **2567**
- **Weather**, Today and past, **1656**
- **Maintenance: turnaround**, When and what done?, **1959**

**Maintenance: routine**, When and what done?, **1839**

- **What should be happening**

**Design and simulation files** (allowances made for fouling, overdesign and uncertainties)

Clarifier, **1203**

Sludge pump, **1456**
Belt filter, **1045**
24-h composite sampler, **636**
Parshall flumes, **954**
**Vendor files**: Parshall flumes, **507**
**Commissioning data, P&ID, internal reports, 61**
**Trouble-shooting files**, **317**

- **What is the current operation**

**Visit control room:** control-room data: values now and from past records,
Torque, **493**
Records of 24 h sampler on feed over the past week, **47**
Records of 24 h sampler on effluent over the past week, **204**
Influent flowrate based on Parshall flume, **403**
Bottoms flowrate from sludge pump to belt filter, **703**
Overflow flowrate based on Parshall flume, **579**
**Process operators**, What was done this shift? **861**
Previous shift, **1310**
**Call to others on-site**, Operators of upstream units about possible upsets, **1154**
**Visit site**, read present values, observe and sense.
Sludge pump: cavitation? **1402**
Sludge pump exit valve wide open or mid-range? **1133**
Rake and skimmer turning at expected rpm, **1056**
Belt filter: flooding? **1752**
**Check diagram and P&ID versus what's out on the plant**, **2373**
**On-site simple tests:**
Clear any stuff out of the sampler lines on 24 h composite sampler of feed and resample via 24-h composite, **1580**
**Sensors: use of temporary instruments**,
Use point gauge on both Parshall flumes and use this information to determine total flowrate, **1516**
**Sensors: calibrate**,
Recheck zero setting for the Parshall flumes, **2204**
**Samples and measurements**,
Grab samples of feed whenever torque starts to increase. Five successively at 20-min intervals. Analyze for suspended solids and compare with 400 ppm reading, **2107**
Feed to the belt filter. Grab sample when torque increases and when increase pump output. Analyze for suspended solids and compare with 3%, **2353**
**Open and inspect**,
Sampler lines; look for plugs in sampler line, **2463**
Stop process; drain clarifier and inspect rake, **2961**

- **Take "corrective" action,**
  Turn off sludge pump and increase rake speed until torque just starts to increase; then start sludge pump, **2622**
  Decrease rake speed; keep sludge pump at usual design flowrate, **2534**
  Reduce the sludge-pump flowrate to 5 L/s, **1819**
  Increase vacuum on the belt filter, **1615**

**Case #38: More trouble on the deprop!** (courtesy of T. E. Marlin, Chemical Engineering, McMaster University) [7, sequence of distillation columns with auxiliaries; depropanizer-debutanizer]

The process is the depropanizer-debutanizer described in **Case #8,** Chapter 2. A P&ID is given in Figure 2-4 accompanying **Case #8.**

There has been a lot of trouble on this unit. So last January, we did a detailed study, and simulation, of the deprop and debut units. Our conclusion was that this unit was operating on spec. for the usual range of feedstocks with all equipment components operating very close to the design values.

It's a hot steamy day in August; you are glad that you are in your air-conditioned office. The phone rings! The laboratory analysis of the vapor product on the deprop indicates a loss of propane to the fuel system that is $2^1/_2$ times the design value. This loss of valuable product to fuel gas is costing lots of money. Fix it fast.

**Case #38: More trouble on the depropanizer**

- **MSDS, 1495**
- **Immediate action for safety and hazard elimination,** Put on safe-park, **315**
  Safety interlock shut down, **1315**
  SIS plus evacuation, **2315**
- **More about the process, 1010**
- **IS and IS NOT,** What, **2978**
  When, **2627**
  Where, **2336**
- ***Why? Why? Why?***, Best goal? To reduce the propane loss to fuel gas, **1360**
- **Weather,** Today and past, **1761**
- **Maintenance: turnaround,** When and what done?, **19**

**Maintenance: routine,** When and what done?, **269**

- **What should be happening,**

**Design and simulation files** (allowances made for fouling, overdesign and uncertainties)

Condenser, E-25, **87**
Distillation column, C8: **478**
Thermosyphon reboiler, E-27, **969**
Reflux pump, F-27, **2486**
Overhead drum, V-30, **2970**

**Vendor files**: Condenser, reboiler and preheater, **1475**
Steam traps, **1110**
**Commissioning data, P&ID, internal reports, 1043**
**Handbook and Google**, Cox charts, **2024**
**Trouble-shooting files, 85**

- **What is current operation**

**Visit control room:** control-room data,
Feed to the C8, depropanizer. FC/1, **2831**
Reflux flowrate, FIC/4, **2036**
Overhead product flowrate of propane, FIC/5, **1633**
Pressure drop $\Delta p$ I/1, **1926**
Level bottoms LIC/2, **387**
Temperature bottoms TI/4, **173**
Temperature mid-column TIC/5, **980**
Temperature top, TI/3, **789**
Pressure on overhead drum, PIC/10, **513**
**Process operators**, This shift, **552**
Previous midnight shift, **1352**
Operating procedures, Column pressure, **1185**
**Call to others on-site**,
Utilities: what is the temperature of the water leaving the cooling tower; what is the flowrate to our unit, **1908**
Operators on downstream propane unit: amount and quality of propane received, **872**
**Visit site**, read present values, observe and sense.
Column pressure, PI- 4, **2405**
Pressure relief to flare PSV-1, **2384**
Look at the flare, **2994**
Temperature mid-column TI- 8, **2123**
Valve position for column feed, FV- 1, **2701**
Valve position for steam to preheater E-24, **2606**
Valve-stem position on PV-10, **2504**
Level in feed drum, V-29; LI- 1, **2942**
Isolation valves around the condenser, **2516**
Feel temperature of water line going into condensers, **2008**
**Check diagram and P&ID versus what's out on the plant, 2769**
**On-site simple tests:** Open vents to fuel gas on each condenser for 10 min; then close, **1512**
**Gather data for key calculations**,
Pressure profile,
Drum V-29 to pumps, F-25, 26, **959**
$\Delta p$ across pump, converted to head, **605**
Pump F-25–26 exit to feed location, **857**
Drum V-30, pump F-27 and reflux into column, **549**

Pump F-27, **1464**
Vapor from top of column to vapor space in V-30, **1966**
Thermosyphon reboiler process fluid side, **1676**
Mass balance, Over column, **1558**
Energy balance: Heat load for condenser. Heat load condensed = heat load picked up in cooling water, **2911**
**Sensors: check response to change**, Temperature at the top of column C8: TI 3, **1955**
Pressure at the top of column C8, PI 4, **1175**
**Sensors: use of temporary instruments**,
Use a surface temperature sensor to measure the temperature of cooling water going into the condensers, **1012**
**Sensors: calibrate**,
Temperature at the top of column C8: TI 3, **1263**
Pressure at the top of column C8, PI 4, **327**
**Samples and measurements**, Sample feed and analyze for the amount of propane, **93**
Sample overhead concentration in gas to flare from top of drum every 10 minutes for 1 hour. Analyze for propane, **2442**
Sample overhead concentration from the tower every 10 minutes for 1 hour. Analyze for propane, **2871**
**Open and inspect**,
Condensers, **2554**

- **Take "corrective" action**,
Direct water from the fire hose onto the outside shell of the overhead condensers, **2174**
Increase column pressure to 1.78 MPa, **424**
Decrease feed to the column to 80% usual flowrate, **774**

**Case #39: The case of the lumpy sunglass display** (from D. R Winter, Universal Gravo-plast, Toronto, 2004) [Difficulty 7; involves feed bin, molding machine, mold and mold design. Context: injection molding of thermoplastics]
The customer wants a display stand for sunglasses. The display required several contoured plates (35 cm by 18 cm) molded of polypropylene. These are mounted vertically and held in place at the edges. The display is to be installed in windows of drugstores so that the sunglasses on display are illuminated by sunshine during the day and by fluorescent tubes behind the molded plate during darker hours. The customer requested a special pearlized pigment be added so that the display had pizzaz.

The feed was a mixture of polypropylene resin, a UV stablizer and the pearlescent pigment. The feed was dried by contacting it with hot air at 93 °C for three hours and was used within 20 minutes of drying. The injection-molding machine was a reciprocating screw; L:D of 20:1; compression ratio 2.5:1 and the shot size for this job was 70% of the machine capacity. The processing temperature was 230 °C. The mold pressure was 4.2 MPa, consistent with the part width and the flow length. A mold was created of aluminum with appropriate gate and vent locations and cooling arrangements for the mold. The nozzle, sprue and cold runner system were carefully designed. The molding cycle was established, a molded sample approved by the customer and production started.

The first 2000 parts had been shipped when the customer called and complained that all the parts looked fine in reflected light but in transmitted light, lumps were evident in the panels. Production was stopped. Your job is to eliminate the lumps that appear when the panels are viewed with transmitted light.

- **MSDS**, **1572**
- **Immediate action for safety and hazard elimination**,
  Put on safe-park, **1799**
  Safety interlock shut down, **1298**
  SIS plus evacuation, **298**
- **More about the process**, **12**
- **More about the product and the mold**, **2238**
- **IS and IS NOT**: (based on given problem statement),
  What, **2481**
  When, **2820**
  Where, **2649**
- **Weather**, Today and past, **2272**

**Maintenance: routine**, When and what done?, **1831**

- **What should be happening**

**Design and simulation files** (allowances made for fouling, overdesign and uncertainties)

Injection-molding machine, **1983**
Mold, **1445**
Feed for this product, **1004**

**Commissioning data, P&ID, internal reports**,

Prototype activity. Check if the prototype had lumps when viewed by transmitted light, **911**

**Handbook**, Data sheets for resin, **667**
**Trouble-shooting files**, **511**

- **What is the current operation**

**Visit control room:** control-room data: values now and from past records,

Fill cycle, **136**
Cool cycle, **2738**
Open cycle, **2521**
Melt feed temperature into mold, **2310**
Injection pressure, **1934**
Injection rate, **1545**
Extruder: rear-barrel temperature, **1857**

Extruder: head temperature, **1183**
Backpressure, **1084**
Mold pressure, **730**
Hold pressure, **594**
Mold temperature, **92**

**Process operators, 425**

Operating procedures, Shutdown procedures used, **279**
Were the feed materials from the same batches of resin, UV stabilizer and pearlized pigment as was used for the prototype?, **1945**

- **Check with colleagues about hypotheses, 2416**

**On-site simple tests:**

Clean possible dirty machine by running a charge of acrylic through the extruder; then try again, **2087**
Use another molding press; run the process under the prototype conditions, **2724**
Redry the resin for three hours at 93 °C and dry additives and use immediately, **2575**
Reduce the screw rpm by 5% and use a pyrometer to measure melt temperature, **2771**
Change melt temperature, increase by 20 °C to 251 °C and keep the cooling time the same, **2187**
Change melt temperature, increase by 20 °C to 251 °C and increase the cooling time, **2009**
Use the standard conditions but try an old batch of dried polypropylene from a different supplier together with the UV stabilizer and pearlized pigment, **1662**
Use the standard conditions, use the polypropylene we have been using and the pearlized pigment but delete the UV stabilizer, **1718**
Use the standard conditions, use the polypropylene we have been using and the UV stabilizer but delete the pearlized pigment, **1624**
Clean the hopper; add a cover over the hopper to prevent atmospheric dirt from falling into the feed, **1285**
Reduce the injection pressure to 7.5 MPa, **1367**

**Redesign of product and mold,**

Increase the diameter of the gates by 10%, **631**

**Sensors: check response to change**

Hot-melt temperature sensor at nozzle, **1337**
Exit cooling-water temperature on top mold, **304**
Exit cooling-water temperature on the bottom mold, **241**
Pressure sensor at nozzle, **1841**

**Sensors: use of temporary instruments,**

Use a pyrometer or laser sensor to measure melt temperature and compare with temperature sensor, **2663**

**Sensors: calibrate,** Hot-melt temperature sensor, **2635**

**Call to vendors, licensee, or suppliers**,
Properties of polypropylene, **77**
Properties of UV stabilizer, **183**
Properties of pearlized pigment, **2359**
**Samples and measurements**,
Sample resin and analyze for moisture, **2206**
**More complicated tests**,
Change mold from a QC-7 aluminum mold to a mold P-20, **2873**
**Open and inspect**, Check the shutoff valve for dirt or contamination, **2592**

- **Take "corrective" action**, Replace the shutoff valve, **2849**

**Case #40: The cool refrigerant** (courtesy T. E. Marlin, Chemical Engineering Department, McMaster University) [7, turbine, compressor, KO pot, refrigeration system; general]
The propylene refrigeration system, given in Figure 8-20, was operated successfully for several years. Since the steam turbine and refrigerant compressor had spare capacity, they modified the process by adding an additional heat exchanger, E101, to cool the process stream in E101, as shown in Figure 8-21. When the process was started up, the design values could not be obtained for either of the streams to be cooled. The temperatures, given in °C, are:

For E100: T5: actual 100, design 101; T6, actual 10; design 5.
For E101: T7: actual 70; design 68; T8, actual 4; design 10.
Into the compressor: T2: actual –5; design –4.

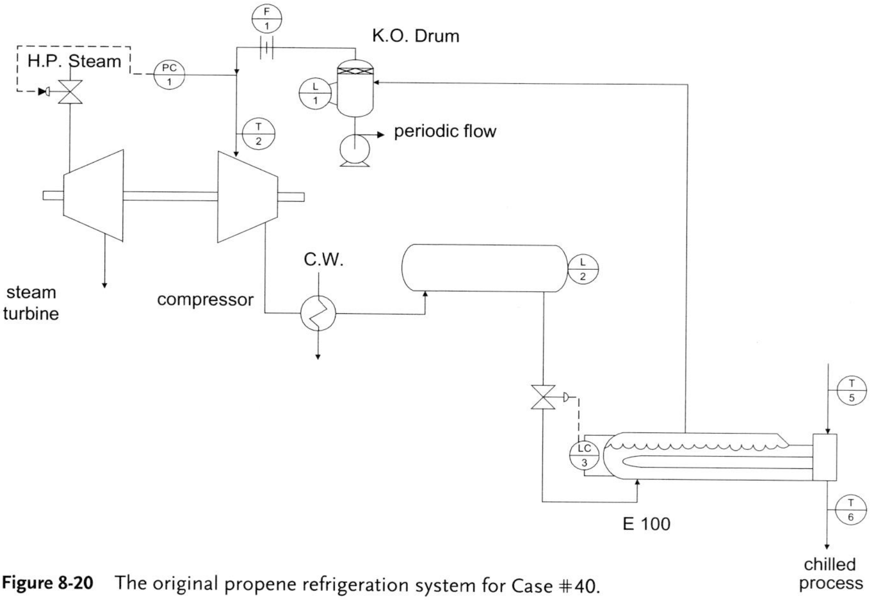


Figure 8-20 The original propene refrigeration system for Case #40.

Because T6 is too high and T8 is too low, the plant cannot produce saleable material. In checking over the design calculations, you realize that the system *should* work. What's going on here?

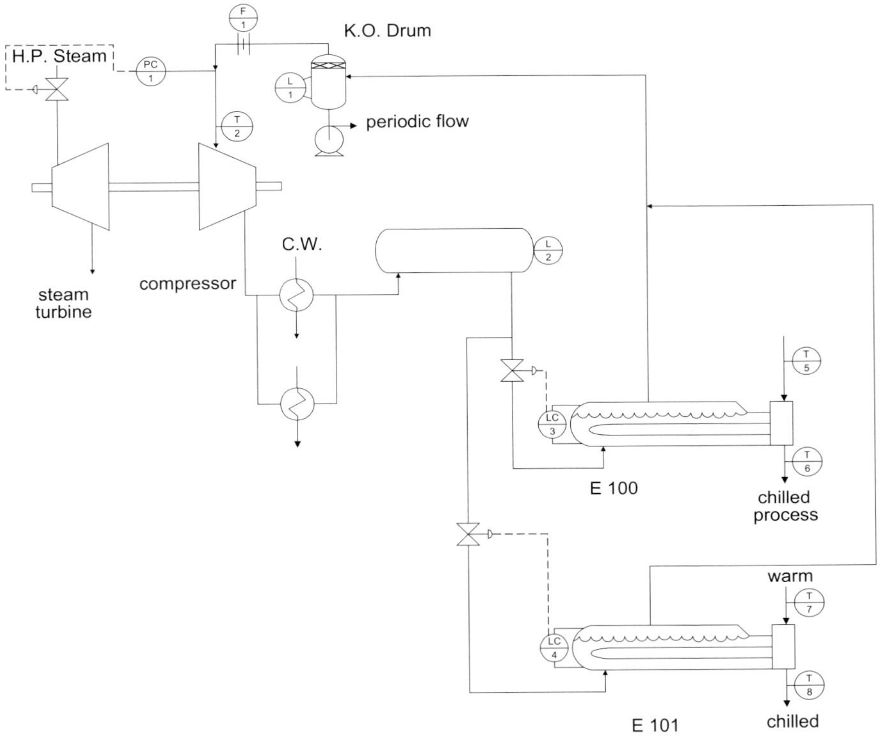


**Figure 8-21** The modified refrigeration system for Case #40.

**Case #40: The cool refrigerant**

- **MSDS, 914**
- **Immediate action for safety and hazard elimination,**
  Put on safe-park, **677**
  Safety interlock shut down, **744**
  SIS plus evacuation, **949**
- **IS and IS NOT**: (based on given problem statement),
  What, **1993**
  When, **1683**
  Where, **1827**
- **Weather**, Today and past, **1918**
- **Maintenance: turnaround**, When and what done?, **1212**

**Maintenance: routine**, When and what done?, **1466**

- **What should be happening**

**Design and simulation files** (allowances made for fouling, overdesign and uncertainties)

E100 chiller, **1031**

E101 chiller, **567**

**Vendor files**: E100 chiller, **883**

E101 chiller, **2485**

**Commissioning data, P&ID, internal reports, 2215**

**Handbook**, Propylene temperature–pressure data and latent heat, **2913**

**Trouble-shooting files, 2544**

- **Calculations and estimations** (that can be done in the office before special tests are done)

Pressure profile, Direction of leak: unable to tell because the pressures of the process streams are not known.

Direction of leak from propylene–cooling water in condensers, **2658**

Energy balance,

Heat load for E100; actual versus design, **2708**

Heat load for E101; actual versus design, **2157**

Propylene flow: design versus actual calculated from heat loads on E100 and E101, **1734**

Rate,

Does propylene boil as nucleate or film form, **2208**

Equipment performance,

E100: does heat flow in the correct direction?, **1707**

E101: does heat flow in correct direction?, **1582**

UA calculations for E100 actual versus design, **190**

UA calculations for E101 actual versus design, **740**

- **What is the current operation**

**Visit control room:** control-room data: values now and from past records,

Temp and pressure of propylene: T2 and P1; are these on pH diagram for saturated propylene gas; if not might it be sensors? or contamination? **1140**

Level: E100, LC/ 3, **69**

Level: E101, LC/4, **362**

Level: KO pot, L/1, **449**

Level: liquid propylene drum, L/2, **944**

Flow of propylene, F/1, **643**

Propylene pressure at compressor suction, PC/1, **713**

**Process operators**, Please tell me what has happened to far, **1369**

Operating procedures, **1125**

**Contact with on-site specialists,**

Operators of unit that provides and receives the process stream from E100: flows and temperatures constant and consistent with what's on this unit?, **1046**

Operators of unit that provides and receives the process stream from E100: pressure in the lines and have you detected any contamination in your process streams, **1809**

Operators of unit that provides and receives the process stream from E101: pressure in the lines and have you detected any contamination in your process streams, **1932**

Operators of unit that provides and receives the process stream from E101: flows and temperatures constant and consistent with what's on this unit?, **2338**

Utilities: steam pressure, flow and degree of superheat, **2144**

Utilities: cooling-water temperature, flow to propylene condensers, **2414**

**Visit site**, read present values, observe and sense.

Steam leave off the top of the steam header?, **2125**

Valve-stem position on valve LC/3, **2783**

Valve-stem position on valve LC/4, **2843**

**Check diagram and P&ID versus what's out on the plant**, **2583**

**On-site simple tests:**

Observe movement of the valve stem on LC/3 when the set point is changed, **1180**

Observe movement of the valve stem on LC/4 when the set point is changed, **615**

Open vent bleed on the top of E100 for 5 minutes, close and observe change, **842**

Open vent bleed on the top of E101 for 5 minutes, close and observe change, **688**

Open vent bleed on the top of propylene drum for 5 minutes, close and observe change, **124**

Consistency check,

Temp and pressure of propylene: T2 and P1; are these on pH diagram for saturated propylene gas; if not might it be sensors? or contamination?, **1140**

**Sensors: check response to change**

LC/3, **394**

LC/4, **247**

**Control system**,

Put LC/3 on manual and adjust level to control temperature T/6, **94**

Put LC/4 on manual and adjust level to control temperature T/8, **2389**

Retune controller LC/3, **2890**

Retune controller LC/4, **2683**

**Sensors: calibrate**, LC/3, **2597**
LC/4, **2072**
**Samples and measurements,**
Sample propylene and analyze for contaminants, **657**
**More complicated tests,**
Reduce the propylene pressure such that the target range of cooling occurs when neither bundle is completely covered with liquid; adjust the levels in E100 and E101 separately until target temperatures are achieved, **130**
**Open and inspect,**
Propylene drum to check the condition of the vortex breaker at the bottom exit nozzle, **265**
E100 and look for fouling inside or outside tubes, **1649**
E100; isolate and pressure test for leaks, **2263**
E101 and look for fouling inside or outside tubes, **2789**
E101; isolate and pressure test for leaks, **2019**
Check for plugs, obstructions or junk in the line from the liquid propylene reservoir and E100, **1638**

- **Take "corrective" action,**
  Replace LC/3 on E100, **1967**
  Replace LC/4 on E101, **1094**
  Replace control valve on propylene to E100, **775**

**Case #41: The ever-increasing column pressure** (courtesy T. E. Marlin, Chemical Engineering Department, McMaster University) [7, distillation plus auxiliaries; general]
The column for the new process was started up for the first time one month ago. From the beginning, the pressure control was not very good; the pressure seemed to deviate from its set point more than in other columns in the plant. However, the product purities seemed to be within specifications, so you didn't worry about it.

Today, the plant operator calls, "We've got a problem on that new column! The pressure controller is not working! The pressure in the column is higher than the set point yet the controller output is 0%! Yes, the production rate and the specifications on all lines are OK. But this pressure is really worrying me." The column is shown in Figure 8-22.

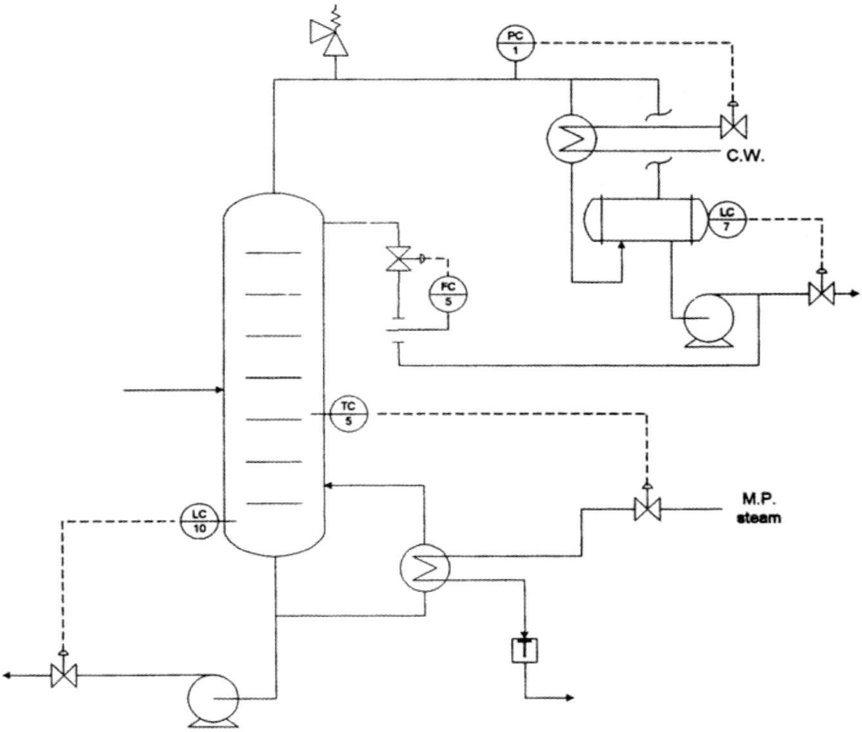


**Figure 8-22** The column discussed in Case #41.

**Case #41: Ever-increasing pressure**

- **MSDS**, 379
- **Immediate action for safety and hazard elimination**,
  Put on safe-park, **180**
  Safety interlock shut down, **479**
  SIS plus evacuation, **58**
- **IS and IS NOT**: (based on given problem statement), What, **758**
  When, **964**
  Where, **679**
- **Weather**, Today and past, **546**
- **Maintenance: turnaround**, When and what done?, **1131**

**Maintenance: routine**, When and what done?, **1483**

- **What should be happening**

**Design and simulation files** (allowances made for fouling, overdesign and uncertainties)

Column, **2706**
Reboiler, **2559**
Control system, **2956**
Condenser, **2867**

**Commissioning data, P&ID, internal reports**, **2695**

**Handbook**, Options for controlling overhead pressure based on Chin's classic article from 1979 *Hydrocarbon Process*, **2184**

Type of hydrocarbon based on overhead pressure and temperature, **2410**

**Trouble-shooting files**, **2423**

- **Calculations and estimations** (that can be done in the office before special tests are done)

Pressure profile,
Direction of flow if there is a leak: condenser, **2069**
Direction of flow if there is a leak: reboiler, **129**
Flow from top of column to the reflux drum, **328**

Equipment performance,
Control system, **445**
Condenser, **205**

- **What is the current operation**

**Visit control room:** control-room data: values now and from past records,
Current overhead temperature, **886**
History of overhead temperature, **736**
History of overhead pressure, **1434**
Current overhead pressure; PC/1, **630**

**Process operators**, When you first started up two months ago, was the control valve on the CW almost closed?, **534**

**Contact with on-site specialists**,
Utilities: cooling water: temperature and flowrate, **1935**
Operators of upstream plant supplying the feed to the column: any change in light ends in composition or change in flowrate, **2385**

**Visit site**, read present values, observe and sense.
Check for consistency between PC/1 and the pressure on the top of the column, **2205**
Valve-stem position on the cooling water, **1109**
Look at the flare, **1247**
Amount of steam flowing to the reboiler, **1313**
Bottom pressure, **1299**
Read the controller output signal to the cooling-water valve, **1619**
Is the control valve on the cooling water fail open or fail closed? does the direction arrow on the valve agree with the direction of flow?, **2938**
Bottom temperature, **1454**

**Check diagram and P&ID versus what's out on the plant**, **1119**

**On-site simple tests:**

Open vent on the condenser for 2 min to bleed off any accumulated inerts, close and check performance, **1753**

Does the valve stem on the cooling water respond to increase in set point, **1543**

Increase the resistance in the vent break line to prevent uncondensed gas from bypassing the condenser, **1888**

Tap the side of the condenser in a vertical line and listen for the change in sound suggesting the location of the vapor-condensate interface, **2631**

**Sensors: check response to change**

Does PC/1 respond to change in set point, **2160**

**Sensors: use of temporary instruments,**

Use a contact surface temperature probe to measure inlet and outlet temperatures of the cooling water to the condenser, **2063**

**Sensors: calibrate,**

Pressure PC/1, **2720**

**Control system,**

Retune the PC/1 control system, **2921**

**Samples and measurements,**

Sample the feed; analyze for light ends and non-condensibles and compare with previous records, **2539**

**More complicated tests,**

Stop the process; crack open the flange on the vent break line and insert a resistance disk with a smaller diameter hole to increase the resistance in the "bypass" line, **2652**

**Open and inspect,**

Control valve on the cooling water; inspect for quality of trim, plugged, **1559**

Condenser and inspect for fouling inside the tubes, **1123**

- **Take "corrective" action,**

  Reduce the feedrate to the column, **1371**

  Direct fire hose water over the condenser, **231**

**Case #42: The weak AN** (courtesy W. K. Taylor, B. Eng. 1966, McMaster University) [7, storage tank, reactor, pump, heat exchanger; ammonia]

Ammonium nitrate is formed by the exothermic neutralization of nitric acid with ammonia. The nitric acid is produced as a 56% solution and is pumped from storage tanks to the reaction leg of the primary neutralizer. In the reaction leg, the acid is mixed with superheated ammonia vapor and also with an ammonia-bearing off-gas stream from the urea plant. The heat of reaction produces a large amount of steam, and the liquid, which is removed to a secondary neutralizer is an 85% solution of ammonium nitrate. Usually about 50% of the ammonia needed to neutralize the acid is provided by the off-gas stream from the urea plant. The usual temperatures in the system are: nitric acid in storage: 40–50 °C; superheated ammonia vapor entering neutralizer: 90–100 °C; the urea plant off-gas: 100–110 °C; the AN product solution: 130–140 °C; the temperature inside the neutralizer: 135 °C. The process is shown in Figure 8-23.

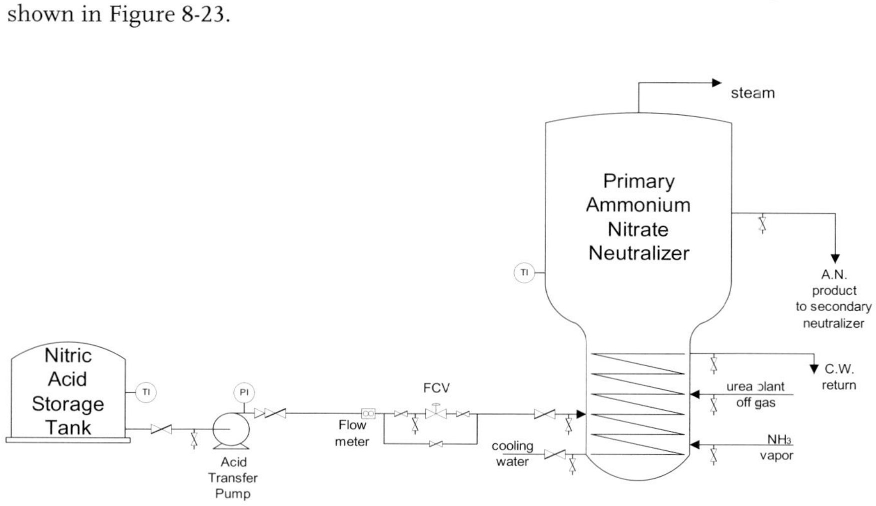


**Figure 8-23** The process for AN for Case #42.

The urgent phone call poses today's problem. The temperature inside the neutralizer-reactor has been falling. The most recent analysis of the AN shows a concentration of about 79%. "This weak AN will cause problems in the downstream processing units where the solution is concentrated in a falling film evaporator to about 99% and then prilled. Let's get this fixed quickly. The evaporator can't concentrate the solution enough for prilling and the whole process will shut down. "

**Case #42: Weak AN**

- **MSDS**, **2716**
- **Immediate action for safety and hazard elimination**,
  Put on safe-park, **2692**
  Safety interlock shut down, **2214**
  SIS plus evacuation, **2406**
- **More about the process**, **2870**
- **IS and IS NOT**: (based on given problem statement), What, **237**
  When, **332**
  Where, **1332**
- **Weather**, Today and past, **1082**
- **Maintenance: turnaround**, When and what done?, **1440**

**Maintenance: routine**, When and what done?, **880**

- **What should be happening**

**Design and simulation files** (allowances made for fouling, overdesign and uncertainties)

Storage tank, **906**
Check the sensors on the off-gas and ammonia line to ensure that there is no mercury used in the sensor, **396**
Design calculations for the orifice plate measuring the nitric acid flowrate, **1940**
Transfer pump, **1796**
Neutralizer, **2826**
Cooling coil, **2628**

**Vendor files**: Transfer pump, **2957**
**Commissioning data, P&ID, internal reports, 1338**
**Handbook**, Heat of formation of ammonium nitrate, **1186**
**Trouble-shooting files, 1116**

- **Calculations and estimations** (that can be done in the office before special tests are done)

Pressure profile,
Direction of leak: cooling-coil reactant, **743**
Direction of leak: steam heating coil to acid in storage tank, **645**
Acid flow from the storage tank into the neutralizer, **142**

Rate,
Of neutralization reaction, **2448**
Of cooling, **2492**

Equipment performance,
Neutralizer and sources of water, **2937**

- **What is the current operation**

**Visit control room:** control-room data: values now and from past records,
Temperature on acid storage tank, **2749**
Pressure at exit of pump, **2638**
Acid flowrate, **2105**
Temperature in neutralizer, **2306**
Temperature of AN leaving the neutralizer, **1306**
Cooling-water temperature out, **1149**
Steam generation from overhead of neutralizer, **1050**

**Process operators**, Please describe what has happened so far, **750**
Has this ever happened before and what did you do, **570**

**Contact with on-site specialists**, Utilities: cooling water, **199**
Operators of downstream plant receiving our AN: check on temperature and concentration of AN received, **448**
Supplier: urea plant off-gas: change in flow, temperature or concentration, **263**
Supplier: ammonia plant supplier: change in flow, temperature or concentration, **1829**
Purchasing; have we changed suppliers for the nitric acid, **1735**

**Visit site**, read present values, observe and sense.

Check that the location of valve stem on acid flow-control valve is mid-range, **1599**
Level of liquid in the neutralizer from the level gauge, **1731**
Do the tabs on the orifice plate, used to measure the acid flowrate, indicate that the orifice is facing the correct direction, **2290**
What sounds do you hear by the neutralizer, **2795**
Are there sounds of cavitation from the transfer pump?, **2532**
Are there any water/steam lines hooked up to any of the process equipment or lines feeding the neutralizer, **1250**
Are there any water/steam lines hooked up to the neutralizer, **1350**
Does the steam off the top of the neutralizer go into the top or bottom of the steam header, **1100**

**Check diagram and P&ID versus what's out on the plant, 698**

**On-site simple tests:**
Check that valve on bypass of acid flow-control valve is shut and block valves fully open, **849**
Shut the valve on exit of transfer pump and compare pressure (head) with head on head-capacity curve for this impeller rpm, **948**

Consistency checks,
Does composition of the off-gas add up to 100%, **250**
Does temperature and composition of AN leaving neutralizer agree with conditions downstream, **296**

Trend checks, Do both the temperature and the concentration of AN decrease in the neutralizer?, **1838**

**Gather data for key calculations**

Mass balance,
Based on the measured flows of reactants, products, **82**
Based on the measured flows of off-gas and pure ammonia, calculate the fraction of the required amount of the ammonia that comes from the off-gas and compare with design, **422**

Energy balance: sink = source,
Heat of reaction = heat removed in cooling water, **2347**

**Sensors: check response to change**
Temperature of AN in the neutralizer, **1697**
Flow-control valve of nitric acid feed, **1548**

**Sensors: use of temporary instruments,**
Contact or laser sensor of temperature of the cooling water entering and leaving the cooling coil, **1877**
Contact or laser sensor of temperature of the contents of the neutralizer, **2841**
Contact or laser sensor of temperature of urea feed to the neutralizer, **2264**
Contact or laser sensor of temperature of the ammonia vapor feed to the neutralizer, **2090**
Contact or laser sensor of temperature of the AN leaving the neutralizer, **1849**

Contact or laser sensor of temperature of the acid in the storage tank, **1749**

**Sensors: calibrate,**

Temperature sensor on the AN neutralizer, **346**

Temperature sensor on the AN at the exit nozzle from the neutralizer, **118**

**Control system,**

Retune the flow control on the nitric acid, **1200**

**Call to vendors, licensee, or suppliers,**

Supplier of nitric acid: concentration and additives, **1293**

**Samples and measurements,**

Product AN at exit nozzle from the neutralizer; analyze concentration of AN, **811**

Nitric acid in storage tank; analyze concentration, **2233**

Nitric acid at nozzle at the entry into the neutralizer; analyze concentration, **2466**

Nitric acid at sampler before acid-transfer pump; analyze concentration, **2775**

Nitric acid at flow-control valve; analyze concentration, **2581**

Ammonia vapor at nozzle into neutralizer; analyze concentration of ammonia, **2321**

Urea off-gas at nozzle into neutralizer: analyze concentration, **1485**

**Open and inspect,**

Cooling coil; shut down; pressure test for leak in the cooling coil in the neutralizer by installing a gauge; pressurizing with air, isolating and note decrease in pressure with time, **838**

Cooling coil; shut down; drain neutralizer, pressure test for leak in the cooling coil in the neutralizer by isolating and hydraulically pressurize coil; look for water leaks, **791**

Cooling-coil fouling: Shut down; drain, make safe for entry. Visually inspect. Look for fouling on outside of coil and at entry and exit on the inside, **355**

Trombone, steam heating coil in acid storage tank. Isolate and statically pressure test with air, gauge and observe pressure decrease with time, **936**

Condition of gas sparge rings on both off-gas and ammonia. Shut down; drain, make safe for entry. Visually inspect, **1472**

Acid transfer pump; isolate, drain, inspect, **2155**

- **Take "corrective" action,**

  Interchange the off-gas and ammonia feeds to nozzles on the neutralizer so that the off-gas is at the lowest nozzle, **2703**

  Relocate the ammonia entry nozzle and sparge so that this is below the cooling coils; install two layers of static mixers above the gas sparge to improve mixing, **2945**

  Install a central draft tube in the neutralizer to improve mixing in the neutralizer, **1399**

**Case #43: High pressure in the debut!** (courtesy of T. E. Marlin, Chemical Engineering, McMaster University) [7, sequence of distillation columns with auxiliaries; depropanizer-debutanizer]

The process is the depropanizer-debutanizer described in **Case #8**, Chapter 2. A P&ID is given in Figure 2-4 accompanying **Case #8**.

This is the startup of the unit after the annual turnaround. During the turnaround several valves were replaced, pumps were dismantled and reassembled, the column internals were inspected, the heat exchangers, condensers and reboilers cleaned and the sensors calibrated. This was done for both the debutanizer, C-9, and the depropanizer, C-8.

Shortly after the unit starts up there is an upset on the depropanizer and we have just resolved that in **Case #41**. This **Case #41** had symptoms that the depropanizer overhead product has far too much C4 and the debutanizer overhead has far too much C3. (You don't have to work **Case #41** before you work on this one.) Now the operator panics and calls you again. This time she thinks that the pressure is so high in the debutanizer that the safety relief valve has popped! That means butane is spewing out; the flare should be flashing! Now what?

#### Case #43: Debutanizer pressure relief

- **MSDS, 1495**
- **Immediate action for safety and hazard elimination,**
  Put on safe-park, **111**
  Safety interlock shut down, **440**
  SIS plus evacuation, **2934**
- **More about the process, 1010**
- **IS and IS NOT**, What, **389**
  When, **1481**
  Who, **1962**
  Where, **1772**
- ***Why? Why? Why?***, Best goal? To manage the purported high pressure on the debutanizer, **1321**
- **Weather**, Today and past, **1065**
- **Maintenance: turnaround**, When and what done?, **2381**
- **What should be happening**

**Design and simulation files** (allowances made for fouling, overdesign and uncertainties)
Condenser, E-28, **2006**
Distillation column, C9: **2367**
Thermosyphon reboiler, E-30, **1382**
Reflux pump, F-29, **1112**
Overhead drum, V-31, **1455**

**Vendor files**: Condenser, reboiler, **1448**
Steam traps, **1244**

**Commissioning data, P&ID, internal reports, 2051**

**Handbook and Google**, Cox charts, **2114**
**Trouble-shooting files**, **2809**

- **What is current operation**

**Visit control room:** control-room data,
Feed to C9, debutanizer, FI/2, **365**
Feed to the C8, depropanizer. FC/1, **601**
Overhead liquid product butane, flowrate FIC/7, **994**
Reflux flowrate, FIC/6, **1491**
Pressure drop Δp I/2, **1102**
Level bottoms LIC/4, **1701**
Level in feed drum, V-31; LIC-5, **1551**
Temperature bottoms TI/12, **2401**
Temperature top, TI/11, **2902**
Analyzer A-1, **2570**
Alarm on PIC/19, **2505**
Pressure on overhead, PIC/19, **2443**
**Process operators**, This shift, **278**
**Call to others on-site**, Utilities: any change in the cooling tower operation that might affect our site, **22**
**Visit site**, read present values, observe and sense.
Column pressure, PI-12, **458**
Pressure relief to flare PSV-3, **202**
Observe flare, **417**
Temperature of the feed to the column TI-10, **505**
Cooling-water temperature in to the condenser, TI-13, **960**
Cooling-water temperature out of the condenser, TI-14, **735**
Valve position for column feed, FV-2, **599**
Valve position on PIC/19; downflow from the condenser to the reflux drum, **853**
Signal to PIC/19, **1301**
Pressure on exit of reflux pump F-29, PI-20, and compare with head-capacity curve at feed flowrate, **1253**
Listen to the reflux pump F29 for sounds of cavitation, **1016**
Observe whether shaft is rotating for the reflux pump F-29. Check that the direction of rotation is consistent with arrow on the housing, **1991**
**Check diagram and P&ID versus what's out on the plant**, **2769**
**On-site simple tests:**
Open bleed valves to fuel on the shell of the condenser for 10 minutes, then shut, **1605**
Tap the side of the condenser along a vertical line to listen for change in sound associated with liquid level in the condenser, **1509**
**Gather data for key calculations**,
Pressure profile,
Drum V-31 to pumps, F-29, **2863**

Δp across reflux pump, converted to head, **2984**
Drum V-31, pump F-29 and reflux into column, **2723**
Pump F-29, **2594**
Thermosyphon reboiler process fluid side, **2531**

Energy balance: sink = source,
Estimate the steam flow to the reboiler based on the reflux rate and the fact that each kg steam boils 5 kg typical organic, **104**

**Sensors: check response to change,**
Temperature sensor at the top of column, TI-11, **1634**
Pressure sensor at top of column PI-12, **2151**
Pressure sensor on overhead line for PIC/19; PT-19, **2319**
Pressure sensor on reflux pump exit line, PI-20, **2031**

**Sensors: use of temporary instruments,**
Measure the surface temperature on the outside of the condenser, **329**

**Sensors: calibrate,**
Temperature sensor at the top of column, TI-11, **464**
Pressure sensor at top of column PI-12, **28**
Pressure sensor on overhead line for PIC/19; PT-19, **702**
Pressure sensor on reflux pump exit line, PI-20, **607**

**Control system,**
Check if alarm on PIC/19 is faulty or if signal to control room is faulty, **813**
Put overhead pressure control, PIC/19; reflux control FIC-6 controllers on manual and try to steady out the column, **901**

**Sample and analyze,** Current feed to the debutanizer for the amount of C3, **2488**

- **Open and inspect,**
  Condenser, **2916**
  Pressure control valve PV-19, **2552**
  Reflux pump, F-29, **2513**
  Line from the top of the column to the condenser, E-28, **2825**
  Sensor tap for pressure on the overhead line, PT-19, **1219**

- **Take "corrective" action**
  Replace the pressure control valve PV-19, **1460**
  Replace the pressure gauge PT-19, **1055**
  Replace high-pressure alarm on PIC/19, **1347**
  Direct water from the fire hose onto the outside shell of the overhead condenser, E-28, **786**
  Reduce the reboiler duty, **2983**

**Case #44: Reactant storage** [8, steam coil, storage tank, controller; general]

Reactant must be stored at 50 ± 1 °C. To achieve this a steam-heated coil is placed inside the open, 2.25-$m^3$ vertical cylindrical storage tank as illustrated in Figure 8-24. The instrument engineers designed a control system that satisfies the temperature requirement. Although the sensor, valve and controller are the fanciest on the market, the best it can do is keep the temperature 50 ±10 °C according to the controller chart. The control engineers have worked on tuning the control loop for a week but find nothing wrong. This reactant ties up $2,000 worth of production per hour. Solve the problem.

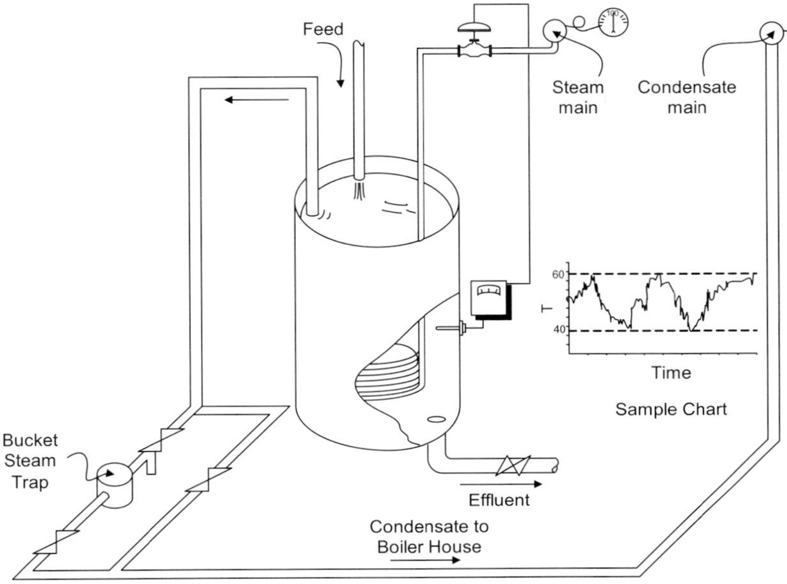


**Figure 8-24** The reactant storage vessel for Case #44.

**Case #44: Reactant storage**

- **Immediate action for safety and hazard elimination**,
  Put on safe-park, **2671**
  Safety interlock shut down, **2145**
  SIS plus evacuation, **32**
- **More about the process, 2805**
- **IS and IS NOT**: (based on given problem statement),
  What, **467**
  When, **1424**
  Who, **1132**
  Where, **1157**
- **Weather**, Today and past, **1858**
- **Maintenance: turnaround**, When and what done?, **2115**
- **What should be happening**

**Design and simulation files** (allowances made for fouling, overdesign and uncertainties)

Controller, **2412**
Steam control valve, **2718**
Heating coil, **2811**
Steam trap, **2696**

**Vendor files**: Steam traps, **2547**
**Commissioning data, P&ID, internal reports**, **2124**
**Handbook**, Steam tables: Temperature for saturated steam at 0. 205 MPa g and 1.5 MPs-g, **2014**
**Trouble-shooting files**, **2254**

- **Calculations and estimations** (that can be done in the office before special tests are done) not sufficient data given.
- **What is the current operation**

**Visit control room:** control-room data: values now and from past records,
Steam pressure in main, **1220**
Feed flowrate into the tank, **1409**
**Process operators**, General comments about the situation, **1492**
Temperature variation in the feed entering, **1037**
**Call to others on-site,** Utilities: steam pressure and quality of steam, **1014**
**Visit site**, read present values, observe and sense.
Pressure in the condensate header, **727**
Check if insulated and quality of the insulation, **1810**
Check that the bypass valve on the condensate trap is closed and that the block valves are open, **935**
**Check diagram and P&ID versus what's out on the plant**, **2287**
**On-site simple tests:** Open bypass on the steam control valve, **517**
Open bypass on condensate trap, **2421**
Open bypass on steam control valve; block off steam valve; open drain, **612**
Give the bucket trap a sharp hit to dislodge any crud that might interfere with the mechanism, **402**
Shut valve on exit of trap for several minutes and then open slowly to "reseal" the trap, **79**
Note position of valve stem on steam valve over "two cycles" of the temperature variation, **2917**
**Gather data for key calculations**
Pressure profile,
Pressure drop across the control valve on inlet side; estimate, **2462**
Estimate the pressure to push the condensate up vertically to the condensate header, **2092**
Mass balance, estimate the mass balance on the steam, **1668**
Energy balance: sink = source, steam required = amount of heat up the reactant, **1903**

**Sensors: use of temporary instruments**,
Use contact or laser pyrometer to measure the temperatures upstream and downstream of the trap; compare with expected 200 and 134 °C, **2777**
Use contact or laser pyrometer to measure feed temperature into the vessel over two cycles, **2577**
Use the stethoscope to listen to the bucket trap over "two cycles" of the temperature variation and compare with expected "loud initially, followed by lower-pitch bubbling and then no noise" for each cycle, **2979**
Use two contact or laser pyrometers to measure the temperatures upstream and downstream of the trap over "two cycles" of the temperature variation and compare with expected 200 and 134 °C, **1702**
**Sensors: calibrate**, Temperature sensor in vessel, **1207**
Pressure sensor on steam main, **705**
**Control system**, Put controller on manual, **2687**
**More complex tests**, Add visible tracer to observe mixing patterns in vessel, **1560**
**Open and inspect**,
Block off steam trap; operate on bypass; open strainer and check for clogging. If clogged, clean and restart, **588**
Steam heating coil, **156**
Bucket steam trap; then clean if necessary and restart, **408**

- **Take "corrective" action**,
Insulate vessel, **666**
Repipe steam connections to coil so that steam comes in the top and condensate to trap is out the bottom of the coil, **1093**
Add a portable, top-entry propeller mixer, **1494**
Relocate the feed pipe entry under the surface, **2898**
Replace bucket trap with a float trap, **2530**
Install a check valve on the line to the condensate header, **1763**

**Case #45: The deprop bottoms and the ISO dilemma** (courtesy of T. E. Marlin, Chemical Engineering Dept., McMaster University) [8, sequence of distillation columns plus auxiliaries; petrochemical, refinery]
The process is the depropanizer-debutanizer described in **Case #8**, Chapter 2. A P&ID is given in Figure 2.4 accompanying **Case #8**.

To be able to sell your products, your plant must obtain ISO certification. As a result, you have established a routine analysis of various streams in the depropanizer-debutanizer system. The initial laboratory analyses indicate too much variability in the mole fraction of propane in the bottoms of the depropanizer, C-8. For the last day, the mole fraction has been about 0.04, while the target is 0.015. Before the new procedures, we never knew that we were operating the plant so poorly, so no one cared! Now everyone is frustrated that you have found this fault.

If you cannot obtain ISO certification, the company will not be able to sell products to key customers. Resolve the problem.

**Case #45: The deprop bottoms and the ISO dilemma**

- **MSDS**, **1495**
- **Immediate action for safety and hazard elimination**,
  Put on safe-park, **1882**
  Safety interlock shut down, **1531**
  SIS plus evacuation, **732**
- **More about the process**, **1010**
- **IS and IS NOT**, What?, **1477**
  When?, **2237**
  Where?, **2537**
- ***Why? Why? Why?***, Best goal? To reduce the propane concentration in the bottoms to within specifications, **2211**
- **Weather**, Today and past, **2303**
- **Maintenance: turnaround**, When and what done?, **1421**

**Maintenance: routine**, When and what done?, **2044**

- **What should be happening**,

**Design and simulation files** (allowances made for fouling, overdesign and uncertainties)
Condenser, E-25, **591**
Distillation column, C8: **515**
Thermosyphon reboiler, E-27, **2380**
Feed pump, F-25; F-26, **16**
Turbine drive, **1193**
Motor drive, **1072**
Reflux pump, F-27, **286**
Feed preheater, E-24, **1989**
Feed drum, V-29, **1478**
Overhead drum, V-30, **1124**
**Vendor files**: Condenser, reboiler and preheater, **10**
Steam traps, **481**
**Commissioning data, P&ID, internal reports**, **245**
**Handbook**, Cox charts, **803**
**Trouble-shooting files**, **1430**

- **Calculations and estimations**

Equipment performance, Nucleate or film boiling in reboiler? check $\Delta T$, **642**
Energy balance: Estimate the steam flow to the reboiler based on the reflux rate and the fact that each kg steam boils 5 kg typical organic, **2821**

- **What is current operation**

**Visit control room:** control-room data,
Control temperature on the C8, depropanizer. TC/5, **1563**
Bottoms temperature. TI/4, **1156**

Steam flow to reboiler. FIC/3, **2078**
Debutanizer overhead analyzer for C3: AC/1, **436**
Level in bottoms, LC/2, **103**
Flow of bottoms to the debut, FI/2, **2251**
Feed flowrate to column. FIC/1, **766**

**Process operators**, This shift; variation in AC/1?, **1334**
Previous shift: variation in AC/1, **433**

**Call to others on site**,
Operators of facility receiving the butane as overheads from the debutanizer: propane contamination in product, **219**

**Visit site**, read present values, observe and sense,
Column temperature on tray #9, TI-8, **1303**
Observe the flare, **2815**
Valve-stem position on LV/2, bottoms flowrate from the deprop, C-8, **2989**
Valve-stem position on FV-3; steam to reboiler, **2614**

**Check diagram and P&ID versus what's out on the plant**, **2769**

**On-site simple tests:**
Put control of steam to reboiler E-27 on manual and control based on TC-5, **2585**

Trend check, do trends in the lab sample results follow the same trends in the readings from AC/1?, **652**

**Control system**,
Retune the control system for the bottoms of the deprop, C-8. TC/5, FC/3 and LC/2, **894**
Operator adjust the set point on TC-5 based on the measurement of A-1, **201**

**Sensors calibrate**,
Calibrate sensor on the top of the debut. A/1, **454**
Calibrate temperature sensor, TC/5 on tray #9 of the column, **952**

**Samples and measurements**,
Sample feed to C-8, the deprop and check for > usual concentration of C3, **1202**
Sample the bottoms from C-8 once per hour for three hours. Flush sample line well before collecting each sample. Analyze in lab for C3, **1107**
Draw three samples of the bottoms from C-8; send one sample to our lab and others to two independent labs. Analyze for C3, **1453**
Are the lab sample results consistent with the readings from AC/1?, **2102**

**Open and inspect**,
Reboiler: pull bundle and check for fouling on both inside the tubes and outside, **1952**
Column at tray #9 and check that TI-8 and TC-5 are "well situated" and the wells are not corroded; that the downcomer is sealed and the tray is level, **1651**
Column at bottoms and check that vortex breaker is present and that thermowell for temperature sensor TI-4 is "well situated" and the well is not corroded. Level sensor is OK?, **1051**

- **Take "corrective" action**
  Install a C3 analyzer on the feed into the column, **2751**
  Replace the thermocouple at TC-5; retune the controller, **1391**
  Replace the steam trap on the condensate leaving reboiler, E-27, **1007**

**Case #46: The not so cool chiller** (courtesy of Scott Lynn, Chemical Engineering Department, University of California, Berkeley) [8, pump, exchanger, ethylene refrigeration; polymerization]
The system below is designed to prepare butene in methyl chloride solvent for polymerization. This reactor feed is to be dried, condensed and cooled to–29 °C with ammonia refrigerant, and finally chilled to –96 °C with boiling $C_2H_4$ in a thermosyphon chiller. The ethylene tube-and-shell chiller will cool the reactor feed to –96 °C at only half the design flowrate or to only –60 °C at the design flowrate. This reduces polymer production by 0.45 Mg/h at an out-of-pocket loss in profit of $1/kg. Get this going correctly. The process is shown in Figure 8-25.

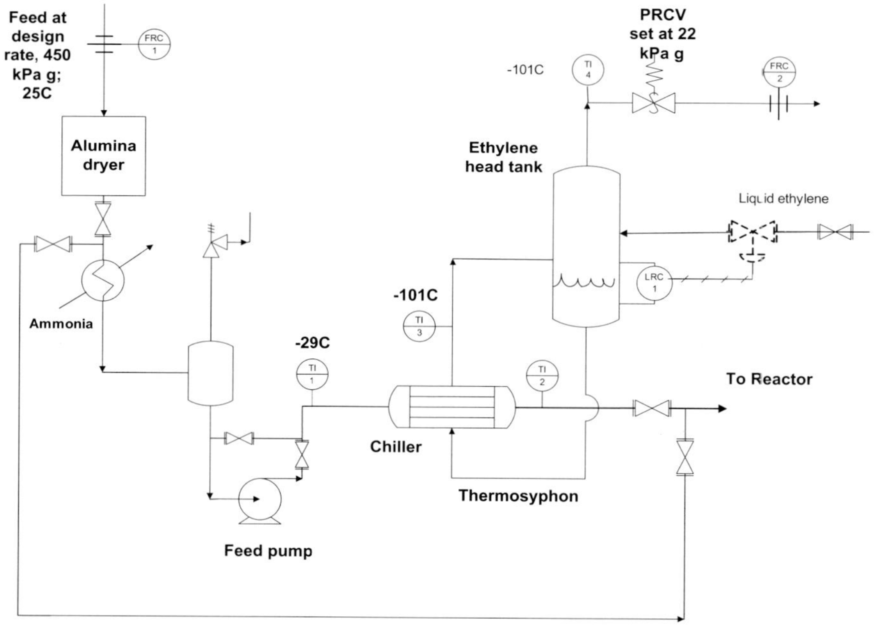


Figure 8-25 The chiller for the reactant feed for Case #46.

**Case #46: Not so cool chiller**

- **MSDS**, 423
- **Immediate action for safety and hazard elimination,**
  Put on safe-park, **148**
  Safety interlock shut down, **1282**
  SIS plus evacuation, **1029**
- **More about the process**, **1136**
- **IS and IS NOT**: (based on given problem statement), What, **2337**
  When, **2096**
  Where, **2896**
- **Weather**, Today and past, **2633**
- **Maintenance: turnaround**, When and what done?, **2332**

**Maintenance: routine**, When and what done?, **1723**

- **What should be happening**

**Design and simulation files** (allowances made for fouling, overdesign and uncertainties)
Chiller, **1616**
Feed pump, **1223**
Drier, **1061**
Ammonia condenser, **647**
Feed drum, **823**
Ethylene head tank, **312**
FRC controller, **1937**
**Vendor files**: Feed pump, **2924**
**Commissioning data, P&ID, internal reports, 2680**
**Handbook**, Methyl chloride: vp at 25 °C and compare with pressure on vapor of 450 kPa g, **2555**
Heat capacities of butene, water, methyl chloride and ammonia, **2640**
**Trouble-shooting files**, 2827

- **Calculations and estimations** (that can be done in the office before special tests are done)

Pressure profile,
From inlet to drier at 450 kPa g to feed drum, **2598**
Direction of leak: ammonia condenser, **343**
Direction of leak: chiller, **134**
From the vapor from the heads tank to the ethylene compressor, **792**
From feed drum to reactor, **613**
Energy balance: sink = source, vaporisation load: design load vs; current full load and half-load, **907**
Rate, In chiller, nucleate vs film boiling on shell side for full versus half-load, **863**
Equipment performance,
Chiller: rating comparison of UA for full load vs half-load, **746**

Chiller: tube side and shell side transfer coefficient, **1429**
Compressor-condenser in the refrigeration loop, **1187**

- **What is the current operation**

**Visit control room:** control-room data: values now and from past records,
Flowrate of vapor to the driers, FRC, **1095**
Feed temperature to chiller, TI, **1302**
Temperature of the process fluid exit of chiller, **1397**
Temperature of the ethylene vapor off the chiller, **1780**
Flow of ethylene off top of head tank, FR, **1942**
Setting on PRCV on ethylene head tank, **2222**
**Process operators**, Please tell me about the operation so far, **2471**
**Contact with on-site specialists**,
Contact operators of reactor; temperature and flowrate to reactor consistent with the temperatures and flowrates from chiller unit, **2193**
Unit supplying and receiving the ethylene, **2788**
**Visit site**, read present values, observe and sense.
Feed pump: sounds like cavitation? **2584**
Valve position for vapor feed at FRC/1, **2697**
Tab on orifice: indicating that the plate has been put in correctly, **2914**
Check that the block valve on bypass around the FRC on the vapor feed to the dryers is CLOSED, **2217**
Liquid level in ethylene head tank, **2084**
Valve position for liquid ethylene feed, **2042**
Temperature on vapor line off top of ethylene head tank, **1625**
Check that the two valves on the vapor bypass line are fully closed. Check that the "flow direction" through the valve is correct, **1817**
Check that the valve on the ethylene thermosyphon line is fully open and that the "direction of flow" through the valve is correct, **714**
Check that the bypass valve on the feed pump is fully closed and that the "direction of flow" through the valve is correct, **875**
Check that the valve on the pump discharge is fully open. Check that "flow direction" on the valve is correct, **1869**
**Check diagram and P&ID versus what's really out on the plant, 1711**
**On-site simple tests:**
Adjust the PRCV to a slightly higher pressure and note temperatures, **1693**
Complete a "turn and seal" test on two valves on the vapor bypass line. Ensure the valves are left shut, **1592**
Complete a "turn and seal" test on the valve on the ethylene thermosyphon line. Ensure the valve is fully open, **1929**
Complete a "turn and seal" test on the bypass around the pump. Ensure the valve is fully closed, **2739**
Tap side of the ethylene head tank to try to detect liquid level. Compare height with level gauge reading, **2616**
Tap side of the feed drum to try to detect liquid level, **2173**

Tap side of the chiller to try to detect the liquid level. Compare with the level of exchanger tubes. Are the tubes completely covered by the ethylene, **2340**

Consistency checks, Check that temperature and pressure measured at top of chiller are consistent for "pure" ethylene, **1821**

Temperature sensors: TI/3 on ethylene chiller agree with TI/4 on top of head tank, **1688**

Process temperature from chiller exit agree with downstream unit, **1529**

**Sensors: check response to change**

Temperature sensor of the ethylene vapor TI/3, **1071**

Temperature sensor of the process liquid out of chiller; TI/2, **1213**

**Sensors: use of temporary instruments**,

Surface or laser temperature sensor for the overhead ethylene from the chiller, **127**

Use clamp-on ammeter to measure the amps on the motor driving the feed pump; compare with specs, **397**

**Sensors: calibrate**,

Temperature sensor of the ethylene vapor TI/3, **756**

Temperature sensor of the process liquid out of chiller; TI/2, **1333**

**Control system**,

Retune the FRC controller on the feed to the drier, **1042**

Retune the LRC controller on the ethylene heads tank, **2298**

**Samples and measurements**,

Sample ethylene and analyze for contamination, **2440**

Sample process liquid from the feed drum. Analyze for water and ammonia, **2875**

**More complex tests**,

Steam regenerator for the drier system. Isolate, once conditions are safe, hydraulically pressure test to identify leaks in the tubes or between the tubesheet and the tubes. Any leak of steam into the drier system might possibly load the adsorbent with water, **2538**

Gamma scan to determine the interface location in the chiller, **2947**

**Open and inspect**,

Chiller, check for fouling. Isolate, drain, once conditions are safe, pull out the bundle and check for fouling inside and outside tubes; use optic fiber probe as needed, **2091**

Feed pump. Isolate, when conditions are safe, open and inspect, **2256**

Line from the heads tank to the chiller for "thermosyphon operation": isolate, when safe, open and inspect for blockages. Use optic fiber probe as needed, **1161**

- **Take "corrective" action**,

  Replace the two valves on the vapor bypass around the chiller, **776**

  Replace the globe valve on the thermosyphon line from the head tank with a ball valve, **214**

**Case #47: The fluctuating production of acetic anhydride** [8, reactor, vaporizer, condensers, liquid ring vacuum pump, control; petrochemical]

To produce acetic anhydride from acetic acid, via the Wacker process, about 0.6 kg/s of acetic acid is vaporized and sent into a series of three cracking coils that are operated under vacuum. The three cracking coils are inside a furnace as is illustrated in Figure 8-26. But producing acetic anhydride is not a simple process. The acetic acid is cracked in the furnace to produce ketene and water. The reaction is reversible. If the water is not condensed and removed rapidly from the reactor effluent, the reverse reaction could occur and the product from the reactor would be acetic acid – the reactant you started with. If the removal of the water is successful, then gaseous ketene leaves the reactor condenser system and reacts with fresh liquid acetic acid in an absorber to produce acetic anhydride. The condensers for the water are a series of Liebig condensers using first water and then brine as the coolant. A further complication is that ketene dimerizes and forms a gunk that plays havoc with the operation of the reciprocating vacuum pumps.

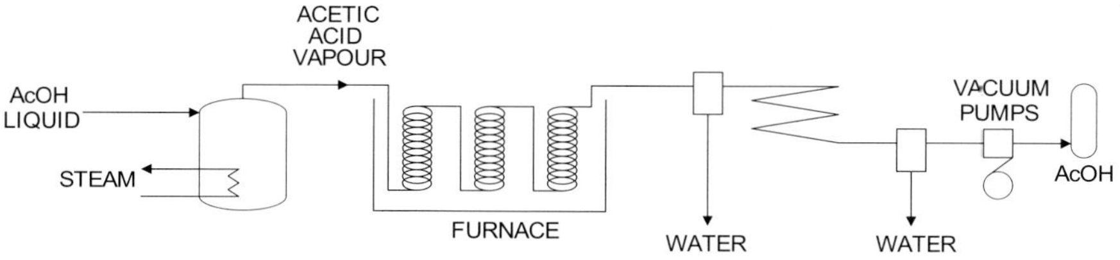


**Figure 8-26** The furnace reactor and condensers to produce acetic anhydride in Case #47.

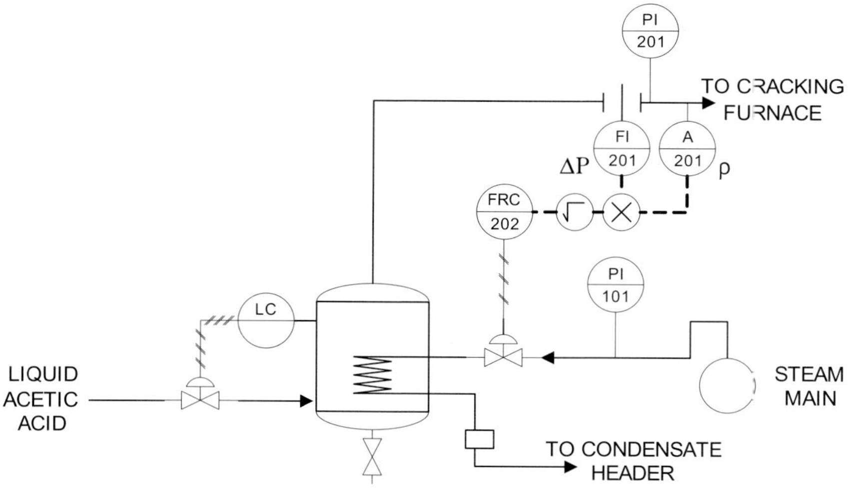


**Figure 8-27** The vaporizer for the feed to the furnace in Case #47.

The system has experienced serious upsets. The amount of acetic anhydride product fluctuates greatly. Some suspect the vaporizer that is shown in Figure 8-27. Some think that the condensation of the water from the ketene is the source of the fluctuations. Still others point to the vacuum pump. Perhaps it is the absorber. Sort it out! and sort it out fast!

**Case #47: Fluctuating production of acetic anhydride**

- **MSDS**, 339
- **Immediate action for safety and hazard elimination**,
  Put on safe-park, **203**
  Safety interlock shut down, **2653**
  SIS plus evacuation, **2080**
- **IS and IS NOT**: (based on given problem statement),
  What, **2361**
  When, **2061**
  Where, **1654**
- **Weather**, Today and past, **1542**
- **Maintenance: turnaround**, When and what done?, **1204**

**Maintenance: routine**, When and what done?, **1021**

- **What should be happening**

**Design and simulation files** (allowances made for fouling, overdesign and uncertainties)
Cracking reactor, **46**
Condenser, **117**
Vacuum pumps, **363**
Orifice meter FRC/201, **474**
Absorber-reactor, **985**
Vaporizer, **622**
Steam trap, **541**
Gas-fired furnace for the reactor coils, **1433**
**Vendor files**: Vacuum pump, **1038**
**Commissioning data, P&ID, internal reports**, **1023**
**Handbook**, Properties of acetic acid, **1267**
**Trouble-shooting files**, **1363**

- **Calculations and estimations** (that can be done in the office before special tests are done)

Pressure profile,
Pressure drop through reaction coil, **1681**
Direction of flow if there is a leak: in the condenser, **1981**
Mass balance, on acetic anhydride produced, **2104**

Energy balance: sink = source,

On condensers: heat removed in water/brine = heat to be removed from condensing reactant, **2454**

Thermodynamics,

Vapor-liquid equilibrium in KO pots between condensers, **2963**

Rate,

Rate and catalyst addition in reaction coils, **2670**

Boiling characteristics in vaporizer: nucleate or film, **2553**

Equipment performance,

Condensers: heat-transfer coefficients, **2515**

- **What is the current operation**

**Visit control room:** control-room data: values now and from past records,

Steam pressure PI 101, **2973**

Acetic acid flowrate FRC 201, **2590**

**Process operators**, Please describe the operation so far, **2153**

**Contact with on-site specialists**,

Steam plant: pressure and steadiness of steam supplied to battery limits, **1607**

Steam plant: condensate return header, **1927**

Source of acetic acid; specs agree with those used in the design, **229**

Cooling water to condensers, **56**

**Visit site**, read present values, observe and sense.

Inspect the tab on the orifice plate to determine if the plate is in backwards and to note the size of the orifice, **370**

Does condensate discharge into the top of the condensate header, **931**

Variation in the pressure and flowrate of the fuel to the cracking furnace, **768**

Upstream and downstream distance from the orifice plate to see if distance meet usual design specs, **598**

Duration of the cycle of the variation in production, **1164**

**Check diagram and P&ID versus what's out on the plant**, 2787

**On-site simple tests:**

Glove test on the steam trap to estimate the temperature of the trap upstream and downstream, **1706**

Stethoscope on trap to listen is discharge of condensate, **2203**

Tap the side of the evaporator to try to discern the level of acid in the vaporizer, **2057**

Consistency checks,

Pressure on steam header on-site and the pressure at the steam drum, **2402**

**Sensors: calibrate**, LIC on vaporizer, **855**

FRC 201, **1256**

Pressure in acetic acid vapor line PI 201, **1872**

Steam pressure to coil PI 101, **2220**

**Control system**,

Controller on manual and check the FRC 201, **2604**

Controller on manual and read PI 201, **2933**
Controller on manual and check for steadiness in the downstream production, **560**
**Call to vendors, licensee**, Vacuum pump, **2905**
**More complicated tests,**
Gamma scan on vaporizer to locate level relative to the coil, **715**
**Open and inspect,**
Evaporator and look for fouling, **2493**
Isolate coil in vaporizer and pressure test for leaks, **2148**
Orifice plate FRC 101 to see if it is put in backwards, **2702**
Isolate, inspect cracking coils for carbon formation inside the tubes, **2548**

- **Take "corrective" action,**
Retune the control system on the vaporizer, **253**
Replace the bucket trap with a float trap, **889**

**Case #48: The column that just wouldn't work** (courtesy of T. E. Marlin, Chemical Engineering, McMaster University) [8, sequence of distillation columns with auxiliaries; depropanizer-debutanizer]

The process is the depropanizer-debutanizer described in **Case #8**, Chapter 2. A P&ID is given in Figure 2.4 accompanying **Case #8**.

This is the startup of the unit after the annual turnaround. During the turnaround several valves were replaced, pumps were dismantled and reassembled, the column internals were inspected, the heat exchangers, condensers and reboilers cleaned and the sensors calibrated. This was done for both the debutanizer, C-9, and the depropanizer, C-8.

Shortly after the unit starts up the operator is in a panic because, according to the laboratory analysis, the depropanizer overhead product has far too much C4 and the debutanizer overhead has far too much C3. The plant is losing $1000s per hour and the plant manager is furious. Fix the problem.

**Case #48: The column that just wouldn't work**

- **MSDS, 1495**
- **Immediate action for safety and hazard elimination,**
Put on safe-park, **227**
Safety interlock shut down, **2874**
SIS plus evacuation, **2619**
- **More about the process, 1010**
- **IS and IS NOT,**
What, **2907**
When, **2240**
Where, **1680**
- ***Why? Why? Why?***, Best goal? To meet specs on overhead and bottoms, **1366**
- **Weather**, Today and past, **1065**

- **Maintenance: turnaround**, When and what done?, **181**

  Was the vortex breaker in the reflux drum V-30 in "good" shape, **24**

- **What should be happening**

**Design and simulation files** (allowances made for fouling, overdesign and uncertainties)

Condenser, E-25, **725**
Distillation column, C8: **519**
Thermosyphon reboiler, E-27, **1673**
Reflux pump, F-27, **1524**
Overhead drum, V-30, **2324**

**Vendor files**: Condenser, reboiler and preheater, **2915**

Reflux pump, F-27, **1755**
Steam traps, **1535**

**Commissioning data, P&ID, internal reports, 1863**
**Handbook and Google**, Cox charts, **1162**
**Trouble-shooting files, 664**

- **What is current operation**

**Visit control room:** control-room data,

Feed to the C8, depropanizer. FC/1, **1820**
Reflux flowrate, FIC/4, **1712**
Overhead product flowrate of propane, FIC/5, **1628**
Pressure drop $\Delta p$ I/1, **2309**
Level bottoms LIC/2, **2026**
Level in reflux drum: LIC/3, **2705**
Steam to reboiler, FIC/3, **2760**
Temperature of feed into the column, TIC/2, **2974**
Analyzer A-1; concentration of C3 in bottoms of C8 and feed to C9, **1573**
Temperature bottoms TI/4, **1515**
Temperature mid-column TIC/5, **1077**
Temperature top, TI/3, **1374**
Pressure on overhead drum, PIC/10, **673**

**Process operators**, This startup shift, **943**

Any change to the feedstock to this unit?, **974**

Operating procedures, Column temperature too high; then increase the reflux, **321**

**Call to others on-site**, Utilities: is the steam-production rate and steam temperature and pressure the usual values?, **123**

Utilities: is the cooling-water flowrate and temperature to our unit what we expect?, **39**
Operators of plants downstream receiving the butane, propane and pentane, **2731**
Operators of upstream plant providing feed, **1732**

**Visit site**, read present values, observe and sense.

Column pressure, PI-4, **412**
Pressure relief to flare PSV-1, **2425**
Look at the flare, **882**
Temperature mid-column TI-8, **2183**
Valve position for column feed, FV-1, **1766**
Valve position for reflux, FV-4, **1913**
Valve position for propane overhead product to downstream processing, FIC-5, **1695**
Pressure on the reflux pump exit, F-27 as shown on PI-11, **2277**
Is there noise of cavitation around reflux pump F-27, **2127**
Are the leads to the motor drive on reflux pump F-27 correct so that the motor turns in the "correct" direction, **2925**
Check that the direction of rotation of the pump shaft is correct, **2593**
Valve position for steam to preheater E-24, **2138**
Valve-stem position on PV-10, **2366**
Level in feed drum, V-29; LI-1, **2106**
Isolation valves around the reflux control valve, FIC/4, **1655**
Isolation valves around the reflux pump, F-27, **1747**

**Check diagram and P&ID versus what's out on the plant, 2769**

**On-site simple tests:**

Open bypass around the reflux control valve, FIC/4; note any change in overhead temperature TI-3 and flow of reflux FIC/4, **1217**

**Gather data for key calculations**,

Pressure profile,

Drum V-29 to pumps, F-25, 26, **529**
$\Delta$p across pump, F-25, 26 converted to head, **96**
Pump F-25–26 exit to feed location, **463**
Drum V-30, pump F-27 and reflux into column, **164**
Pump F-27, **868**
Vapor from top of column to vapor space in V-30, **582**
Thermosyphon reboiler process fluid side, **1405**

Mass balance, Over column, **1287**

Energy balance:

Heat load for condenser. Heat load condensed = heat load picked up in cooling water, **1015**
Steam load to reboiler based on 1 kg steam boils 5 kg organic, **1088**
Water flowrate to overhead condenser consistent with steam flowrate at bottoms. 15 L water / kg steam, **1426**
With data from hand-held voltmeter, clamp-on ammeter and power-factor meters calculate the power used by pumps F-25,–26 and compare with specs, **890**

**Sensors: check response to change**,

Temperature at the top of column C8: TI 3, **572**
Analyzer A-1, **76**

Flowmeter for reflux, FE-4, **375**
Pressure at the top of column C8, PI 4, **747**

**Sensors: calibrate**,
Temperature at the top of column C8: TI 3, **1690**
Analyzer A-1, **2257**
Flowmeter for reflux, FE-4, **2270**
Pressure at the top of column C8, PI 4, **2312**

**Control systems**,
Output signal from controller to reflux valve FIC/ 4, **1246**
Put reflux control system, FIV/4, on manual; increase the flowrate, **1795**

**Samples and measurements**,
Sample feed and analyze composition and compare with usual, **2158**
Sample overhead concentration in vapor from column every 10 minutes for 1 hour. Analyze for composition, **2196**
Sample bottoms concentration from the tower every 10 minutes for 1 hour. Analyze for composition, **2088**

- **More ambitious tests**, Gamma scan on column to look for collapsed tray(s), **2730**
- **Open and inspect**,
  Condensers; E-25, check baffles and for fouling, **2694**
  Access hole to column and check if trays are level and sealed; for those trays that can be seen from access hole, **2855**
  Reflux pump: check for damage to impeller, wear rings are not worn, F-27, **2891**
  Reflux control valve: check stem, trim, FV/4, **2803**
- **Take "corrective" action**,
  Shut down the column; take out all the trays and add weirs to "seal" the self-sealed downcomers, **2284**
  Replace the check valve on the exit line of the reflux pump, F-27, **2349**
  Replace the control valve on the reflux, FIC/4, **2039**
  Reduce feedrate to the column to 1/2, **1647**
  Put on safe hold, **1604**

**Case #49: The case of the faulty stretcher pedal** (from D.R. Winter, Universal Gravo-plast, Toronto, 2004) [Difficulty 8; involves feed bin, molding machine, mold and mold design. Context: injection molding of thermoplastics]

The client manufactures hospital stretchers, sometimes called gurneys. The plastic component the client wishes molded is a foot pedal that would stop the stretcher from rolling. The peal is 24 cm long with a central hub about which the pedal functions like a teeter-totter so that the pedal can be pushed from either direction. The foot part of the pedal is 6.5 cm across and ribbed. The pedal is 2.6 cm thick with webs of 3 mm thickness. The central hub is 3.2 cm in diameter with wall thicknesses of 1 cm. Ribs support the under part of the pedal because the torque created is substantial. A metal core was used in the mold to create the hole for the axle, over which the pedal was fitted.

The resin selected was an alloy of polycarbonate and ABS that provides an impact strength of 640 J/m. A mold was designed, tests were done and the customer approved the prototype. Production was started.

After several months in the field, some pedals were breaking in Hong Kong, California, British Columbia. The breaks seemed to be independent of the hospital location. An analysis of the broken pedals suggested that the breakage occurred at the hub; inspection showed weld or knits lines. Solve the problem! Figure 8-28 is a sketch of the pedal.

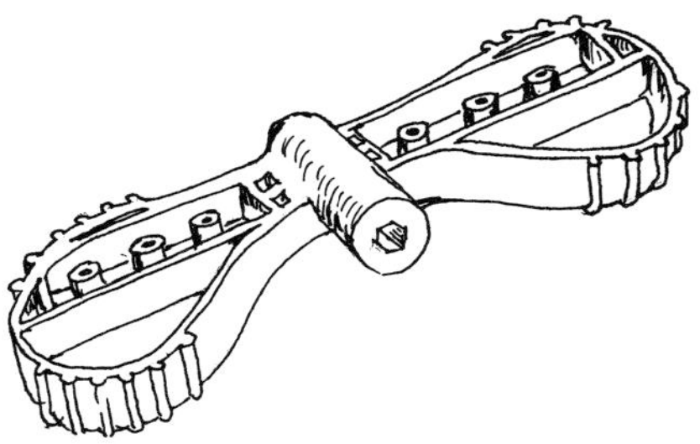

**Figure 8-28** The pedal.

**Case #49: The case of the faulty stretcher pedal**

- **MSDS,300**
- **Immediate action for safety and hazard elimination,**
  Put on safe-park, **947**
  Safety interlock shut down, **523**
  SIS plus evacuation, **1584**
- **More about the process, 2282**
- **More about the product and the mold, 2093**
- **IS and IS NOT**: (based on given problem statement),
  What, **2729**
  When, **2523**
  Who, **191**
  Where, **1222**
- **Weather**, Today and past, **1343**

**Maintenance: routine**, When and what done?, **1791**

- **What should be happening**

**Design and simulation files** (allowances made for fouling, overdesign and uncertainties)

Injection molding machine, **2473**
Mold, **2704**

Feed for this product, **1862**
**Commissioning data, P&ID, internal reports**, Prototype development, **1522**
**Handbook**, Data sheets for resin, **1691**
**Trouble-shooting files**, **2118**

- **What is the current operation**

**Visit control room:** control-room data: values now and from past records,
Fill cycle, **2314**
Cool cycle, **2892**
Open cycle, **2629**
Total cycle, **384**
Feed temperature into mould, **721**
Injection pressure, **904**
Extruder: rear-barrel temperature, **1076**
Extruder: head temperature, **1419**
Shot to cylinder size, 1939
Concentration of foamer, **1622**
Backpressure, **1980**
Mold temperature, **2447**
**Process operators**, **2007**
Operating procedures, Shutdown procedures used, **2591**
Cleaning procedures used, **2673**
Were the feed materials from the same batches of resin, coloring agent and blowing agent as was used for the prototype?, **149**

- **Check with colleagues about hypotheses**, **344**

**On-site simple tests:**
Increase injection speed by 10%, **623**
Decrease injection speed by 10%, **873**
Change injection temperature, increase to 243 °C and keep the cooling time the same, **990**
Change feed temperature, increase to 243 °C and increase the cooling time, **1022**
Clean the mold and extruder before the run by sending through a charge of acrylic. Then use standard conditions, **1308**
Increase the foamer from 0.5 to 2.1% and maintain the same conditions otherwise, **1479**
Increase the foamer from 0.5 to 0.8% and maintain the same conditions otherwise, **1608**
Change mold to 1 feed gate near the hub with a fill cycle of 8 s; all the rest the same, **1741**
Increase the mold temperature to 65 °C, **1965**
Increase the mold temperature to 90 °C, **2267**
Increase the cold water flow to the mold, **2053**
Decrease the cold water flow to the mold, **2693**

Increase the backpressure from 400 to 500 kPa, **2556**
Change the foamer from an exothermic to an endothermic type, **95**
Increase rpm from 160 rpm to 190 rpm, **373**
Use the same mold, same resin, same operator but move from a different molding machine, **782**
Use the same mold, same resin, same machine, same conditions but different operator, **965**
Use the same mold, same operator, same machine, same conditions but resin from a different supplier, **1159**
Use the same mold, same resin, same machine, same conditions, same operator but delete the coloring agent, **1126**
Separately dry the resin pellets just before using and maintain the same conditions, **1754**

**Gather data for key calculations,**
Calculations of the heat loss through the walls for various parts of the mold, **160**

**Redesign of product and mold**, One feed gate instead of two, **653**
Longer plastic neck around the axle from 2.5 to 3.5 cm, **807**
Realignment of axle so that plastic over the axle is centered over the pressure on the pedal, **1388**

**Sensors: check response to change**
Temperature sensor at nozzle, **1049**
Temperature sensor at first stage, **1726**

**Sensors: use of temporary instruments,**
Use a pyrometer or laser sensor to measure melt temperature and compare with temperature sensor, **1597**

**Call to vendors, licensee, or suppliers,**
Properties of feed polycarbonate ABS blend: **2249**
Properties of foamer, **2784**
Properties of coloring agent, **2172**

**Samples and measurements,**
Measure moisture in resin, **1783**

**More complicated tests**, Change the mold from a aluminum to P-40, **1557**

**Case #50: The cleanup column** (courtesy W. F. Taylor, B. Eng. 1966, McMaster University) [9, vacuum distillation column plus auxiliaries; ammonia]
The solvent sulfinol is stripped of unwanted *degradation product heavies* in a solvent recovery column. The solvent is then recycled for reuse. The column, 0.75 m diam × 10.5 m high, treats a small sidestream of organic solvent to remove impurities as the bottoms. The feed flowrate is 0.06 to 1. L/s. Feed enters at tray #5 and the recovered solvent is taken off at the tray above, #4, and pumped, via a positive displacement pump, 114-J, to the sulfinol storage tank, about 15 m away.

The top four trays act as a water wash to prevent overhead losses. The bottom ten trays do the separation. Live stripping steam is injected into the bottoms via valve HCV-13. Most of the steam eventually goes overhead. The wash water fed to tray #1 is sufficient to condense all the organic product and some water; the product contamination is about 20 to 50% water. However, most of the overhead stream going to the vacuum equipment is water. The column is shown in Figure 8-29.

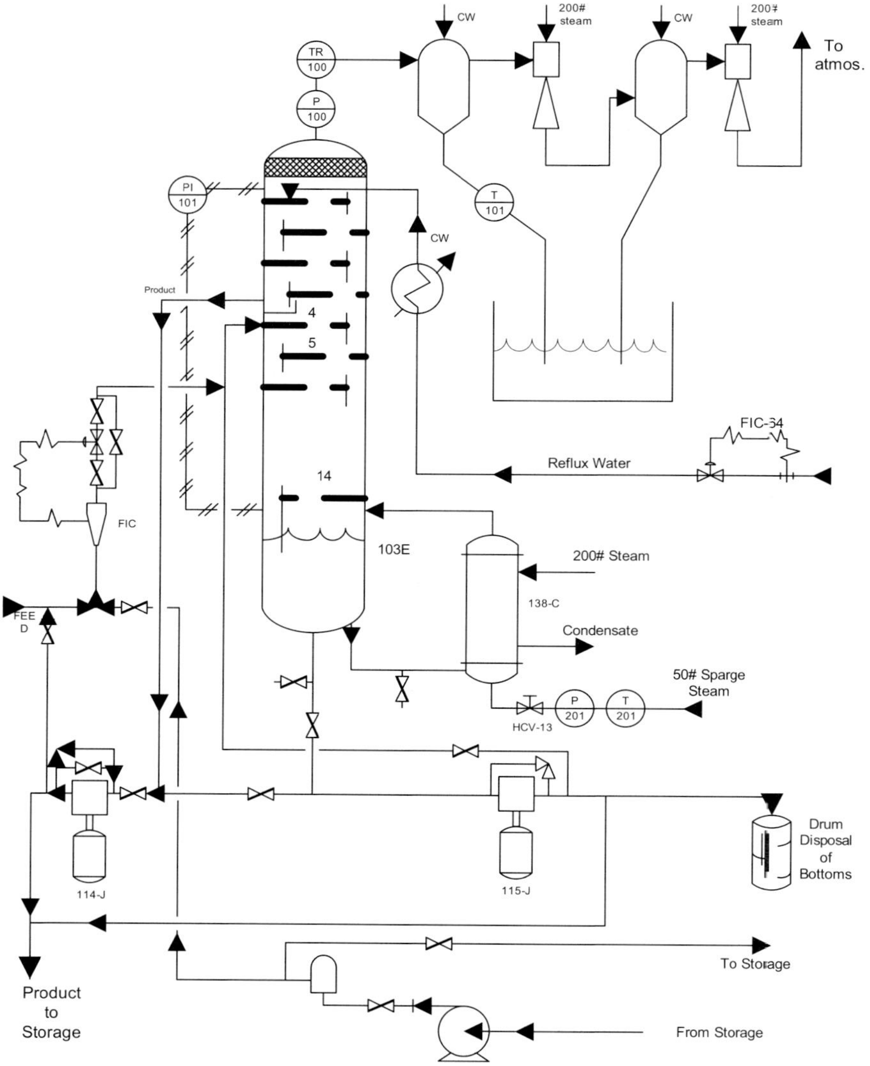


**Figure 8-29** The vacuum distillation unit for Cases #50 and 51.

The vapor pressure of the steam is about 100 times that of the organic product at the tower conditions. The vapor pressure of the organic product, in turn, is about 10 times that of the bottoms. The tower operates under vacuum developed by a two-stage steam ejector system. This particular column is operated for a day once a week so that production downtime costs are not a significant contribution to any problems on this plant.

*The problem:*
The tower has never worked properly. Overhead losses of organic into the hot well are probably high, but they have not been monitored. Operating conditions are rarely stable. No standard operating procedure has been developed. The bottoms product typically contains 10% organic solvent, whereas the column was designed to produce a bottoms concentration of < 2% sulfinol.

But today is particularly frustrating. It was a cold and blustery night and product pump 114-J is just not delivering. The amount of "overhead product" being pumped to storage is almost zip!

**Case #50: Cleanup column**

- **MSDS, 719**
- **Immediate action for safety and hazard elimination,**
  Put on safe-park, **67**
  Safety interlock shut down, **2715**
  SIS plus evacuation, **2068**
- **More about the process, 589**
- **IS and IS NOT**: (based on given problem statement),
  What, **2968**
  When, **986**
  Where, **574**
- **Weather**, Today and past, **753**
- **Maintenance: turnaround**, When and what done?, **157**

**Maintenance: routine**, When and what done?, **1410**

- **What should be happening**

**Design and simulation files** (allowances made for fouling, overdesign and uncertainties)
Column, **1122**
Feed pumps system, **1205**
Product pump system, 114-J, **1305**
Vacuum system, **1354**
**Vendor files**: Pump 114-J, **1486**
**Commissioning data, P&ID, internal reports, 1964**
**Handbook**, Sulfinol composition, **1900**
Expected density of liquid product, **1750**
Steam tables, **1650**
**Trouble-shooting files, 1540**

- **Calculations and estimations** (that can be done in the office before special tests are done)

Rate,

Use ΔT to estimate whether boiling in reboiler is nucleate or film, **2908**

Equipment performance, insufficient data available to do calculations

- **What is the current operation**

**Visit control room:** control-room data: values now and from past records,

Pressure of sparge steam P/ 201, **2605**
Temperature of sparge steam, T/201, **2230**
Temperature of water in the barometric leg, **1850**
Temperature of the overhead TR/ 100, **1950**
Pressure at the top of the column PI/100, **1674**
Pressure in the overhead line to the ejectors. PI/ 101, **847**
Flowrate of feed to column PI/101, **946**

**Process operators**,

When the process was shut down the last time, was the pump 114-J drained to prevent possible freezing in the idle pump, **695**
Please describe what has been done so far, **650**

**Contact with on-site specialists**,

Utilities: steam pressure and flow to unit stable, **132**
Utilities: cooling water to reflux cooler and to the barometric condenser: temperature and flow, **419**
Utilities: reflux water; temperature, flowrate and composition, **1973**

**Visit site**, read present values, observe and sense.

Is the steam tracing on, **2374**
Compare speed of pump 114-J with specs, **2475**
Compare pump stroke on 114-J with specs, **2868**
Pressure drop across trays 1 to 14, **797**
Pressure drop across trays 1 to 4, **2517**
Pressure drop across trays 4 to 14, **1393**
Flowrate of reflux water, FIC/64, **2030**
Temperature at the bottom of the column, **2413**
Temperature of the feed to the column, **1612**
Listen to pump 114-J for sounds of cavitation, **1854**
Read level glass on the sump at tray 4, **1525**

**Check diagram and P&ID versus what's out on the plant**, **1336**

**On-site simple tests:**

Close bypass valve on 114-J, **1443**
Shut-off steam tracing, wait for h and note performance, **1039**
Block off pump 114-J and open bypass direct to storage. Observe liquid level on tray 4 and the "bulls eye" on the line to product storage, **761**
Open suction line on 114-J; drain, **524**
Reduce the flowrate of reflux water, check bottoms temperature and sample/ analyze bottoms, **211**

Install a pressure gauge on the suction of 114-J and note readings under present, usual operation and compare with past, **393**
Pressure at top of the column; compare P/100 with PI/101, **282**

**Gather data for key calculations**
Pressure profile,
Calculated NPSH supplied to pump 114 J, **2141**
Pressure profile from tray 4 to storage, **2281**
Mass balance,
On sulfinol over the still, **2802**

**Samples and measurements,**
Measure flowrate of water from the hot well and compare with design value, **2895**
Sample water from the hot well and analyze concentration of organic, **2691**

**More complicated tests,**
Stop feed to column. Maintain vacuum conditions. Unhook suction line from pump 114-J; hook up water line to the suction line and blow water backward up the suction line and watch the level via the sight glass on tray 4, **2785**
Install a pressure gauge on the suction of 114-J, disconnect the dP and measure the pressure in the column at tray 4. Hook up a portable liquid pump; start pump, measure liquid flowrate and compare the $\Delta p$ measured with an estimate $\Delta p$ for flowrate, **2857**
Gamma scan around tray #4 to locate liquid level in on tray 4 and in the product sump, **1420**

**Open and inspect,**
Inspect foot valve on 114-J and compare valve size with the size of suction line. Correct as needed, **2117**
Check for plugs, obstructions in the suction pipe from Tray 4 to pump 114-J, **2386**
Check the strainer on the suction line to pump 114-J for plugs and obstructions, **1227**
Open column and look for blockage in the nozzle leaving product sump and the condition of the vortex breaker, **221**

- **Take "corrective" action,**
  Replace two valves on the steam to both ejectors, **834**
  Replace bypass valve on 114-J, **1880**
  Replace faulty pressure relief valve on pump 114-J, **2177**

**Case #51: The cleanup column revisited** (courtesy W. K. Taylor, B. Eng. 1966, McMaster University) [9, vacuum distillation column plus auxiliaries; ammonia]
The diagram of this unit is given in **Case #50.**

The solvent sulfinol is stripped of unwanted *degradation product heavies* in a solvent-recovery column. The solvent is then recycled for reuse. The column, 0.75 m diam × 10.5 m high, treats a small sidestream of organic solvent to remove impurities as the bottoms. The feed flowrate is 0.06 to 1.0 L/s. Feed enters at tray #5 and the recovered solvent is taken off at the tray above, #4, and pumped, via a positive displacement pump, 114-J, to the sulfinol storage tank, about 15 m away.

The top four trays act as a water wash to prevent overhead losses. The bottom ten trays do the separation. Live stripping steam is injected into the bottoms via valve HCV-13. Most of the steam eventually goes overhead. The wash water fed to tray #1 is sufficient to condense all the organic product and some water; the product contamination is about 20 to 50% water. However, most of the overhead stream going to the vacuum equipment is water.

The vapor pressure of the steam is about 100 times that of the organic product at the tower conditions. The vapor pressure of the organic product, in turn, is about 10 times that of the bottoms. The tower operates under vacuum developed by a two-stage steam-ejector system. This particular column is operated for a day once a week so that production downtime costs are not a significant contribution to any problems on this plant.

*The problem:*
The tower has never worked properly. Overhead losses of organic into the hot well are probably high, but they have not been monitored. Operating conditions are rarely stable. No standard operating procedure has been developed. The bottoms product typically contains 10% organic solvent, whereas the column was designed to produce a bottoms concentration of $< 2\%$ sulfinol.

Get the bottom composition within design specs. And, by the way, the product pump 114-J is sporadic; just not delivering the amount of product we expect, consistently.

**Case #51: More about the cleanup column**

- **MSDS, 719**
- **Immediate action for safety and hazard elimination**,
  Put on safe-park, **67**
  Safety interlock shut down, **2715**
  SIS plus evacuation, **2068**
- **More about the process, 589**
- **IS and IS NOT**: (based on given problem statement),
  What, **2968**
  When, **986**
  Where, **574**
- **Weather**, Today and past, **2753**
- **Maintenance: turnaround**, When and what done?, **157**

**Maintenance: routine**, When and what done?, **1410**

- **What should be happening**

**Design and simulation files** (allowances made for fouling, overdesign and uncertainties)

Column, **1122**
Feed pumps system, **1205**
Product pump system, 114-J, **1305**
Vacuum system, **1354**
**Vendor files**: Pump 114-J, **1486**
**Commissioning data, P&ID, internal reports, 1964**
**Handbook**, Sulfinol composition, **1900**
Expected density of liquid product, **1750**
Steam tables, **1650**
**Trouble-shooting files**, **1540**

- **Calculations and estimations** (that can be done in the office before special tests are done)

Rate,
Use $\Delta T$ to estimate whether boiling in reboiler is nucleate or film, **2908**
Equipment performance, insufficient data available to do calculations.

- **What is the current operation**

**Visit control room:** control-room data: values now and from past records.
Pressure of sparge steam P/201, **2605**
Temperature of sparge steam, T/201, **2230**
Temperature of water in the barometric leg, **1850**
Temperature of the overhead TR/100, **1950**
Pressure at the top of the column PI/100, **1674**
Pressure in the overhead line to the ejectors. PI/101, **847**
Flowrate of feed to column PI/101, **946**
**Process operators**, Please describe what has been done so far, **2650**
**Contact with on-site specialists**,
Utilities: steam pressure and flow to unit stable, **132**
Utilities: cooling water to reflux cooler and to the barometric condenser: temperature and flow, **2419**
Utilities: reflux water; temperature, flowrate and composition, **1973**
**Visit site**, read present values, observe and sense.
Is the steam tracing on?, **1376**
Compare speed of pump 114-J with specs, **2475**
Compare pump stroke on 114-J with specs, **2868**
Pressure drop across trays 1 to 14, **2644**
Pressure drop across trays 1 to 4, **2517**
Pressure drop across trays 4 to 14, **2133**
Flowrate of reflux water, FIC/64, **2030**
Temperature at the bottom of the column, **2413**
Temperature of the feed to the column, **1612**
Listen to pump 114-J for sounds of cavitation, **2854**
Read level glass on the sump at tray 4, **2225**
**Check diagram and P&ID versus what's out on the plant, 1336**

**On-site simple tests:**

Close bypass valve on 114-J, **1443**

Block off pump 114-J and open bypass direct to storage. Observe liquid level on tray 4 and the "bulls eye" on the line to product storage, **761**

Open suction line on 114-J; drain, **524**

Reduce the flowrate of reflux water, check bottoms temperature and sample/ analyze bottoms, **2221**

Install a pressure gauge on the suction of 114-J and note readings under present, usual operation and compare with past, **393**

Pressure at top of the column; compare P/100 with PI/101, **282**

**Gather data for key calculations**

Pressure profile,

Calculated NPSH supplied to pump 114 J, **2141**

Pressure profile from tray 4 to storage, **2281**

Mass balance,

On sulfinol over the still, **2802**

**Samples and measurements**

Measure flowrate of water from the hot well and compare with design value, **2895**

Sample water from the hot well and analyze concentration of organic, **2691**

**More complicated tests**,

Stop feed and reflux to column. Maintain vacuum conditions. Unhook suction line from pump 114-J; hook up water line to the suction line and blow water backward up the suction line and watch the level via the sight glass on tray 4. When it reaches the top of the sight glass, stop the flow and observe the level over the next 10 minutes, **785**

Gamma scan around tray #4 and in the stripping section to locate liquid level in on tray 4 and possible tray collapsed, **404**

**Open and inspect**,

Inspect foot valve on 114-J and compare valve size with the size of suction line. Correct as needed, **2117**

Check for plugs, obstructions in the suction pipe from Tray 4 to pump 114-J, **2386**

Check the strainer on the suction line to pump 114-J for plugs and obstructions, **1227**

Open column and look at the product sump and the trays in the stripping section, **1120**

- **Take "corrective" action**,

  Replace two valves on the steam to both ejectors, **834**

  Replace bypass valve on 114-J, **1880**

  Replace faulty pressure relief valve on pump 114-J, **2177**

**Case #52: The case of the swinging loops** (courtesy of W. K. Taylor, B. Eng. McMaster, 1966) [9, reactor, compressor, separator; ammonia]
Ammonia is produced on two interconnected reactor loops for the production of ammonia are given in Figure 8-11, the figure in **Case #29**. Feed gas consists of hydrogen and nitrogen in the proper 3:1 ratio with about 1% methane as an inert. In this ammonia synthesis reaction about 10% conversion occurs per pass through the reactor. Feed gas is compressed to 34.5 MPa abs and fed to a common header that feeds two reactor loops. Liquid product is condensed and removed from the system; gas is recycled back to the loop via the recycle stage compression. The reactor operates at 500 °C. There is an internal gas-gas heat exchanger within the reactor. The $\Delta T$ due to exothermic reaction is about 50 °C.

Each compressor is a multistage reciprocating constant-speed machine rated at about 3000 kW. Bypass valve B is operated to control the $\Delta p$ across the recycle stage that must not exceed 3.5 MPa. Opening valve B lowers the $\Delta p$ and the flow of recycle gas to the loop. The recycle flow is about five times the flow of fresh feed.

Bypass valve A is operated to trim the loop: closing this valve forces more gas over to the reactor; opening the valve causes gas to bypass the loop. Valve A is used to control the reaction temperature. If too much gas is fed to the reactor and the catalyst is inactive, the high flow might extinguish the reaction. Similarly, if the flow to the reactor is too low, the reaction will go further because of the longer reaction time; the reactor will overhead because there is not enough flow to carry away the heat of reaction. Normally valve A is open slightly during plant operation.

Methane is an inert coming in with the feed. The methane concentration is kept about 15% in the loop gas to the reactors by maintaining a small purge.

The pressures, levels in the separators and the temperature profile in the reactor are shown in the control room.

The design provides operating flexibility. If one compressor breaks down the other machine can feed both loops thus keeping the reactors at operating temperatures while repairs are done. This avoids costly startup expense. Isolation block valves on the compressors are not shown on the figure. Furthermore, both loops are equalized in pressure thus evening out any slight variations introduced by the compressors.

*The problem:*
The plant has been in operation with everything apparently running smoothly. Production rates, however, are only 80% of design. When the rates are increased by increasing the front-end feedrate and closing the compressor kickbacks, then the whole process becomes very hard to control. Operating logs report the following "... all the gas goes to the North loop for a while and then it swings and all the gas goes to the South loop ..." "... the reactor is overheated ... and the loops flip flop again from South to North ... or sometimes from North to South. Usually it starts with the South". You are brought in to sort out what the operators call "The Swinging Loops". This costs us $40 000/ day.

**Case #52: The swinging loops**

- **MSDS, 186**
- **Immediate action for safety and hazard elimination**,
  Put on safe-park, **2665**
  Safety interlock shut down, **2578**
  SIS plus evacuation, **2962**
- **More about the process, 483**
- **IS and IS NOT**: (based on given problem statement),
  What, **1976**
  When, **1583**
  Who, **1704**
  Where, **1803**
- **Weather**, Today and past, **1264**
- **Maintenance: turnaround**, When and what done?, **1080**
- **What should be happening**

**Design and simulation files** (allowances made for fouling, overdesign and uncertainties)
Reactor, **608**
Internal heat exchanger, **903**
Reciprocating compressor, **153**
Refrigeration system, **335**
Condensers: refrigerant, **1928**
Condensers: water, **2929**
Gas-liquid separators, **2278**
Valves A and B, **1756**
**Vendor files**: Reciprocating compressors, **1907**
Refrigeration system, **1353**
**Commissioning data, P&ID, internal reports, 1158**
**Handbook**, Thermal properties of hydrogen and ammonia, **1053**
**Trouble-shooting files, 1467**

- **Calculations and estimations** (that can be done in the office before special tests are done)

Pressure profile,
Direction of leak: condensers with CW, **107**
Direction of leak: condensers with refrigerant, **64**
Around the loop and fresh feed into the loop, **260**
Mass balance,
Bleed to maintain the inerts in the recycle at 15%, **406**
Thermodynamics, **489**
Equipment performance,
Controlling any hot spots, **217**
Maximum temperature before the catalyst is damaged, **1309**

- **What is the current operation**

**Visit control room:** control-room data: values now and from past records.

Valve A in South loop, **1024**
Valve A in North loop, **1074**
Valve B in South loop, **1113**
Valve B in North loop, **1208**
Temperatures in bed in South reactor, **1414**
Temperatures in bed in North reactor, **1496**
Production of liquid ammonia South loop, FR, **2417**
Production of liquid ammonia North loop, FR, **2258**
Δp across catalyst bed South reactor, **2113**
Δp across catalyst bed North reactor, **2060**
Hydrogen concentration in feed gas to the reactor: South loop, **2021**
Hydrogen concentration in feed gas to the reactor: North loop, **2156**
Temperature at the exit of the cooling water exchanger, South loop, **2804**
Temperature at the exit of the cooling water exchanger, North loop, **2607**
Temperature at the exit of the refrigeration exchanger, South loop, **2778**
Temperature at the exit of the refrigeration exchanger, North loop, **2958**
Calrod startup heaters to both South and North loops, **2333**

**Visit control room:** control-room data:

As soon as one of the loops starts to swing record the values and note the "usual values". Here are the data when the South loop swings taken 4 min after the swinging was first noted by the operators, Valve A in South loop, **2354**
Valve A in North loop, **1852**
Valve B in South loop, **1982**
Valve B in North loop, **1845**
Temperatures in bed in South reactor, **1602**
Temperatures in bed in North reactor, **920**
Production of liquid ammonia South loop, FR, **683**
Production of liquid ammonia North loop, FR, **556**
Δp across catalyst bed South reactor, **131**
Δp across catalyst bed North reactor, **369**
Hydrogen concentration in feed gas to the reactor: South loop, **289**
Hydrogen concentration in feed gas to the reactor: North loop, **1824**
Calrod startup heaters to both South and North loops, **1884**

Process operators,

What is the frequency of the swinging loops, **2318**
What is the cycle time of the swinging loops, **2392**
What do you do when you first notice a hot spot, **2872**
Can you predict when one loop will start swinging, **2903**
What do you think is the most noteworthy observation when swinging is occurring, **2623**

Is it always the South loop that gets the hot spot that seems to trigger the swinging loops or does it happen with equal frequency for the North loop, **1620**

Operating procedures,

What to do if there is a hot spot, **1657**

New procedure; if reactor "hot spot" is 10 to 20 °C; immediately close valve A in that loop, **1181**

**Contact with on-site specialists**,

Supervisor for construction crew for the North loop and the tie-ins: precautions that were taken to ensure that lubricating oil and or water was not left in any of the lines especially the tie-in lines, **706**

Any possible preservative oil or residual valve stem lubricant left on any of the tie-in lines/valves for the recent construction, **806**

**Visit site**, read present values, observe and sense.

Change set point on valve A, South and North loops, and observe whether the valve stem moves, **988**

Change set point on valve B, South and North loops and observe whether the valve stem moves, **80**

Observe the "kickback" behavior of the compressor on the South loop, **206**

Observe the "kickback" behavior of the compressor on the North loop, **655**

Observe-listen for surge in compressors on South loop, **887**

Observe-listen for surge in compressors on North loop, **800**

**Check diagram and the P&ID versus what's actually out on the plant**, **1216**

**On-site simple tests:**

Record evidence and actions over a three hour period whenever the swinging loop behavior "just starts", **1384**

Consistency checks,

For temperature sensors in catalyst bed; do they make sense relative to each other for Loop South reactor, **1145**

For temperature sensors in catalyst bed; do they make sense relative to each other for Loop North reactor, **1725**

Trend checks,

Cycle time: duration and frequency, **1268**

**Sensors: check response to change**

Temperatures North and South loop, **720**

Ammonia production flowrate: North and South loop, **580**

Level on liquid receivers: North and South loop, **1059**

Temperatures on exit of condensers: both water and refrigerated condensers. North and South loop, **1437**

Hydrogen analyzer in loop circuit. North and South loop, **1714**

**Sensors: calibrate**,

Eight temperature sensors North loop reactor, **2244**

Eight Temperature sensors South loop reactor, **2146**

Ammonia production flowrate: North and South loop, **2444**

Level on liquid receivers: North and South loop, **1686**

Temperatures on exit of condensers: both water and refrigerated condensers. North and South loop, **1534**
Hydrogen analyzer in loop circuit. North and South loop, **1239**
Kickback relief on North loop compressor, **1979**
Kickback relief on South loop compressor, **1678**

**Call to vendors, licensee,**
Supplier of the catalyst: anything to watch for that might poison the catalyst, **1873**
Was the same batch of catalyst shipped for the new reactor in the North loop as was used in the previously commissioned South loop system, **2375**

**Samples and measurements**
Shut down the system; sample the "reduced" catalyst from both reactors and compare the activity of each, **1295**
Sample the feed to South reactor when it has the hot spot; the samples are to be every 2 minutes for the first 10 minutes of the swinging loop cycle. Analyze for inerts, **1340**
Sample the feed to North reactor when it has the hot spot; the samples are to be every 2 minutes for the first 10 minutes of the swinging loop cycle. Analyze for inerts, **633**
Fresh feed to South loop; analyze for inerts. Sample once per hour for 24 hours and once every two minutes for ten minutes during the swinging loop, **940**
Fresh feed to North loop; analyze for inerts. Sample once per hour for 24 hours and especially during the swinging loop, **2430**

**More complicated tests,**
Isolate the North loop and run commissioning tests the same way we had done before turnaround for the South loop alone, **2756**

**Open and inspect,**
Open the reactors and note the height of catalyst in the two reactors, **1710**
Valve A; South loop and look for chatter anything that might vibrate or oscillate to cause fluctuations, **1949**
Valve A; North loop and look for chatter anything that might vibrate or oscillate to cause fluctuations, **2949**
Valve A from both North and South loops and pressure test for leak across the butterfly valves, **350**
Valve B from both North and South loops and pressure test for leak across the butterfly valves, **100**

- **Take "corrective" action,**
  Increase the bleed rate in the South loop to decrease the inert concentration in the recycle from 15% to 14%, **867**
  Replace valve on purge line in the South loop, **1867**
  Replace the butterfly valves (used for valves A and B on both North and South loops) with full size globe valves, **2865**
  Replace the temperature sensors in the South loop reactor, **2737**
  Replace kickback valve system on the South loop reactor, **763**

## 8.3 Summary

I hope you enjoyed your journey to improve your skill and confidence as a trouble shooter. I hope you had a chance to use the triad method; this provides a rich experience for you to see how others handle a situation, to spend the time to really understand a process so that you can respond quickly, as the *expert system*, to any questions the *trouble shooter* might pose.

Working the cases on your own is a great experience as well. I trust that the range of cases gave you enriched insights about processes and trouble shooting. I welcome your feedback on how to improve the cases and, of course, cases and answers of cases you solved.

# 9
# What Next?

Here we summarize the highlights of this book so that we can define where we are now. Then we emphasize why reflection and self-assessment are key – yet time-consuming, challenging and frustrating – activities to develop confidence. Ideas are given in Section 9.3 on how to use goal setting as a neat way to develop skill beyond that discussed in this book. Sections 9.4 and 9.5 focus on how to go beyond this book for knowledge about process equipment (as given in Chapter 3) and for interesting and challenging trouble-shooting cases (as given mainly in Chapter 8).

## 9.1
## Summary of Highlights

This book summarizes research about what is known about trouble shooting and what you can do to improve your skills. This is not a series of anecdotes of my trouble-shooting experience or how I personally solve problems. This book is flexibly designed to address your interests and needs. This book is filled with activities!

- **Five key skill areas needed by trouble shooters:** The five areas of skills needed in trouble shooting are 1) problem solving, 2) knowledge of process equipment, 3) process safety and properties of materials, 4) "systems thinking", and 5) people skills. Although research has shown that knowledge of process equipment (common faults, typical symptoms) is of vital importance in determining success in trouble shooting, equal emphasis is given in this book to develop your skill in the other areas and to give you practice in using all the skills to trouble shoot. Section 1.3 gives an inventory for you to reflect on and rate your skills in these five areas. Five cases, **Cases #3–7**, follow in Section 1.6 on which you can try out your skills.

  *Where were your five strengths?*
  *What were the areas you wanted to work on?*
  *How did you rate your skill after you had worked on some of the five cases in Section 1.6?*

*Successful Trouble Shooting for Process Engineers.* Don Woods

ISBN: 3-527-31163-7

- **In problem solving:** the four key components of the process are: 1) the overall problem-solving process as characterized by frequent monitoring, emphasis on checking and double checking, being organized and systematic and keeping the problem in perspective. 2) the elements of data handling and critical thinking: data gathering and resolution, based on fundamentals, with valid reasoning and being complete. 3) how well we synthesize and put all the ideas together: having five to seven working hypotheses and being flexible. 4) decision-making: based on criteria, priorities and avoiding bias. This provides, in Table 2-1, target skills for us to emulate. To help quantitatively see progress in skill development an evaluation form (Worksheet 2-2) and a Trouble-Shooter's Worksheet (Worksheets 2-1 and 2-4) are given. The example use of the Trouble-Shooter's Worksheet was illustrated for **Case #8**: "the depropanizer: the temperatures go crazy".

  *The target skills are described in Table 2-1. Reconsider your rating in Chapter 1 based on your better understanding of these problem-solving skills.*
  *Did the Trouble-Shooter's Worksheet help you? If not, develop your own guide to help you to be systematic.*

- **Knowledge of process equipment:** symptoms–probable causes. Scattered through the literature and one's experience are details of symptoms (and cause) from malfunctioning processes. Sometimes the information is confounded because the cause is not the root cause. Chapter 3 attempts to organize this information for ease in use with cross referencing for systems of equipment. Consistent SI units of measurement, terminology and organization are used, following the book *Process Design and Engineering Practice.*

  *How does this collection of practical details match yours?*
  *How might you combine this collection with yours?*
  *How might you expand and keep it up-to-date?*

- **Want feedback about trouble shooting?** ... then Chapter 4 gives story time describing the adventures of five engineers as they trouble shoot problems that range in complexity. But these aren't your usual stories. The story is interrupted. Each of the five scripts consists of about three parts with each part concluding with a few questions for you to consider. This question break was introduced to give you a chance to reflect on how you would have handled the case, and to decide what you should do next. At the end of each case an assessment is given of the problem-solving processes used by each of the trouble shooters.

  *How was your approach similar to, different from Michelle, Pierre, Dave, Saadia or Frank?*
  *How does your rating of their approach agree with my rating of their approach?*
  *How well could you describe the approach you took?*
  *How would you rate your approach?*

- **If you want to improve your confidence and skill in *problem solving*** ... then Chapter 5 provides activities, with feedback, to develop skills and confidence in
  - awareness or the ability to describe your problem-solving processes;
  - strategies or the ability to see patterns in the process;
  - exploring the context using the *Why? Why? Why?* process;
  - being creative;
  - being skilled in performance assessment.

  For each, target skills were listed, some examples were provided, activities to develop the skills were described and forms of evidence were given. **Cases #9** and **#10** (the bleaching plant and to dry and not to dry) were introduced.

  *What problem solving strategy do you use? How does it compare with the MPS 6-step?*
  *Is the Why? Why? Why? approach effective for you?*
  *When are the best times to apply it?*
  *For creativity, are you willing to spend the extra time to write down a wide range of ridiculous ideas?*
  *For creativity, what triggers work best for you?*

- **If you want to improve your confidence and skill in *data collection and critical thinking*** ... then Chapter 6 provides activities, with feedback, to develop skills and confidence in
  - gathering information, selecting diagnostic actions, testing hypotheses/causes. Criteria are given for selecting tests. Biases and mistakes made in gathering and interpreting data are confirmation bias, over-interpretation, under-interpretation and mis-interpretation, availability bias, premature closure, anchoring. The Jungian typology dimension P-J provides a useful indicator of one characteristic of your personal style for gathering and interpreting data.
  - being consistent in our use of words, and our use of knowledge about processes and process equipment. Specifically the focus was on identifying fact versus opinion, gathering accurate cause–symptom data and astutely reversing the connection to link symptom–cause. Our usage should be consistent with the rules of English, mathematics, the fundamentals of science and engineering and practical experience.
  - classification; the process of dividing large sets of information into meaningful parts. In making the division we should use a single basis/criteria per level, use no single entries and avoid faulty coordination or subordination. This was illustrated for the task of classifying the starting information into "symptoms" and "triggering" events.
  - identifying patterns.
  - reasoning. A systematic 9-step process is suggested and illustrated in the context of part of Michelle's reasoning in **Case #3**. **Case #11** (the lazy twin) was introduced.

*Did the list of actions given in Section 6.1 agree with yours?*
*What types of actions do you prefer to select when you trouble shoot?*
*What is your style in matching hypotheses with symptoms?*
*What methods do you use to select tests that will correctly test hypotheses?*
*Did the two inventories about bias provide useful insight? If not, what else might you do to identify your preference and style?*
*How easy was it for you to separate facts from opinions?*
*How easy was it matching hypotheses consistent with symptoms?*
*What special techniques do you use to identify "symptoms"?*
*What methods do you use to spot patterns in data?*
*Did the diagramming of an argument help you?*

- **If you want to improve your confidence and skills in *interpersonal skills* ...** then visit Chapter 7 that provides activities, with feedback, to develop skills and confidence in:
  - communicating,
  - listening,
  - applying the basic fundamental principles of interpersonal relationships,
  - building trust and
  - understanding our own uniqueness and the uniqueness of others.
  - Factors that affect our performance and the performance of others include:
  - unwillingness to admit having made a mistake,
  - stress,
  - level of frustration and lack of motivation.
  - preference to follow their own approach even though it may result in incorrect and even unsafe operation of the process.
  - preference to infer and interpret what we see or hear.

  A questionnaire is included to give us a chance to evaluate the environment in which we trouble shoot. **Cases #12–14** (the drop boxes, the lousy control system, the condenser that was just too big!) were introduced. **Cases #15–18** were included as exercises.

  *Which activities were most useful to you?*
  *How strong is your tendency to infer?*
  *How does your environment rate?*
  *What was the best new idea you learned from this Chapter?*

- **If you want to improve your skill and confidence in *trouble shooting*** ... then work the cases in Chapter 8, reflect on the process you used, check on the number of target skills you displayed and set goals for improvement. You can experience the cases as triads or as an individual. If you get the chance select the triad experience! The cases have been carefully selected to provide varying levels of difficulty, different types of situations (from startup to usual operations) and different varieties of equipment. Start looking at cases at Level 3 and 4. The lower ratings start with **Case #19**.

*Which case did you enjoy the most?*
*From which case did you learn the most?*
*When you experienced the triad activity, which role did you prefer?*
*What was the biggest challenge in reflecting on and evaluating the approach you used?*
*Did you complete a Trouble-Shooter's Worksheet when you started? If not, why not?*

This book was designed as a challenging and enjoyable educational experience.

## 9.2 Reflection and Self-Assessment are Vital for the Development of Confidence

Research has shown that the quality of the answer to a problem and the problem-solving skill and confidence improves if we:

- pause during the process and write down reflections of what you have done so far and where you are going next.
- have clear goals describing the skill, have measurable criteria about how you will know when you have the skill and have opportunities to collect evidence.

The design for self-assessment and confidence building is illustrated in Chapter 5.

- **The target skills are listed**. These are the proven behaviors of successful trouble shooters. Write these in terms that can be observed. Try to remove any ambiguity. Include some quantitative criterion to measure achievement of the target skill. *This was done, for example in Sections 5.1.1, 5.2.1, 5.4.1.* In the context of the overall process of trouble shooting, this was presented in Table 2.1.
- **An activity is posed**. The activity gives you a chance to display the behavior. *In Chapter 5 the tasks were described in Sections 5.1.2, 5.2.2, with the materials to be used given in Sections 5.1.3, 5.2.3.* For the overall process of trouble shooting, the overall process was illustrated first through the completed Trouble-Shooter's Worksheets (for **Case #8** in Chapter 2, and the **Cases #2–4** in Appendix C) and through the examples given in Chapter 4 of Michelle, Pierre, Dave, Saadia and Frank. For your approach, two options were presented in Section 8.1: the triad and the individual activity. Cases were scattered through the various chapters with a list of options given in Section 8.1.
- **Feedback forms and forms of evidence** are created. *In Chapter 5 the forms of evidence were listed in Sections 5.1.4, 5.2.4.* For the overall trouble-shooting process, the forms of evidence will include: the problem statement (and the marks, underlines and notations you made directly on this); your trouble shooter's notebook or worksheets that you used; the pauses and reflections you wrote as you worked through the case, the Trouble-Shooter's Worksheet (and especially the hypotheses–symptoms chart), the written requests for information (if you worked in triads) or the sequence of codes for the ques-

tions (if you worked as an individual). The most important form of evidence is Worksheet 2-2. This is significant because it relates directly to the target behaviors of successful trouble shooters, noted in Table 2.1. Use this to give yourself-feedback about the process you used.

## 9.3 Going Beyond this Book: Setting Goals for Improvement

Prepare yourself for success and use reflection and self-assessment effectively.

### 9.3.1 Prepare Yourself for Success

- Continually update your knowledge of process equipment. Research has shown that the key is a broad knowledge of process equipment. Continually draw on your experience and the experience of colleagues to enrich that experience.
- Praise yourself for where you are now. Too often we focus only on the negative and things we can't do well. Throughout this book you have noted that the feedback was always five strengths and two areas to work on. At this time, write down your five strengths
  1. ________________________________________
  2. ________________________________________
  3. ________________________________________
  4. ________________________________________
  5. ________________________________________
- Set achievable goals. These should be expressed in terms of the target skills of successful trouble shooters, described in Table 2.1. These should be consistent with the two areas you want to work on. You might identify your goals as:
  1. to maintain the five strengths noted above and

  to shift one of the following "areas to work on" to a strength within the next year.

  My two areas to work on are:
  1. ________________________________________
  2. ________________________________________
- Arm yourself for success.

  In Section 6.1.2b I listed the stuff I had available in my office related to trouble shooting. A notebook, paper or hand-held electronic, is essential. Gloves, tape measure and string and always good additions. The first time you appear with a stethoscope will require courage! But after you have solved a bunch of tricky problems because of the stethoscope, others will want one too.

### 9.3.2 Use Reflection and Self-Assessment Effectively

Follow the principles of self-assessment, given in Sections 5.5 and 9.2. Assessment or performance review is not a dirty word! Conscientious self-assessment is the guide to growth.

## 9.4 Going Beyond this Book: Updating your Rules of Thumb and Symptom ← Cause Data for Process Equipment

Rules of thumb are generalized, usually numerical, values of "usual" practice. Although we all generate such "experience" numbers intuitively, it helps to organize and write these down. You might create files for different equipment, following, for example, the titles used in Chapter 3. As you read articles in such journals as *Chemical Engineering Progress, Chemical Engineering* and *Hydrocarbon Processing* record the rules of thumb in your paper or electronic files. In creating the files, decide on a system of units that you will consistently use. Take the time to rework information from other systems of units. Create your own set of rules of thumb for the processes and unit operations with which you work.

Extend the rules of thumb to include cause → effect data. Check through the vendor files on the web or that are in the equipment files. Some sources in the literature include Bloch and Geitner (1983), and McNally Institute at http://www.mcnallyinstitute.com (2003).

## 9.5 Beyond this Book: Sources of Other Cases

Some computer simulation games for trouble shooting have been developed by Doig and colleagues for the SYSCHEM process (1977, 1980). Trouble-shooting cases reported in the literature tend to describe the problem, the process used to discover the fault and the corrective action taken. Although this format is not easy to use to help polish your skill, these cases broaden our perspective, provide more cause → effect data and can be converted into the format used in this book to aid in skill development.

Some sources include Liberman (1985), Saletan (1994), the Riance series (1983 ff), the articles by Henry Kister that appear in Hydrocarbon Processing or at the AIChE conferences and the Marmaduke series in Power Magazine 1950 ff with some reprinted by Elonka (1979).

# Literature References


## Preface

Branan, C (1998) "Rules of Thumb for Chemical Engineers," 2nd edition, Gulf Publishing Co.

Gans, M. et al., (1983) "Plant Start-up step by step by step," Chem. Eng. 90, 20, Oct 3 p 74.

Kister, H.Z, series of articles in Hydrocarbon Processing 1979 ff.

Lieberman, N.P. (1985) "Trouble-shooting process Operations" 2nd edition, PennWell Books, Tulsa OK.

Saletan, D, (1994) "Creative Trouble Shooting in the Chemical Process Industries," Chapman and Hall, London.

## Chapter 1

Example articles describing a personal approach to trouble shooting in engineering:

Gans, M. et al., (1983) "Plant Start-up step by step by step," Chem. Eng. 90, 20, Oct 3 p 74.

Laird, D, B. Albert, C. Steiner and D. Little (2002) "Take a Hands-on approach to refinery troubleshooting," Chem. Eng. Prog, June, 68–73.

Smith, K. (2002) "Refine your approach to process trouble shooting and optimization," Hydrocarbon Processing, June, 63–66.

Taylor, W.K. (1980) "Trouble Shooting at Canadian Industries Limited" Chemical Engineering Education, Spring, p 88–89.

- Books to help improve knowledge about safety and hazards.

Kletz, T.A. (1985) "What went Wrong?" Gulf Publishing Co, Houston, TX.

Kletz, T.A. (1983) "Hazop and Hazan–Notes on the Identification and Assessment of Hazards," Institution of Chemical Engineers, London, UK.

Woods, D.R. (1995) "Data for Process Design and Engineering Practice," Prentice Hall.

- Example references summarizing research about trouble shooting:

Dubeau, C.E. et al. (1986) "Premature Conclusions in the Diagnosis of Iron-deficiency Anaemia: cause and effect", Medical Decision Making, **6**, 3, 169–173.

Elstein, A.S, L.S. Shulman and S.A. Sprafka (1978) "Medical Problem Solving: an analysis of clinical reasoning", Harvard University Press, Cambridge MA.

Groen, G.J. and V.L. Patel (1985) "Medical Problem Solving: some questionable assumptions" Medical Education 19, 95–100.

Johnson-Laird, P.N. and W.C. Wasan, eds, (1977) "Thinking: Readings in Cognitive Science", Cambridge University Press, Cambridge, MA.

Kassirer, J.P, and G.A, Gorry (1978) "Clinical Problem Solving: A Behavioral Analysis", Ann. Int. Medicine, 89 245–255.

Kern, L, and M. E. Doherty (1982) "Pseudo-diagnosticity" in an Idealized Medical Problem Solving Environment", J. Medical Education, 57 100–104.

McGuire (1985) "Medical Problem Solving: A Critique of the Literature", J. Med. Education, 60, 587–595.

Nisbett, R. and L. Ross (1980) "Human Inference: strategies and shortcomings of social judgement", Prentice Hall, Englewood Cliffs, NJ.


*Successful Trouble Shooting for Process Engineers.* Don Woods

ISBN: 3-527-31163-7

Spencer, Joanne, (1988) " Critique of Problem solving and trouble shooting approaches", Department of Chemical Engineering, McMaster University, NSERC summer student report.

Tversky, A, and D. Kahneman (1974) "Judgement and uncertainty: heuristics and biases", Science, 185, 1124–1131.

Voltovich, A.E. et al. (1985) "Premature conclusions in diagnostic reasoning", J. Med. Education, 60, 302–307.

Whitman, N. et al. (1986) "Problem Solving in Medical Education: Can it be Taught?", Current Surgery, 43 453–4.

Wolf, F.M, L.D. Grappen and J.E. Billi (1985) "Differential Diagnosis and the Competing-Hypothesis Heuristic", J. AMA, 253, 19, 2858–2861.

## Chapter 2

Covey, S.R. (1990) "Seven Habits of Highly Effective People," Fireside Book, Simon Schuster.

Holmes, T.H. and R.H. Rahe (1967) "The Social Readjustment Rating Scale," J of Psychosomatic Research, Aug, 213–218.

Kepner, C.H. and B.B. Tregoe (1985) "The New Rational Manager," McGraw Hill, New York.

Woods, D. R. (1994) "Problem-solving skills", Chapter 3 in "Problem Based Learning: How to Gain the Most from PBL" Woods, Waterdown, ON, Canada.

Woods, D.R. (2000) "An Evidence-based strategy for problem solving," J. of Engineering Education, Oct, 443–459.

Woods, D.R. et al, (1997) "Developing Problem-solving skills: The McMaster Problem Solving Program," J of Engineering Education, 86, 2, 75–91 and http://www.chemeng.mcmaster.ca/innov1.htm and click on MPS.

Woods, D.R. (1988) "Novice versus Expert Research suggests ideas for implementation," J College Science Teaching, Sept, p 77–79, 66–67; Nov, p 138–141; Dec, p 193–195.

## Chapter 3

Gans, M. et al., (1983) "Plant Start-up step by step by step," Chem. Eng. 90, 20, Oct 3 p 74.

Griff, A. (1968) "Plastics Extrusion Technology", Reinhold Book Co.

Lieberman, N.P. (1985) "Trouble-shooting Process Operations" 2nd edition, PennWell Books, Tulsa OK.

Rauwendall, C. (1986) "Polymer Extrusion", Hanser Publishers, Munich.

Vlachopoulos, J. et al, (2001) "The SPE Guide on extrusion technology and troubleshooting," The Society of Plastics Engineers, Brookfield, CT.

Woods, D.R. (1995) "Process Design and Engineering Practice, Prentice Hall.

Francis, D, and D. Young, "Improving Work Groups: A Practical Manual for Team Building," University Associates, San Diego, CA. (1979).

Woods, D.R, "Group skills," Chapter 5 in "Problem-based Leaning: how to gain the most from PBL," Woods Publisher, Waterdown ON Canada distributed by McMaster University Bookstore, Hamilton, ON (1994).

Fisher, K, et al. (1995) "Tips for Teams," McGraw Hill, New York.

Kirton, M.J. (1976) "Adaptors and innovators: a description and measure," J. Applied Psychology, 61, 622–629.

Keirsey, D. and M. Bates, "Please understand me: character and temperament types," Gnosology Books, Del Mar, Ca (1984) and http://www.keirsey.com.

Schutz, W.C, "FIRO: a three-dimensional theory of interpersonal behavior," Holt Rinehart and Winston, New York, NY, 1958, with the instrument and scoring available from Whetton, D.A, and K.S. Cameron, "Developing Management Skills," Scott Forseman, Glenview, IL,1984, p. 30.

Tannen, D, "You Just Don't Understand: Women and Men in Conversation," Ballantine, New York, 1990.

Johnson, D.W, "Reaching Out," Prentice Hall, Englewood Cliffs, NJ, 1986.

## Chapter 4

Lieberman, N.P. (1985) "Trouble-shooting process Operations" 2nd edition, PennWell Books, Tulsa OK.

Kister, H.Z., Series of articles in Hydrocarbon Processing 1979 ff.

Gans, M. et al., (1983) "Plant Start-up step by step by step," Chem. Eng. 90, 20, Oct 3 p 74.

### Chapter 5

Basadur, M. (1995) "Simplex: a flight to creativity", Creative Education Foundation, Buffalo, N. Y.
Gadsby, R.E. and J.G. Livingstone (1977) "Catalysts and some incidents they have survived," CEP Ammonia Plant Safety, 20, p 70 ff.
Kimbell, R. et al. 1991 "Assessment of Performance in Design and Technology," SEAC report, UK.
Krishnaswamy, R. and N.H. Parker (1984) Chemical Engineering, April 16, p 93–98.
Leifer, L, (1997) "Design team performance: metrics and the impact of technology," in "Evaluating organizational training: models and issues," S.M. Brown and C. Seidner, eds, Kluwer Academic publishers.
Lombard, J.F. and R.A. Culberson (1972) "Defining Reformer Performance," CEP Ammonia Plant Safety, 15, p 29–35.
Woods, D.R. (1988) "Novice versus Expert Research suggests ideas for implementation," J College Science Teaching, Sept, p 77–79, 66–67; Nov, p 138–141; Dec, p 193–195.

### Chapter 6

Branan, C. (1998) "Rules of Thumb for Chemical Engineers," 2nd edition, Gulf Publishing Co.
Halpern, Diane (1996) "Thought and Knowledge; an introduction to critical thinking" 3rd edition, Lawrence Erlbaum.
Scriven, M. (1976) "Reasoning", McGraw Hill.
Walas, S.M (1988) "Chemical Process Equipment," Butterworths.
Woods, D.R. (2001–3) "Rules of Thumb" John Wiley and Sons, forthcoming.
Woods, D.R. (1995) "Process Design and Engineering Practice," Prentice Hall.

### Chapter 7

Johnson, D.W. and F.P. Johnson (1986) "Joining Together," Prentice Hall, Englewood Cliffs, NJ.
Jungian typology or MBTI: see Keirsey, D. and M. Bates (1984) "Please Understand Me" Prometheus Books, Del Mar CA and http://www.keirsey.com.
Kirton KAI see Kirton, M. (1976) "Adaptors and Innovators: a description and measure," J. Applied Psychology, 61, no. 5, 622–629.
Kletz, T.A. (1986) "What went Wrong? Case histories of Process Plant Disasters," Gulf Publishing Co, Houston, TX.
Powers, G.J. and S.A. Lapp (1983) "Fault Tree Analysis," Carnegie Mellon University Short Course, Pittsburgh.
Woods, D.R. (1994) Chapter 5 in "Problem based learning: how to gain the most from PBL," Woods, Waterdown, ON, Canada.

### Chapter 8

Barton, J. and R. Rogers (1997) "Chemical Reaction Hazards", 2nd edition, Gulf Publishing Co, Houston TX.
Krishnaswamy, R. and N.H. Parker (1984) "Corrective Maintenance and Performance Optimization", Chemical Engineering, April 16, p 93–98.
Yokell, S. (1983) "Trouble shooting shell and tube heat exchangers", Chem. Eng. 90 no 15, p 57–75.

### Chapter 9

- Sources of symptom-cause information:

Bloch, H.P. and F.K. Geitner (1983) "Practical Machinery Management for Process Plants, part 2: Machinery Failure analysis and troubleshooting," Gulf Publishing.

- Simulation:

Doig I.D. (1977) "Training of Process Plant Malfunction Analysis," Chemeca 77, Canberra, 14–16 Sept, p 144–148 and (1980) "Trouble-shooting systems and experiences at New South Wales," Chemical Engineering Education, Summer 1980, p 130.

- Sources of other cases:

Drew, J.W. (1983) 'Distillation Column Startup," Chem. Eng. Nov 14, p 221.
Elonka, S.M. (1979) "Marmaduke Surfaceblow's Salty Technical Romances" R. Krei-

ger Publishing, New York. Collection of articles from Power Magazine.
Karassik, I.J. (1981) "Centrifugal Pump Clinic," Marcel Dekker Inc. New York.
Kister, H.Z., Series of articles in Hydrocarbon Processing 1979 ff.
Lieberman, N.P. (1985) Trouble-shooting process Operations, 2nd edition, PennWell Books, Tulsa, OK.
Riance, X.P. (1983 ff) "Learning–the hard way" series of articles in Chemical Engineering Magazine, in 1985: Oct 3, Oct 31, In 1985: May 13, June 10; In 1986, Feb 17. In 1987, Jan19. In 1988, Feb 15 1986.
Saletan, D, (1994) "Creative Trouble Shooting in the Chemical Process Industries," Chapman and Hall, London.
Shah, G.C. (1979) Trouble shooting reboiler systems," Chem. Eng. Prog., July 53–58.
Yokell, S. (1983) "Trouble shooting shell and tube heat exchangers", Chem. Eng. 90 no 15, p 57–75.

- Other articles for rules of thumb and trouble shooting:

Eckert, J.S. (1979) "Design of Packed Distillation Columns" Section 1.7 in "Handbook of Separation Techniques for Chemical Engineers", P.A, Schweitzer, ed, McGraw Hill, New York. p 1.221–1.240.
Ellerbe, R.W. (1979) "Batch Distillation" Section 1.3 in "Handbook of Separation Techniques for Chemical Engineers", P.A, Schweitzer, ed, McGraw Hill, New York. p. 1.164–1.167.
Farminer, K.W. (1988) "Defoaming" in "Encyclopedia of Chemical Technology and Design", J McKetta, ed, Marcel Dekker, NY.
Gans, M. and Fitzgerald, F.A. (1966) "Plant Startup" Chapter 12 in "The Chemical Plant" R. Landau, ed. Reinhold Publishing Co. New York.
Gardner, K.A. (1974) "Anticipation of Operating Problems in the Design of Heat Transfer Equipment" in "Heat Exchangers: Design and Theory Sourcebook" N. Afgan and E.U. Schlunder, eds, Scripta Book Co, McGraw Hill, New York.
Godard, K.E. (1973) "Gas Plant Startup Problems" Hydrocarbon Process. Sept. p 151–155.
Griffith, S. and Keister, R.G. (1970) "This Butadiene Unit Exploded" Hydrocarbon Process, **49**, no. 9, p 323.
Kister, H.Z. (1979) "When Tower Startup has Problems" Hydrocarbon Process Feb p 89–94.
Kletz, T. (1979) "Learn from these HPI fires" Hydrocarbon Processing Jan p 243–250.
Kletz, T.A. (1986) "Hazop and Hazan" 2nd edition, The Institution of Chemical Engineers, London, UK.
Kletz, T.A. (1985) "What Went Wrong? Case Histories of Process Plant Disasters" Gulf Publishing Co, Houston, TX.
Lapp, S.A. and Powers, G.J. (1977) "Computer Aided Synthesis of Fault-trees" IEEE Transactions on Reliability, April p 2–13.
Lefevre, L.J. (1986) "Ion Exchange: Problems and Troubleshooting" Chem Eng, July 7, p 73–75.
Lieberman, N.P. (1983) "Process Design for Reliable Operations" Gulf Publishing Co. Houston, TX.
Lieberman, N.P. (1985) "Trouble-shooting process Operations" 2nd edition, PennWell Books, Tulsa OK .
McLaren, D.B. and Upchurch, J.C. (1970) "Guide to Trouble-free Distillation" Chem. Eng. **77**, No 12, June 1, p 139–152.
Penny, W.R. (1970) "Guide to Trouble-free Mixers", Chem. Eng. **77**, No 12, June 1, p 171–180.
Powers, G. J. and Lapp, S.A. (1983) "Fault Tree Analysis" Carnegie Mellon University Short Course, Pittsburgh.
Powers, G.J. and Lapp, S.A. (1981) 'A Short Course on Risk and Reliability Assessment by Fault tree Analysis" Carnegie Mellon University.
Powers, G.J. and Lapp, S.A. (1976) "Computer-aided Fault Tree Synthesis" Chem. Eng. Prog. April p 89–93.
Reason, J. and Mycielska, K. (1982) "Absent Minded? The Psychology of Mental Lapses and Everyday Errors" Prentice Hall, Englewood Cliffs, NJ.
Reginald, S. and Gupta, J.P. (1986) "A Note on 1.2 Heat Exchanger Trouble Shooting" Int. Comm. Heat Mass Transfer, 13, p 235–243.
Riance, X.P. (1983) "Learning- the Hard Way" Chem. Eng. Oct 3, p115–116.

Riance, X.P. (1987) "More Learning–the Hard Way: part 4, Chem. Eng. Jan 19, p 131–133.
Shah, G.C. (1978) "Trouble Shooting Distillation Columns" Chem. Eng. July 31 p 70–78.
Swartz, A. (1988) "Evaporator Operation: Trouble shooting" in Encyclopedia of Chemical Technology and Design", J McKetta, ed, Marcel Dekker, NY.
Talley, D.L. (1976) "Startup of a Sour Gas Plant" Hydrocarbon Process. April p 90–92.
Troyan, J.E. (1960) "Trouble Shooting New Processes" Chem. Eng. Nov 14, p 223–226.
Troyan, J.E. (1961) "Trouble Shooting New Equipment" Chem. Eng. March 20, p 147–150.
Troyan, J.E. (1961) "More on Trouble Shooting New Equipment: Pumps, compressors and agitators" Chem. Eng. May 1, p 91–94.
Wetherhorn, D. (1970) "Guide–Trouble-free Evaporators" Chem. Eng. 77, No. 12, June 1, p 187–192.

**Appendix A**

Chin, T.G. (1979) "Guide to distillation pressure control methods", Hydrocarbon Processing, October, p 145–153.
Goyal, O.P. (2000) "Evaluating troubleshooting skills", Hydrocarbon Processing, October, p 100C.

# Index



*Successful Trouble Shooting for Process Engineers.* Don Woods

ISBN: 3-527-31163-7